PRACTICAL ASPECTS
OF
ION TRAP MASS
SPECTROMETRY

Volume III

*Chemical, Environmental,
and Biomedical Applications*

MODERN MASS SPECTROMETRY

*A Series of Monographs on
Mass Spectrometry and Its Applications*

EDITOR-IN-CHIEF
Thomas Cairns, Ph.D., D.Sc.

Forensic Applications of Mass Spectrometry
Edited by Jehuda Yinon

*Practical Aspects of Ion Trap Mass Spectrometry, Volume I
Fundamentals of Ion Trap Mass Spectrometry*
Edited by Raymond E. March and John F. J. Todd

*Practical Aspects of Ion Trap Mass Spectrometry, Volume II
Ion Trap Instrumentation*
Edited by Raymond E. March and John F. J. Todd

*Practical Aspects of Ion Trap Mass Spectrometry, Volume III
Chemical, Environmental, and Biomedical Applications*
Edited by Raymond E. March and John F. J. Todd

PRACTICAL ASPECTS OF ION TRAP MASS SPECTROMETRY

Volume III

Chemical, Environmental, and Biomedical Applications

Edited by
Raymond E. March
John F. J. Todd

CRC Press
Boca Raton   New York   London   Tokyo

LIBRARY OF CONGRESS CATALOGING-IN-PUBLICATION DATA

Practical aspects of ion trap mass spectrometry / edited by Raymond E. March, John F. J. Todd

    p. cm. — (Modern mass spectrometry)
    Includes bibliographical references and index.
    ISBN 0-8493-4452-2 (vol. 1)
    ISBN 0-8493-8253-X (vol. 2)
    ISBN 0-8493-8251-3 (vol. 3)
    1. Mass spectrometry. I. March, Raymond E. II. Todd, John F. J. III. Series
QD96.M3P715 1995
539.7′028′7—dc20   95-14146
                                                       CIP

© 1995 by CRC Press, Inc.

No claim to original U.S. Government works
International Standard Book Number 0-8493-4452-2 (vol. 1)
International Standard Book Number 0-8493-8253-X (vol. 2)
International Standard Book Number 0-8493-8251-3 (vol. 3)
Library of Congress Card Number 95-14146
Printed in the United States of America 1 2 3 4 5 6 7 8 9 0
Printed on acid-free paper

EDITORIAL ADVISORY BOARD

FOREWORD

Publication of the volumes in this new series, *Modern Mass Spectrometry*, represents a milestone for scientists intimately involved with the practice of mass spectrometry in all its various forms. Forthcoming monographs in this series will focus on various selected topics within the rapidly expanding realm of mass spectrometry. Individual volumes will provide in-depth reports on mainstream developments where there is an urgent need for a specific mass spectrometry treatise in an active and popular area.

While mass spectrometry as a field is quite well-served by several publications and a number of societies, the application of mass spectrometric techniques across the basic scientific disciplines has not yet been recognized by existing journals. The present distribution of research and application papers in the scientific literature is widespread. There is a multidisciplinary audience requiring access to concise reports illustrating the latest successful approaches to difficult analytical problems.

The distinguished members of the Editorial Advisory Board all agreed that a platform exists for a premier book series with high standards to cover comprehensively general aspects of developing mass spectrometry. Contributing authors to the series will provide concise reports together with a bibliography of publications of importance selecting worthy examples for inclusion. Due to the rapid and extensive growth of the literature in mass spectrometry, there is a need for such reports by authorities who critique the entire subject area. There is an increasing urgency to provide readers with timely, informative and cogent reviews stripped of outdated material.

I believe that the decision to publish the series *Modern Mass Spectrometry* reflects the realization that increasing numbers of mass spectrometrists are applying nascent state of the art approaches to some fascinating problems. Our challenge is to develop the best forum in which to present these emerging issues to encourage and stimulate other scientists to comprehend and adopt similar strategies for other projects.

Lofty ideals aside, our immediate goal is to build a reputable series with scientific authority and credibility. To this end, we have assembled an excellent Editorial Advisory Board that mirrors the prerequisite multidisciplinary exposure required for success.

I am confident that the mass spectrometry community will be pleased with this "new publication approach" and will welcome the opportunities it presents to foster the development of interactions between the various scientific disciplines. No doubt there is a long and difficult road ahead of us to ensure the series grows into a position of leadership, but I am convinced that the hard work of our outstanding Editorial Advisory Board, and the enthusiasm of our authors and readers will achieve the degree of success for which we all seek.

Thomas Cairns
Editor-in-Chief

THE EDITORS

Raymond E. March, Ph.D., is presently Professor of Chemistry at Trent University in Peterborough, Ontario, Canada. He obtained his B.Sc. degree from the University of Leeds in 1957; his Ph.D. degree, which he received in 1961 from the University of Toronto, was supervised by Professor John C. Polanyi.

Dr. March has conducted independent research for over 28 years and has directed research in gas phase kinetics, optical spectroscopy, gaseous ion kinetics, analytical chemistry, and mass spectrometry.

He has published and/or co-authored over 130 scientific papers in the above areas of research with emphasis on mass spectrometry, both with sector instruments and quadrupole ion traps. Dr. March is a coauthor with Dr. Richard J. Hughes of *Quadrupole Storage Mass Spectrometry*, published in 1989.

Professor March is actively engaged in the supervision of graduate student research and is an Adjunct Professor of Chemistry at Queen's and York Universities in Ontario; he is the Associate Director of the Trent/Queen's Cooperative Graduate Program. Professor March is a Fellow of the Chemical Institute of Canada and a member of the American, British, and Canadian Societies for Mass Spectrometry. Research in Dr. March's laboratory is supported by the Natural Sciences and Engineering Research Council of Canada, the Ontario Ministry of the Environment, and Varian Associates. Dr. March has enjoyed long-term collaborations with the coeditor, John Todd, and with colleagues at the University of Provence and Pierre and Marie Curie University of France, and with colleagues in Italy.

John F. J. Todd, Ph.D., is currently Professor of Mass Spectroscopy and was, until recently, Director of the Chemical Laboratory at the University of Kent, Canterbury, England. He obtained his B.Sc. degree in 1959 from the University of Leeds, from whence he also gained his Ph.D. degree and was awarded the J. B. Cohen Prize in 1963, working in the radiation chemistry group led by Professor F. S. (now Lord) Dainton, FRS. He was a postdoctoral research fellow in the laboratory of the late Professor Richard Wolfgang at Yale University, Connecticut, from 1963 through 1965 and was one of the first appointees to the academic staff of the, then new, University of Kent in 1965.

Since arriving in Canterbury Professor Todd's research interests have encompassed mass spectral fragmentation studies, gas discharge chemistry, ion mobility spectroscopy, analytical chemistry, and ion trap mass spectrometry. His work on ion traps commenced in 1968, and he first coined the name QUISTOR for quadrupole ion store, to describe a trap coupled to a quadrupole mass filter for external mass analysis. Acting as a consultant to Finnegan MAT, he was a member of the original team that developed the ion trap commercially. Research in Professor Todd's laboratory has been supported by the Science and Engineering Research Council, the Engineering and Physical Sciences Research Council, the Defense Research Agency, the Chemical and Biological Defense Establishment, and Finnegan MAT, Ltd.

Dr. Todd has published and/or co-authored over 100 scientific papers, concentrating on various aspects of mass spectrometry. With Dr. Dennis Price, he co-edited several volumes of *Dynamic Mass Spectrometry,*; he edited *Advances in Mass Spectrometry 1985*; and he is currently editor of the *International Journal of Mass Spectrometry* and *Ion Processes*.

Professor Todd is actively engaged in the supervision of full-time and industrial-based part-time graduate students, all working in the field of mass spectrometry. He is a Chartered Chemist and a Chartered Engineer and is currently an elected member of the Council of the Royal Society of Chemistry. He recently completed a four-year term as Treasurer of the British Mass Spectrometry Society. Outside the immediate confines of academic work, he was for ten years Master of Rutherford College at the University of Kent and also for four years was appointed as the Chairman of the Canterbury and Thanet Health Authority. He is presently a Governor of the Clergy Orphan Corporation and of Canterbury Christ Church College. He has enjoyed long-term collaborations with co-editor Professor Raymond March, with colleagues at Finnigan MAT in the United Kingdom and the United States and with groups in Nice (France) and Padova and Torino (Italy).

DEDICATION

To ion trappers, young and old, everywhere.

PREFACE

This monograph is Volume 3 in the series *Modern Mass Spectrometry*, published by CRC Press, and it is the third volume of a mini-series of three volumes on *Practical Aspects of Ion Trap Mass Spectrometry.* Volume 3, "Chemical, Environmental and Biomedical Applications," is a companion to Volumes 1 and 2, subtitled "Fundamentals of Ion Trap Mass Spectrometry" and "Ion Trap Instrumentation," respectively.

Volume 3 is composed of 14 chapters that have been arranged into four parts:

Part I: Fundamentals;
Part II: Practical Ion Trap Technology;
Part III: Applications Involving Small Molecules; and
Part IV: Environmental and Biomedical Applications.

Part 1. Fundamentals. Chapter 1 is devoted to a discussion of the theory, operating principles, and recent improvements in ion trap mass spectrometry. The reader is introduced to the basic operations of the ion trap for ion isolation, chemical ionization, mass range extension, high mass-resolution, and tandem mass spectrometry. The themes of ion isolation, chemical ionization and tandem mass spectrometry are taken up and expanded upon in Chapter 2. The practical aspects of tandem mass spectrometry are examined and illustrated with case histories. Chapter 2 concludes with an examination of practical aspects of laser desorption and photodissociation. In Chapter 3 is presented a review of modern ion trap research and includes an examination of ionization techniques, the injection of externally-generated ions, the operation of the ion trap and revisits tandem mass spectrometry.

Part II. Practical Ion Trap Technology. Chapter 4 deals at length with practical ion trap technology for gas chromatography/mass spectrometry and gas chromatography combined with tandem mass spectrometry, and reviews recent advances in these areas. Chapter 5 deals with the combination of high performance liquid chromatography with mass spectrometry which is well recognized as possessing enormous potential for analyzing a wide variety of compounds. It should be noted that gas chromatography is limited to some 20% of all organic compounds. This chapter is most appropriate given recent instrumental developments. Chapter 6 continues in the same vein by discussing the utilization of the ion trap for ion spray liquid chromatography/mass spectrometry and capillary electrophoresis/mass spectrometry. Some excellent applications are discussed in this chapter. Part II is rounded off with a detailed discussion in Chapter 7 of chemical ionization in the ion trap together with some illustrative examples.

Part III. Applications Involving Small Molecules. Chapter 8 is concerned with the energetics and efficiencies of collision-induced dissociation in the ion trap. Ion dissociation in collision with helium buffer gas is induced by resonance excitation which is therefore, an essential element of tandem mass spectrometry. Resonance excitation is discussed in detail together with the experimental determination of resonantly excited ion energies. Chapter 9 deals with the determination of ion structures in the ion trap. Following a brief introduction to daughter ion mass spectra, this chapter reviews a number of case histories concerning the identification of specific ion structures. In Chapters 10 and 11 there is a change of pace. In Chapter 10, the reader is introduced to atmospheric glow discharge/ion trap mass spectrometry; in this chapter, a theoretical treatment of the glow discharge mechanism is followed by a discussion of the energetics of charged particles in a discharge and the formation of negative ions. Chapter 11 deals with the operation of dynamically-programmed scans by which data may be acquired from hundreds of experiments when experimental parameters are varied incrementally; in this chapter are given a number of applications of dynamically-programmed scans which include automated tandem mass spectrometry, mapping of the stability diagram, and investigation of "Black Canyons."

Part IV. Environmental and Biomedical Applications. Chapter 12 is devoted to discussions of the Clean Water Act of 1977, the Safe Drinking Water Act of 1986, and the performance of the ion trap mass spectrometer for environmental analyses. A number of field applications are examined, together with the application of the ion trap to the determination of volatile and less-volatile organic air pollutants. Chapter 13 is concerned specifically with the application of the ion trap for multi-residue pesticide analysis; this chapter includes examination of the driving force behind this application, the use of chemical ionization, and the analytical approach to pesticide analysis. Chapter 14 returns to the theme of liquid chromatography/mass spectrometry using an electrospray ion source. Applications of electrospray/ion trap mass spectrometry to environmental analysis and to the determination of neuropeptides are presented.

The principal objective of this monograph is to convey to the reader an appreciation of the ion trap as an instrument of such high sensitivity and enormous versatility that it can be used in many routine, though sophisticated, analytical procedures and in many avenues of research often in tandem with other instruments or components, such as external ion sources and lasers.

The ion trapping field continues to be very active and is growing. The criteria by which we come to this conclusion are *the number of ion trap presentations, the number of ion trap manufacturers, and the number of new commercial ion trap devices introduced to the market.* Let us examine each criterion in turn. In each succeeding year for the past eleven years, the number of ion trap presentations at the American Society for Mass Spectrometry Conference on Mass Spectrometry and Allied Topics has exceeded that in the previous year. The number of manufacturers of ion trap devices now numbers four, that is, Finnigan MAT, Varian Associates, Bruker-Franzen and Teledyne Electronic Technologies. In addition to the relatively new ion trap instruments introduced by Bruker-Franzen and Teledyne Electronic Technologies, we have seen the introduction of Varian Associates' Saturn 4D gas chromatograph/tandem mass spectrometer (GC/MS/MS) and Finnigan MAT's two new instruments, the GC-Q gas chromatograph/tandem mass spectrometer and the LC-Q liquid chromatograph/tandem mass spectrometer (LC/MS/MS) with a high mass-resolution zoom feature. The exhaustive reader will note that several contributors to this monograph have forecast or called for the introduction of benchtop GC/MS/MS and LC/MS/MS instruments. In the ion trapping field, the pace of instrumental development is outstripping that of publishing!

As this monograph is expected to appear almost simultaneously with its companion monographs, Volumes 1 and 2, it is clear that the effort required by all to bring about simultaneous publication of the three monographs has been quite substantial. As editors, we are extremely grateful to all of the principal players in the ion trapping field who, in contributing to this set of three volumes, have enabled us to realize an extensive yet coherent account of the state of ion trap mass spectrometry. While it was a great privilege for us to be invited, in 1991, to undertake the preparation of an initial monograph on the quadrupole ion trap for CRC Press, it is to the enormous credit of our contributors that, as they have responded so positively and promptly to our invitations to participate in this endeavor, we have been able to prepare three volumes in all. Despite the torturous route by which the three monographs came into being, the cooperation, good humor and dedication of all of the contributors have enabled us to complete this task in a reasonable period.

Finally, we wish to thank the many people who have assisted us in one way or another with the many tasks that must be carried out in order to arrive at the publication of a monograph from a collection of manuscripts. First of all, to our contributors without whom this monograph would not have appeared. We give thanks for their individual inspiration; we thank them for the fruits of their labors, and for their patient toleration. At CRC Press, we thank Joel Claypool, Julie Spadaro and Al Starkweather for their ready co-operation, encouragement and skill in bringing this monograph into being; we thank especially Nora Konopka for her cheerfulness and devotion to the task at hand in maintaining communication throughout. We thank the Series Editors, Dr. Thomas Cairns and Dr. M. Allen Northrup for their guidance and assistance. We would be remiss if we were not to thank the people in our laboratories and to crave their indulgence for the time which we have devoted to this and earlier monographs, time which might have been devoted to them. However, it is all part of the general laboratory experience as so many of our contributors will testify. In particular, we wish to thank Dr. Marian Langford and Tony Franklin in Canterbury, Nathalie Mechin, Laurent Kirsch, and Philippe Liere in Professor Jean-Claude Tabet's laboratory in Paris, and Jeff Plomley, Frank Londry, Dr. Mila Lausevic, Dr. Zoran Lausevic, Xuewu Jiang, Dr. Maurizio Splendore and Pierre Perrier (from Professor Jacques André's laboratory in Marseille) at Trent.

Raymond E. March
John F. J. Todd

CONTRIBUTORS

Cecilia Basic
Division of Immunology
Beckman Research institute
City of Hope
Duarte, CA

Matthew M. Booth
Department of Chemistry
University of Florida
Gainesville, FL

William L. Budde
U.S. Environmental Protection Agency
Office of Research and Development
Cincinnati, OH

Thomas Cairns
Psychemedics
Culver City, CA

Silvia Catinella
Consiglio Nazionale Ricerche
Servizio Spettrometria Massa
Padova, Italy

M. Judith Charles
Department of Environmental Science
and Engineering
University of North Carolina
Chapel Hill, NC

Kin S. Chiu
U.S. Food and Drug Administration
Office of Regulatory Affairs
Los Angeles, CA

Colin S. Creaser
Department of Chemistry & Physics
Nottingham Trent University
Nottingham, England

Donald M. Eades
Lawrence Livermore National Laboratory
Livermore, CA

Anthony M. Franklin
Chemical Laboratory
University of Kent
Canterbury, Kent, England

Gary L. Glish
Department of Chemistry
University of North Carolina
Chapel Hill, NC

Jack D. Henion
College of Veterinary Medicine
Cornell University
Ithaca, NY

Jodie V. Johnson
Chemistry Department
University of Florida
Gainesville, FL

Jon A. Jones
Department of Chemistry
University of Florida
Gainesville, FL

Brent L. Kleintop
Wyeth-Ayerst Research
Princeton, NJ

Marian L. Langford
DERA Security Systems
Fort Halstead
Sevenoaks, Kent, England

H. K. Lim
Wyeth-Ayerst Research
Princeton, NJ

Hung-Yu Lin
Sterling Winthrop Inc.
Malvern, PA

Raymond E. March
Department of Chemistry
Trent University
Peterborough, Ontario, Canada

A. Mordehai
Varian Associates
Palo Alto, CA

David Navarro
U.S. Food and Drug Administration
Office of Regulatory Affairs
Los Angeles, CA

Randall E. Pedder
Extrel Mass Spectrometry
Monroeville, PA

Emil Siegmund
U.S. Food and Drug Administration
Office of Regulatory Affairs
Los Angeles, CA

James L. Stephenson, Jr.
Department of Chemistry
University of Florida
Gainesville, FL

Robert J. Strife
Proctor & Gamble
Cincinnati, OH

John F. J. Todd
Chemical Laboratory
University of Kent
Canterbury, Kent, England

Pietro Traldi
Consigli Nazionale Ricerche
Servizio Spettrometria Massa
Padova, Italy

Robert D. Voyksner
Research Triangle Institute
Research Triangle Park, NC

Nathan A. Yates
Chemistry Department
University of Virginia
Charlottesville, VA

Richard A. Yost
Department of Chemistry
University of Florida
Gainesville, FL

TABLE OF CONTENTS

Part 3.   Applications Involving Small Molecules

Part 4.   Environmental and Biomedical Applications

Part 1

FUNDAMENTALS

Part 1

FUNDAMENTALS

Chapter 1

ION TRAP THEORY, DESIGN, AND OPERATION

John F. J. Todd

CONTENTS

0-8493-8251-3/95/$0.00+$.50
© 1995 by CRC Press, Inc.

3

I. INTRODUCTION

Mass spectrometers are instruments in which positive or negative gas-phase ions formed from a sample are analyzed according to their mass/charge (m/z) ratios and the relative abundances of these ions recorded. A *mass spectrum* is therefore either a listing or a graphical plot of the relative abundances (or intensities) of the ions *versus* the m/z-values; usually the intensity axis is normalized so that the most intense ion signal in the spectrum is given a value of 100 or 1000.

While mass spectroscopy may yield structural information about the sample, as well as (hopefully) its relative molar mass, it may also be used as a means of quantitatively assaying an analyte. However, mass spectroscopy should not be confused with the optical spectroscopies such as infrared, ultra violet or even nuclear magnetic resonance spectroscopy, where the signal observed arises from the absorption or emission of radiation from energy states within the atoms or molecules according to quantum mechanical selection rules. Rather the acquisition of a mass spectrum should be seen as conducting a chemical reaction in which the sample is reacted to form the ions, which are then subjected to analysis and detection after a period of time. As with any chemical reaction, the precise nature of the products observed will depend upon the conditions under which the reaction is carried out (the reagents, energetics (temperature), etc.), the means of analysis employed, and the time-scale of the process.

The mass spectrometer consists, therefore, of three basic components: the ion source, the analyzer, and the detector, together with supplementary devices such as the sample inlet system and the means for data recording (see Fig. 1.1). In most instruments, the analyzer and the detector regions are generally maintained at a sufficiently low pressure as to minimize collisions between the ions and background gas molecules. Depending upon the means of ionization being employed, the pressure within the ion source may range from atmospheric to *ca.* 10^{-4} Pa and this, in turn, determines the degree of sophistication of the pumping system which must be employed. The sampling system may be a simple *batch* inlet arrangement for introducing pure materials, but frequently the mass spec-

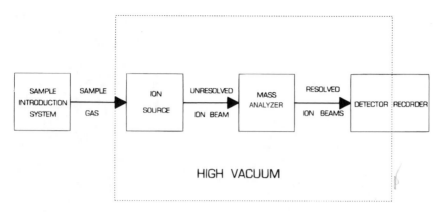

FIGURE 1.1
Block diagram representing the operational details of a mass spectrometer.

trometer is coupled to a separation instrument such as a gas or liquid chromatograph, and this, together with the nature of the sample being analyzed, determines the most appropriate method of ionization, and often the type of mass analyzer to be employed.

The commercially-available mass analyzers may be classified either as *beam transport* or as *ion trapping* devices. In the former category fall the magnetic sector, quadrupole, and time-of-flight instruments: all these instruments involve the movement of ions through space, over which electric and/or magnetic fields are applied, in order to effect a separation according to the m/z ratio. On the other hand, ion trapping analyzers, such as the Fourier transform ion cyclotron resonance mass spectrometer and the ion trap itself, involve containment of the ions within a confined region where they are subjected to time-dependent fields. By their very nature, the ion trapping instruments operate with pulsed ionization, and the same is true of time-of-flight analyzers; both the magnetic sector and the quadrupole spectrometers run with continuous ion beams. Quite often additional analytical benefits may be obtained by combining two or more different kinds of analyzer together to form *hybrid* machines. It is beyond the scope of this chapter to provide a critical analysis of the relative merits and demerits of the various types of analyzers available, and the reader is referred elsewhere for such a discussion.[1]

II. THEORY AND OPERATING PRINCIPLES OF THE ION TRAP MASS SPECTROMETER

A. General Description

The ion trap mass spectrometer is a member of the quadrupole family of instruments, indeed, its origins go back to the pioneering work of

FIGURE 1.2

Photograph of an *exploded* ion trap revealing details of the three electrodes, the filament, and the detector. Reproduced by courtesy of Finnigan Corporation.

Wolfgang Paul, and it was described in the same patent as that which first described the quadrupole mass filter. Its essential features are its simplicity and compactness, as can be seen from the *exploded* view of a trap shown in Fig. 1.2. The ion trap consists of three cylindrically-symmetric electrodes, two end-caps, and a ring. Each of these electrodes has accurately-machined hyperbolic internal surfaces and, in the normal mode of use, the end-cap electrodes are connected to earth potential while a radiofrequency (RF) *drive* potential oscillating, typically, around 1 MHz, is applied to the ring electrode; for certain applications (see below), a direct current (DC) potential may also be applied to the ring electrode. Ions are created within the ion trap by the gated injection of electrons emitted from a filament, shown on the right of Fig. 1.2 (or ions may be injected into the trap from an external source), and a range of m/z-values is held in bound or *stable* orbits by virtue of the RF potential. Upon increasing the amplitude of this potential, the motion of the ions becomes progressively more energetic such that their trajectories become *unstable* along the axis of symmetry (the z-axis) of the trap. As a result, the ions are ejected through holes in the left hand end-cap and are collected on the *channeltron* electron multiplier shown on the extreme left of the picture. The ejection of the ions occurs in the order of increasing m/z-value and the resulting signal represents, therefore, the mass spectrum of the ions which were originally trapped. This mode of operation of the ion trap, called *mass-selective ejection*, is to be distinguished from two earlier means

FIGURE 1.3

Photograph of an assembled ion trap together with a view along the axis of the trap with the ion exit end-cap electrode removed. Sample eluted from a capillary GC column is transferred to the trap via the cylindrical insulator supported from the upper end-cap electrode. Reproduced by courtesy of Finnegan Corporation.

of recording mass spectra with the trap, namely *mass-selective detection*, and *mass-selective storage;* the principles involved in the latter two modes are described elsewhere.[3]

Fig. 1.3 shows that the assembled trap incorporates teflon or quartz *spacers*. These spacers are employed to ensure that the structure is reasonably *gas-tight* since, as we shall see below, the mass spectral performance of the ion is improved greatly by maintaining a pressure of *ca.* 10^{-1} Pa of helium buffer gas during operation. A schematic diagram of the Ion Trap Detector (ITD™) is shown in Fig. 1.4.

B. Theory of the *Perfect* Quadrupole Trap

The theory behind the operation of the ion trap[4] is best considered by first examining the equations both for the electric field within the ion trap of *perfect* quadrupolar geometry and for the resulting motion of the ions. The shape of the potential, ϕ_{xyz}, developed within the trap when the electrodes are coupled to the RF and DC potentials as indicated above is described by

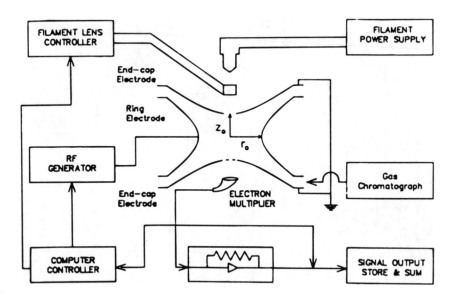

$$\phi_{xyz} = (U - V\cos\Omega t)\frac{(x^2 + y^2 - 2z^2)}{(r_0^2 + 2z_0^2)} + \frac{U - V\cos\Omega t}{2} \qquad (1.1)$$

where U represents the maximum DC potential and V the maximum RF potential applied between the ring and the end-cap electrodes, Ω is the angular frequency of the RF drive potential, r_0 is the internal radius of the ring electrode and $2z_0$ is the closest distance between the two end-cap electrodes. For the perfect quadrupole field, the arrangement of the electrodes corresponds to the case in which $r_0^2 = 2z_0^2$, although it has been found that a different relationship may give better performance in practice (see below). The oscillation of the RF potential causes the field to reverse in direction periodically so that the ions are alternately focused and defocused along the z-axis, and *vice versa* in the radial plane.

The resulting motion of the ions within the trap may be visualized with the aid of a mechanical model, designed by Paul,[5] in which the potential within the trap is represented by a saddle-shaped surface mounted on a rotating turntable (see Fig. 1.5). When the surface is stationary and a ball bearing is placed in the center as shown, clearly it will roll off down one of the sloping sides, representing an *unstable* ion trajectory. However, if the surface is first rotated at the correct angular frequency then, before the ball has had a chance to roll off along one of the *downward* surfaces, an *upward* surface will have rotated to a point where the ball is

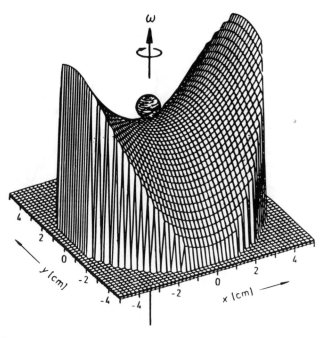

FIGURE 1.5

Diagram of a mechanical model representing the instantaneous saddle-shaped potential surface acting within an ion trap, together with an ion shown as a solid sphere. The field is focusing in the x-direction and de-focusing in the y-direction. If the solid surface is rotated about the vertical axis at the correct frequency then the ball remains trapped on the surface and describes a trajectory similar to that shown in Fig. 1.6. Reproduced with permission from Ref. 5.

arrested and focused back to the center. In this way, the ball remains trapped on the surface in a *stable* trajectory, which has the shape of a Lissajous' figure. A photograph of two circuits of a charged aluminium microparticle contained in a quadrupole ion trap[6] is reproduced in Fig. 1.6.

Returning to our more rigorous treatment, the force, $\vec{F}$, acting upon an ion of mass m and charge ze is given by

$$\vec{F} = -ze \cdot \nabla\phi = m\vec{A} \qquad (1.2)$$

from which, with Eq. (1.1), the forces acting upon the ion in each of the perpendicular directions are given by

FIGURE 1.6
A Lissajous figure created by an
illuminated electrically-charged
aluminum microparticle suspended in
an RF quadrupole trap. Reproduced
with permission from Ref. 6.

$$\left(\frac{m}{e}\right)\ddot{x} + (U - V\cos\Omega t)\,\frac{x}{(r_0^2 + 2z_0^2)} = 0 \qquad (1.3)$$

$$\left(\frac{m}{e}\right)\ddot{y} + (U - 1\ V\cos\Omega t)\,\frac{y}{(r_0^2 + 2z_0^2)} = 0 \qquad (1.4)$$

$$\left(\frac{m}{e}\right)\ddot{z} - 2(U - V\cos\Omega t)\,\frac{z}{(r_0^2 + 2z_0^2)} = 0 \qquad (1.5)$$

It will be noted that none of these expressions contains cross-terms between x, y, and z, with the result that the motion may be resolved into each of the perpendicular coordinates, respectively; furthermore, the forces acting upon the ion depend linearly upon the displacement, x, y, or z. The x- and y-components are identical and may be treated independently provided that we ignore any angular momentum which the ions may have around the z-axis. Because of the cylindrical symmetry, the x- and y-components are often combined to give a single radial r-component using $x^2 + y^2 = r^2$.

The z-component of motion is out-of-phase by half a cycle with respect to the x- and y-motion (hence the minus sign), and the factor of two arises because of the asymmetry of the device brought about by the need to observe the Laplace condition $\nabla^2\,\phi = 0$ when applied to Eq. (1.1). These equations are all examples of the Mathieu equation, which has the generalized form

$$\frac{d^2u}{d\xi^2} + (a_u - 2q_u\cos2\xi)u = 0 \qquad (1.6)$$

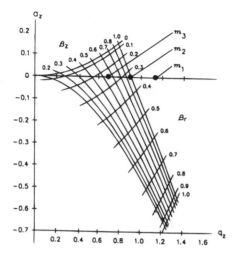

FIGURE 1.7
Stability diagram for the ion trap plotted in (a_z, q_z) space (see text for definitions). The points marked m_1, m_2, and m_3 ($m_1 < m_2 < m_3$) refer to the (a_z, q_z) coordinates of three ions: m_1 has already been ejected and detected, m_2 is on the point of ejection and the species m_3 is still trapped.

where

$$u = x, y, z \tag{1.7}$$

$$\xi = \frac{\Omega t}{2} \tag{1.8}$$

$$a_z = -2a_x = -2a_y = -\frac{16eU}{m(r_0^2 + 2z_0^2)\Omega^2} \tag{1.9}$$

and

$$q_z = -2q_x = -2q_y = \frac{8eV}{m(r_0^2 + 2z_0^2)\Omega^2} \tag{1.10}$$

Thus the transformations, Eqs. (1.7)–(1.10), relate the Mathieu parameters a_u and q_u to the experimental variables and also to the *time* variables Ω and t. The a_u and q_u parameters are quite fundamental to the operation of the ion trap since they determine whether the ion motion is stable or unstable. The diagram shown in Fig. 1.7 (which is actually only a small portion of a much larger family of curves) defines the areas within which the axial (z) and radial (r) components of motion are stable; the region of overlap indicates the (a_z, q_z) coordinates corresponding to those ions which are held in the ion trap. Clearly, when the values of U and V are held constant and the value of the drive frequency Ω is gradually reduced, both a_z and q_z will increase to such an extent that the (a_z, q_z) *working coordinate* of the ion will cross one of the boundaries so that the ion

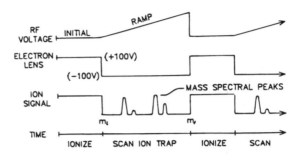

FIGURE 1.8
Simple timing sequence (the *scan function*) for the ion trap mass spectrometer. Reproduced with permission from Ref. 3. Copyright 1984 Finnigan Corporation. All rights reserved.

trajectory becomes unstable. In terms of the mechanical model described earlier, this reduction of the drive frequency Ω corresponds to slowing down the frequency of rotation of the surface to such an extent that the ball is no longer prevented from falling out along one of the downward surfaces.

A similar effect can be achieved with the ion trap when the values of U and/or V are increased while Ω is held constant. For the acquisition of a simple mass spectrum with the commercial ion trap, as indicated in Section IIA, U is set at 0 and V is increased linearly so that the ion trajectories become unstable successively in order of increasing m/z. When q_z reaches the value of $q_{eject} = 0.908$ along the $a_z = 0$ line, the ion motion remains stable in the radial plane whereas the axial component of the trajectory becomes unstable so that the ions are ejected from the trap. The timing sequence for the operation of the ion trap is represented by the *scan function* shown in Fig. 1.8, which is a profile of the variation of V with time. First, there is a period of *ca.* 1 ms when the amplitude of the RF potential is reduced to 0 to *clean out* the trap, after which the trap is filled with ions (either by internal or external ionization). During this process, the value of V is held constant for *ca.* 3 ms, generally at a value where ions of m/z 15 and below have values of $q_z > q_{eject}$ and are, therefore, not trapped. The RF potential is then ramped rapidly up to the *start mass*, i.e., the lowest value of m/z selected for the spectrum, after which it is increased at a slower rate, usually corresponding to a scan speed of 180 μs per atomic mass unit for a singly-charged ion. Each sweep of the scan function is called a *microscan* and, usually, a number of microscans (e.g., 10) are summed to give a *macroscan* which is then displayed or stored in a data file.

Before leaving the theory of the trap, we note that superimposed upon the stability diagram shown in Fig. 1.7 are the so-called iso-β lines. The parameter β appears in the solution to the Mathieu equation (1.6), and defines the long-period *secular* frequency of oscillation of the ion, as

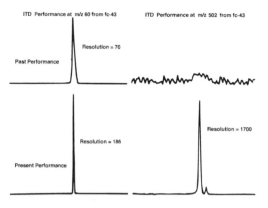

FIGURE 1.9
The effect of approximately 10^{-1} Pa Torr of helium pressure within the ion trap upon three mass spectral peaks formed in the electron ionization of perfluorotributylamine (*FC-43*): past performance = no helium; present performance = added helium. Reproduced with permission from Ref. 7.

exemplified by Fig. 1.6. The use of this parameter becomes evident in Section VI, and it is especially important when utilizing the ion trap for collision-induced decomposition experiments in tandem mass spectrometry applications (see Chapter 2).

III. SOME PRACTICAL IMPROVEMENTS IN ION TRAP OPERATION

The preceding section described the underlying theory of operation of an ideal ion trap having perfect quadrupolar geometry. In developing the practical, commercial instruments several important modifications have been made to this basic mode of operation in order to improve the performance. This section considers each of these refinements in turn.

A. The Use of Helium *Buffer* Gas

It was stated in Section II.A that the pressure within the mass spectrometer analyzer is generally maintained at a sufficiently low value as to minimize collisions between the ions and background gas molecules. It comes initially as a surprise, therefore to learn that the optimum conditions for operating an ion trap mass spectrometer occur with a residual pressure of helium in the range 10^{-2} to 10^{-1} Pa. In fact, under these conditions both the sensitivity and the resolution of the instrument are improved, as shown in Fig. 1.9. The explanation for these effects (which in spectrometry generally tend to work in opposition to each other) is that the light helium atoms arrest the motion of the heavier ions, removing their kinetic energy and causing them to migrate towards the center of the trap. Consequently, when being ejected, the ions tend to be starting from essentially the same position along the axis of the trap (leading to

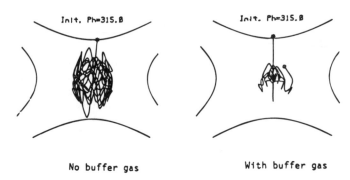

No buffer gas   With buffer gas

FIGURE 1.10
Computer-modelled trajectory plots of the motion of m/z 69 ions in the $(r - z)$ plane of
the ion trap with and without added helium buffer gas. Reproduced courtesy of W. J. Fies,
Jr., Finnigan Corporation.

improved resolution) and with minimum radial dispersion, forming a
tightly focused beam which passes cleanly through the holes in the end-
cap electrode (resulting in greater sensitivity). This behavior can be seen
from the computer simulations of an ion trajectory both with and without
collisional damping reproduced in Fig. 1.10.

B.   The Use of a *Stretched* Geometry

In presenting the theory of operation of the ion trap in Section II.B,
we noted that for a *perfect* quadrupole field $r_0^2 = 2z_0^2$. However, it has
been announced[8] that the commercial traps have a *stretched* geometry, in
which the shapes of the end-caps and the ring electrodes are retained,
but the distance between the end-caps has been extended by some 11%
beyond the *ideal* value (see Fig. 1.11). This change was made because,
during the early stages in the development of the ion trap mass spectrometer, it was discovered that the assignment of m/z-values to the ejected
ions could, apparently, be compound-specific, and an intensive search for
a solution led to the empirical optimization of the dimensions as above.
This modification to the design remained a *trade secret* until the announcement, and further details of this aspect of the evolution of the trap are to
be found in Chapter 4 of Volume 1.[4]

A more detailed discussion of the effect which *stretching* has upon
the nature of the electric field is beyond the scope of this introductory
chapter. Essentially, departure from the pure quadrupole geometry leads
to the presence of higher order nonlinear field components which influence, therefore, the motion of the ions in quite significant ways. This effect
has been exploited very successfully by Franzen, who has developed a
number of ion traps in which the shapes of the end-cap electrodes have

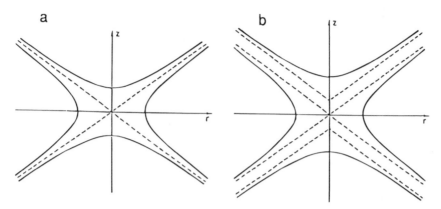

FIGURE 1.11
Section through the $(r - z)$ plane showing the electrode arrangement in a trap with (a) *pure* quadrupolar geometry and (b) the *stretched* geometry.

been modified in order to superimpose closely-specified contributions of the nonlinear fields upon the fundamental quadrupole component. A full account of this work is presented in Volume 1, Chapter 3.[4]

C. The *Segmented* Scan Function

We have seen from Eq. (1.10) that at a constant amplitude of the RF potential, V, the q_z coordinate of an ion depends inversely on the m/z-value; thus, after ion creation or injection but before analysis, the different types of ions are trapped at different values of q_z. Furthermore, it is found experimentally that the efficiency of trapping depends upon the value of q_z. As a result, in a mass spectrum recorded with an ion trap there could be considerable instrumental distortion of the spectral intensities, especially when the spectrum encompasses a wide range of mass/charge ratios. For this reason, the standard ion trap software supplied for analytical applications utilizes a *segmented* scan function in which the maximum range of m/z (10 to 650 u) is split into four separate portions, as illustrated in Fig. 1.12. Here we see that there are up to four separate ion creation/ion ejection sequences which are coupled together sequentially; the data are recorded in such a way that in the visual display of the complete mass spectrum the *joins* are invisible, and one continuous trace is observed. The various ranges of m/z covered by each segment are shown in the diagram; if a shorter scan range is required, then the scan function is simply truncated at the appropriate point. For example, if the scan range desired was m/z 100 to 300, then the scan function would start at $V = 0$ (initial cleanout) for segment II and finish one quarter of the way through the ramped portion of segment III.

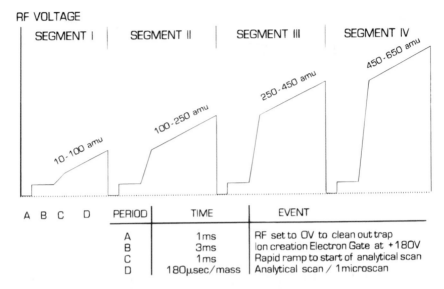

PERIOD	TIME	EVENT
A	1 ms	RF set to 0V to clean out trap
B	3 ms	Ion creation Electron Gate at +180V
C	1 ms	Rapid ramp to start of analytical scan
D	180μsec/mass	Analytical scan / 1 microscan

FIGURE 1.12

Diagram showing how the scan function for the complete mass range of the ion trap mass spectrometer (10 to 650 *u*) comprises four segments, each with its own ion creation/ion ejection sequence.

Clearly a vital feature of any mass spectrometer, especially one used for routine chemical analysis, is that the spectra recorded should be reproducible and should correspond to standard, library mass spectra of the same compounds recorded on different types of instruments under essentially the same conditions of ionization. Mass spectrometers are calibrated, therefore, with standard reference compounds, both to ensure that the m/z scale is set accurately and to give ion peak relative intensities which accord with the accepted values under the specified conditions. The reference compound generally employed for the ion trap is perfluorotributylamine (*FC-43*) since this generates ions over the appropriate range of m/z and is both clean and easy to handle within the vacuum system. With the segmented scan, tuning the ion trap involves optimizing the values of V for each of the four ionization periods so as to achieve the same relative intensities of four selected reference peaks (m/z 69, 131, 264, and 502) as is observed in the generally-accepted standard electron ionization mass spectrum of the compound. An alternative approach to tuning the segmented scan function is described in the following section.

D.   Automatic Gain Control (AGC)

Despite the improvements brought about by the use of the buffer gas, the *stretched* geometry and segmented scan functions, the early mass spectra reported with the ion trap showed significant distortions as the sample concentration was changed. Clearly, these distortions were unsatisfactory for analytical applications such as GC/MS, where quantitative integrity of the output as the analyte peaks are being eluted is essential. In particular, there could be a significant mismatch between the electron ionization (EI) mass spectra recorded with the ion trap compared to those obtained with *standard* instruments and held in mass spectral library collections.

The problems are essentially twofold. On the one hand, high sample concentrations combined with significant trapping times of, say, several milliseconds, lead to the occurrence of ion/molecule reactions, thus changing the identities of the ions being analyzed and also causing a loss of quantitative response. Secondly, the buildup of ion density within the ion trap can lead to space-charge effects, substantially modifying the electric fields to which the ions are being subjected (thereby causing, for example, shifts in the positions of the boundaries of the stability diagram), resulting in changes in the mass/charge ratio assignments of the ions.

The situation may be viewed rather more quantitatively as follows. The number of ions (N) formed in the trap by electron ionization is given by

$$N = k[S]it \qquad (1.11)$$

where k is a constant, from which we see that there should be a linear dependence of N upon the sample concentration, $[S]$, the electron beam current, i, and the duration of the ionization period, t. Thus in a GC/MS application, for example, as the value of $[S]$ changes during the elution of a component of the analyte, there will be linearity of response between the ion trap signal and the sample concentration only when the ideal behavior described by Eq. (1.11) is maintained. The limits of ideality are governed by the considerations represented by Fig. 1.13, where it is evident that outside the clear rectangle (the *ideal* region) there are shaded areas where space-charge effects and ion/molecule reactions will become appreciable. The easiest way in which to remain within the ideal region as the value of $[S]$ changes is to alter the ionization time in a controlled manner using the method of AGC. The idea is to incorporate two ionization stages into the scan function, as indicated in Fig. 1.14. The first ionization time is of fixed duration, for example, 200 μs, after which ions formed from the background gases (typically up to m/z 45) are removed by mass-selective ejection, and the remaining analyte ions are detected without further mass analysis by, for example, reducing the value of V to 0. The *total ion* signal measured in this *pre-scan* is then used to calculate the optimum ionization time for the second stage in order to avoid the effects noted above. This procedure occurs each time the scan function is

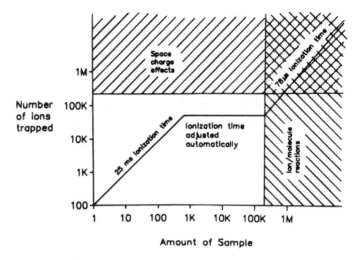

Amount of Sample

FIGURE 1.13

Ion trap response with automatic gain control (AGC). Reproduced with permission from Ref. 3. Copyright 1984 Finnigan Corporation. All rights reserved.

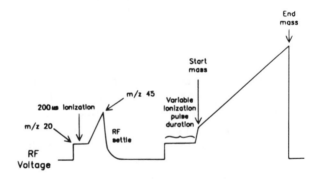

FIGURE 1.14

The AGC scan function and representations of the associated ion signals. Reproduced with permission from Ref. 3. Copyright 1984 Finnigan Corporation. All rights reserved.

repeated, and the resulting ionization times are recorded along with spectral intensities in order to normalize the data before retrieval.

In the case of the segmented scan function operated with AGC, tuning of the instrument with the reference compound is effected by adjusting the length of each of the ionization periods in the different segments so as to optimize the match between the relative intensities of the selected ion peaks and the standard spectrum. The ratios between the different ionization times in the different segments are then kept constant but their magnitudes adjusted according to the total ion signal observed during the pre-scan. This approach is clearly an extremely elegant method, in which a degree of *machine intelligence* is employed, and has established ion trap mass spectrometry as a standard quantitative analytical method.

E. Axial Modulation

A further substantial improvement in performance has been obtained through the technique of *axial modulation*. One of the inherent features of the ion trap in this mode of operation is that while the ions of lower mass/charge ratio are being *scanned* out of the ion trap into the detector, the ions of higher mass/charge ratio are still in the trap, and the space-charge potential which they contribute causes a broadening of the peaks arising from the ions being ejected. This deleterious effect on peak shape can be reduced dramatically by applying a supplementary oscillating field of *ca.* 6 $V_{(p-p)}$ at a frequency of about half that of the RF drive potential between the end-cap electrodes during the *analytical* portion of the scan function, that is, as the value of V is being ramped up. Under these conditions, just as the ions are being ejected, their secular motion enters into resonance with the supplementary field so that the ions are energized as they suddenly *come into step* and are, therefore, much more tightly bunched as they are ejected. This technique of axial modulation has been employed with spectacular success as a means of extending the mass/charge range of the trap (see below and Volume 1, Chapter 9);[4] the application of a supplementary oscillating field is also the basis of studying the collision-induced dissociation of ions in the trap (see below and Chapter 2).

IV. ION ISOLATION

The discussion up to this point has been concerned with the analysis of ions contained within the trap by successively ejecting ions in order of increasing mass/charge ratio by increasing the amplitude of the RF *drive* potential applied to the ring electrode. However, one very important additional feature of the ion trap is that there are several ways in which ions with specific m/z values may be stored selectively while the remaining ions are rejected. Such a facility is essential for tandem mass spectrometry experiments (see Chapter 2), but is also a very valuable tool in further improving the sensitivity of the trap towards low abundance species present in a larger analyte matrix (by eliminating space-charge saturation effects due to the presence of excessive numbers of *background* ions), analyses by selective chemical ionization, and the study of ion/molecule reaction kinetics in general.

Chapter 2 contains a more detailed exposition of the methods available for ion isolation, and only the simplest approach will be mentioned here.

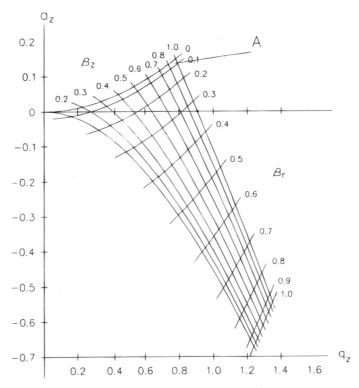

FIGURE 1.15

Stability diagram for the ion trap plotted in (a_z, q_z) space (see text for definitions). If the amplitudes of the DC and RF voltages are adjusted so that the (a_z, q_z) coordinates for a selected ion fall within the region marked A, then only this species will remain trapped.

So far we have considered only experiments in which the magnitude of the DC potential, U, applied to the ring electrode is 0. However, if we now consider the superposition of a negative DC voltage on top of the RF potential, the effect is for the (a_z, q_z) coordinate for the ion to move into the upper region of the stability diagram (Fig. 1.7). If we choose values of U and V so that the working coordinate lies at the point marked A on Fig. 1.15, then only the ions with specified values of m/z will retain stable trajectories, those with lower mass/charge ratios will be lost through axial instability beyond the right-hand ($\beta_z = 1$) boundary, and the higher m/z-value ion trajectories will become unstable in the radial direction beyond the left-hand ($\beta_r = 0$) boundary. The basis of this method is exactly the same as the principle underlying the operation of the quadrupole mass filter. As with the quadrupole, provided the value of V has been selected properly, the resolution of the isolation process depends upon the ratio U/V. This new method of operating the ion trap may be

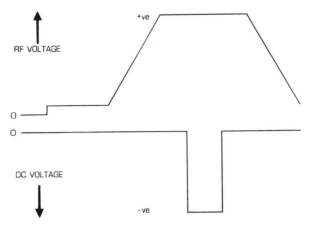

FIGURE 1.16
Partial scan function showing the application of DC and RF voltages in order to trap a single (or narrow range of values of) m/z (see also Fig. 1.15).

represented by modifying the scan function in the manner shown in Fig. 1.16.

V. CHEMICAL IONIZATION

We have already seen in our discussion on automatic gain control (Section III.D) that conditions can exist in the trap where the ions may undergo reaction with any molecular species which may be present. While this effect may be undesirable when trying to perform quantitative conventional electron ionization analyses, like many phenomena in mass spectrometry, it can also be turned to advantage.

As a result, under the correct conditions, chemical ionization (CI) mass spectrometry may be performed readily with the ion trap. We may take as a simple example methane-chemical ionization, in which the reagent ions $CH_5^{+\bullet}$ are formed by reaction between $CH_4^{+\bullet}$ and neutral CH_4. In the CI sources of conventional beam transport mass spectrometers, high pressure (*ca*, 10^2 Pa) conditions are necessary for there to be sufficient collisions between the reactants for the reagent ions to be formed during the ion residence time of the source, typically 1 μs. However, in the ion trap, where residence times are of the order of milliseconds, one can achieve the same outcome at significantly lower pressures of methane. The formation of CI reagent ions is accomplished through alteration of the scan function, by incorporating an additional *reaction* step, as represented by the dashed line in Fig. 1.17. Thus, if methane is present at a partial pressure of *ca*. 10^{-3} Pa then, during the portion of the scan labeled A' (where the amplitude of the RF potential V is held at a low value),

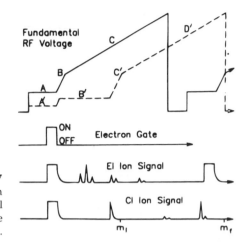

FIGURE 1.17

Scan functions for effecting electron ionization (EI = solid line) and chemical ionization (CI = dashed line) in the ion trap.

primary electron ionization of the methane occurs leading to the formation and trapping of CH_5^+ ions (Eq. (1.12))

$$CH_4^{+\cdot} + CH_4 \rightarrow CH_5^+ + CH_3^\cdot \tag{1.12}$$

but essentially no trapping occurs of ions formed directly from the analyte. The next stage (B') is to increase the RF potential so as to trap efficiently any product ions, for example, proton adducts, resulting from the reaction of CH_5^+ and the analyte (which is present at a partial pressure of *ca.* 10^{-4} Pa), followed by analysis of the resulting mass spectrum (portion D'). As with electron ionization in the ion trap, more reproducible CI mass spectra may be obtained using an analogous approach to AGC called automatic reaction control (ARC).

Further detailed discussion of CI mass spectrometry with the ion trap is to be found in Part II of this volume. However, it is worth noting at this point one important advantage which this method has over conventional high pressure chemical ionization sources, i.e., the ability of the trap to be operated mass-selectively. The reason for so doing is that a parallel ion/molecule reaction, between primary CH_3^+ ions and CH_4 to yield $C_2H_5^+$ ions, can occur concurrently under the conditions described above. These secondary ions may also react with the analyte and cause alternative, more extensive, fragmentation processes by a more energetic charge-transfer reaction, thus giving rise to a mixed CI mass spectrum. Such a complication can easily be avoided by inserting an ion isolation step, for example using the approach described in Section IV, between portions A' and B' of the scan function shown in Fig. 1.17.

VI. EXTENSION OF THE MASS/CHARGE RANGE

As currently marketed commercially, the ion trap instruments have an upper mass/charge limit ($(m/z)_{max}$) of 650 Da e^{-1}, which is determined by rearrangement of Eq. (1.10) to give

$$(m/z)_{max} = \frac{4eV_{max}}{q_{ej}(r_0^2 + 2z_0^2)\Omega^2} \qquad (1.13)$$

where q_{eject} ($= 0.908$) is taken as being the value of q_z at which ion ejection occurs and V_{max} is the maximum value (zero–peak) of the RF drive potential (*ca.* 8 kV). Thus, in order to increase the value of $(m/z)_{max}$ for a given maximum RF amplitude one may reduce the values of r_0 and/or Ω, or reduce the value of q_{eject}. Each of these approaches has been explored (see Volume 1, Chapter 1 for details);[4] the most significant results[9] have been obtained through the use of axial modulation (see Section III.E), but utilizing supplementary frequencies corresponding to ion ejection at very low values of β_z, for example, $\beta_z = 0.01$, rather than $\beta_z = 1$. This form of resonance ejection has the effect of increasing $(m/z)_{max}$ by 100 times, and the method has been verified by recording mass spectra of clusters of cesium iodide, generated using an external Cs^+ ion bombardment source, up to values in excess of m/z 45,000 Da e^{-1}. This topic is considered further in Chapter 1 of Volume 2, and Chapter 3 of this volume of this series.[4]

VII. HIGH RESOLUTION MASS SPECTRA WITH THE ION TRAP

Extension of the mass/charge range of the ion trap has led naturally to a search for means of increasing the resolution of the mass spectra obtained. With the conventional mass-selective ejection method described earlier, the commercial ion traps have a resolving power ($m/\Delta m$) of *ca.* three times the value of m/z on the *full peak width at half-maximum* (FWHM) definition. However, it has been discovered[10] that, by slowing down the speed with which the ions are *scanned out* of the trap, e.g., by a factor of 2000, and applying the axial modulation technique, resolutions in excess of 10^6 may be obtained. A detailed account of this work, together with a discussion of the associated problems of accurate mass calibration and measurement, are given in Volume 2, Chapter 1.[4]

VIII. TANDEM MASS SPECTROMETRY

Tandem mass spectrometry involves the use of successive activation/ analysis sequences in order to explore the chemistry of precursor ions which have been selected in an initial analysis stage. Since the ion trap operates in the time domain and the scan function may be readily adapted to include steps such as ion isolation, much work has been devoted to developing the trap as a tandem mass spectrometer. The essential stage is that in which collisional activation of the ions may be effected, and a number of means of achieving this have now been characterized. The first of these is the use of resonant *tickle* excitation through the application of a supplementary oscillating potential applied across the end-cap electrodes; this means of activation is analogous to axial modulation but with lower amplitude potentials so as not to eject the ions from the trap. Gas phase collisions between the ions and the helium buffer gas then lead to internal excitation and subsequent dissociation of the former into product ions which are then mass-analyzed, or mass-selected for a second activation step. A similar effect may be achieved through *boundary excitation* in which the (a_z, q_z) coordinates of the mass-selected species are held near to the $\beta_z = 0$ or $\beta_r = 0$ boundaries just within the envelope of the stability diagram so that they gain translational energy from the RF drive potential. A third method which has been reported is the use of surface-induced dissociation through impacting the ions on one of the electrodes by means of a rapid DC voltage pulse applied to one of the electrodes.

A more detailed account of tandem mass spectrometry with the ion trap is to be found in Chapter 2.

IX. APPLICATIONS

As originally conceived, the ion trap mass spectrometer was developed commercially for combination with capillary gas chromatography as a *bench-top* system (the ion trap detector, ITD™) and, indeed, at the present time this remains the principal market for the device. However, as is often the case with innovative mass spectral techniques, it is only a short time before manufacturers and/or users are extending the range of applications into other fields. Perhaps the most impressive feature of the ion trap is the speed with which it has evolved: it is just ten years since the first GC/MS systems based on this technology hit the marketplace.

Reference to the Tables of Contents in each chapter of this volume reveals how extensive the range of applications now is. Thus we see that ion traps are routine analytical tools in drug testing (Chapter 6), environmental monitoring (Chapter 12) and pesticide analysis (Chapter 13). The ease with which chemical ionization (possibly with selection of the reagent ion) and alternate chemical ionization/electron ionization (ACE, Chapter 7) may be effected, with minimal changes to the hardware, make the instrument extremely attractive for the user who is facing an unpredictable variety of GC/MS problems to solve.

As with other types of analyzers, application of mass spectrometry to analyzing the effluents from liquid chromatographs (Chapter 5), biological materials (Chapters 13 and 14) and to the direct analysis of atmosphere-borne materials (Chapter 10) involves alternative means of ionization: for the trap this requires *external* ion-creation, followed by pulsed injection into the trapping field. At the same time higher mass/charge range operation, high mass-accuracy and MS/MS capability, possibly with automated operation (Chapter 11) are becoming important. Commercial systems which incorporate such developments are still at the *engineering* stage, but the underlying technology already exists and it is only a matter of time before we see ion trap-based instruments being employed across the whole range of applications of mass spectrometry.

X. CONCLUSION

In this introductory chapter the aim has been to provide a general appreciation of the theory, design, and operation of the ion trap without overburdening the reader with rigorous mathematical details or excessive reference to the rapidly expanding primary literature on the subject.

Perhaps the most impressive features of the ion trap are its overall simplicity of operation and high sensitivity, while encompassing the potential for high mass/charge ratio range, high resolution and multistage tandem mass spectrometry normally associated with much more complex and expensive instruments. Although some of the technology needed to transform these high-performance research-level systems into products with commercially-acceptable reliability is still evolving, there is no doubt about the clear advantages to society of having such an economically-priced, versatile addition to the range of tools available to the analyst.

REFERENCES

1. See, for example, Busch, K. L.; Glish, G. L.; McLuckey, S. A. *Mass Spectrometry/Mass Spectrometry; Techniques and Applications of Tandem Mass Spectrometry.* VCH Publishers, New York, 1988; Chapman, J. R. *Practical Organic Mass Spectrometry,* 2nd Ed., John Wiley & Sons, Chichester, 1993.
2. Paul, W.; Steinwedel, H. *U.S. Patent* 1960, 2,939,952.
3. Todd J. F. J. *Mass Spec. Rev.* 1991, *10*, 3.
4. R. E. March and J. F. J. Todd (Eds.) *Practical Aspects of Ion Trap Mass Spectrometry,* Vols. 1 and 2, "Fundamentals and Instrumentation", Modern Mass Spectrometry Ion Trap series, CRC Press, Boca Raton, FL, 1995.
5. Paul, W. *Angew. Chemie.* 1990, *29*, 739.

6. Wuerker, R. F.; Shelton, H.; Langmuir, R. V. *J. Appl. Phys.* 1959, *30*, 342.
7. Stafford, Jr, G. C.; Kelley, P. E.; Syka, J. E. P.; Reynolds, W. E., Todd, J. F. J. *Int. J. Mass Spectrom. Ion Processes.* 1984, *60*, 85.
8. Louris, J.; Schwartz, J.; Stafford, G.; Syka, J.; Taylor, D. *Proc. 40th ASMS Conf. Mass Spectrometry and Allied Topics.* Washington, May/June 1992, p. 1003.
9. Kaiser, Jr., R. E.; Louris, J. N.; Amy, J. W.; Cooks, R. G. *Rapid Commun. Mass Spectrom.* 1989, *3*, 225.
10. Schwartz, J. C.; Syka, J. E. P.; Jardine, I. *J. Am. Soc. Mass Spectrom.* 1991, *2*, 198.

Chapter 2

ION TRAPS AS TANDEM MASS SPECTROMETERS

Raymond E. March, Robert J. Strife, and Colin S. Creaser

CONTENTS

0-8493-8251-3/95/$0.00+$.50

I. INTRODUCTION

Recent developments in quadrupole ion trap mass spectrometry, combined with a broadening of commercial interests in instrumental development, have led to the appearance of ion trap instruments with enhanced operational versatility. Much of the impetus for these changes is derived from the demands of gas chromatographers for enhanced mass spectrometric performance. The ready facility with which chemical ionization (CI) on the gas-chromatographic time-scale has been incorporated into ion trap scan functions appears merely to have whetted the appetite of the analytical community for further access to ion trap operational modes which, up to this time, have been restricted largely to the research laboratory. While research efforts are being devoted to improvement of ion trap performance particularly, for example, with respect to enhanced mass resolution, the realization of commercial ion trap tandem mass spectrometry combined with gas chromatography is incipient.

The practice of tandem mass spectrometry with commercial quadrupole ion traps is relatively facile, once the appropriate software is made available and minor instrumental modifications are carried out. The present methods of ionization, that is, electron impact (EI) and CI, are largely unchanged in ion trap tandem mass spectrometry compared with normal operation of the device; similarly, the final analytical scan of the instrument is unchanged. The essential new elements in the tandem mass spectrometric operation of the ion trap are those of mass selection (or ion isolation) and ion activation of the mass-selected ion species. Therefore, in this chapter, emphasis has been placed primarily on the discussion of the various methods by which ion isolation and activation can be effected and which have been investigated up to this time.

The practice of tandem mass spectrometry has been pursued very effectively over the past several decades using sector instruments and, more recently, triple stage quadrupole (mass filters) instruments. However, as the tandem mass spectrometric operation of the ion trap differs substantially from that of the earlier instruments, some discussion of the instrumental differences is presented.

A. Historical Background

The pioneering researches which J. J. Thomson carried out near the end of the 19th century are of fundamental importance to the study of mass spectrometry and of gas-phase ion/molecule chemistry. In 1897, Thomson showed[1,2] that cations and anions in the gas phase at low pressure could be separated, and that the behavior and reactions of each could be studied. In addition, Thomson built the first tandem mass spectrometer and demonstrated clearly the veracity of the statement, "When two mass spectrometers are coupled together, the whole is greater than the sum of the parts."[3] In this instrument, a beam of cations passed successively between the poles of two electromagnets which were arranged perpendicularly. Once ions had been formed and accelerated, they passed through a first field-free region which preceded the first magnet then, after passing through the first magnetic field they entered a second field-free region between the two magnets. It was this second field-free region which presented such a novel opportunity for a mass-selected beam of ions to undergo collisions with background gas molecules. The ionic products of these encounters could then be momentum-analyzed in the field of the second magnet. Virtually the entire range of possible charge permutation reactions,[4] such as charge transfer, partial charge transfer, charge inversion, and charge stripping, were demonstrated by Thomson at the turn of the century. Although Thomson's early papers are not readily available, a recent historical account by Beynon and Morgan[5] of the contributions of Thomson and others to mass spectrometry is recommended to the reader.

Thus tandem mass spectrometry came into being; yet it was some 60 years later that this technique was first used for the analysis of mixtures[6] and as a probe for structure determination of gas-phase ions.[7]

B. Elements of Tandem Mass Spectrometry

The essential three elements of tandem mass spectrometry, MS/MS, in the quadrupole ion trap are ion isolation followed, in time, by a period during which ion activation and/or ion reaction takes place, followed by a mass-selective instability scan of the charged contents of the ion trap. The first element, *ion isolation,* may be carried out in a variety of ways as discussed below. The purpose of ion isolation is to isolate within the ion trap a potential reactant ion species, or a narrow mass range of potential reactant ion species. The second element, the *ion activation and/or ion reaction period,* is a period of time when the amplitude of the RF drive potential is held constant and the isolated reactant ion species is allowed to undergo unimolecular and/or bimolecular reactions which, in the widest sense, will include cluster formation, photodissociation, collisions with surfaces, resonance excitation, and collision-induced dissociation (CID). It is during the third element, the *mass-selective instability scan,* that the secondary ionic products of such reactions of the initially isolated reactant ions are mass-analyzed. Thus tandem mass spectrometry presents an opportunity for selected-reactant ion chemistry to occur in its various forms, followed by analysis of the reaction products.

The process of tandem mass spectrometry in the quadrupole ion trap is clearly analogous to that carried out with sector mass spectrometers. For example, with a reverse geometry instrument, reactant ions are momentum-selected by the magnetic sector and pass into a field-free region in which ion chemistry can occur; the products of these reactions are energy-selected by an electrostatic sector. Sector instruments and quadrupole instruments are discussed in more detail in Section II, below.

C. Chemical Noise

In tandem mass spectrometry, which is commonly referred to by the acronym MS/MS,[8,9] the sensitivity in terms of signal/noise ratio which can be achieved is enhanced in comparison with that achievable with single-stage mass spectrometry. The signal intensity in MS/MS is reduced in the collision-activated dissociation or fragmentation process following the first mass-selective stage, yet the reduction in noise level is greater. The separate contributions of chemical noise and electronic noise are more clearly differentiated in MS/MS. The considerable reduction in chemical noise by a mass-selective stage prior to the mass-selective analytical stage affords a lower detection limit for tandem mass spectrometry despite the accompanying loss of signal intensity. In tandem mass spectrometry with the quadrupole ion trap, the loss of signal intensity may be compensated partly by prolongation of the ionization period.

D. Tandem Mass Spectrometry Literature

The first comprehensive review of the state of the art of tandem mass spectrometry appeared in 1983;[10] some 50 of the leading MS/MS practitioners contributed to this publication which is highly recommended to the reader. The review, which appeared concurrently with the announcement of the first commercial version of the quadrupole ion trap, the Finnigan MAT Ion Trap Detector, describes the general types of MS/MS applications which could be pursued with the instrumentation which was available at that time.

More recently, a further comprehensive review of tandem mass spectrometry has appeared.[11] In addition, there have been a number of reviews of ion trap mass spectrometry which have included, in varying degrees, discussions of tandem mass spectrometry.[12–20]

E. Gas Chromatography

The number of gas chromatographs in the world is probably some ten times the number of mass spectrometers and mass filters combined. Thus there is a market of considerable magnitude for relatively inexpensive mass-selective detectors for gas chromatographs. The ready acceptance of the ion trap as a mass detector for gas chromatographs has been due to its relatively low cost, simplicity of operation, high performance characteristics, ruggedness, and low maintenance costs. In addition to the observation of reproducible standard EI mass spectra of compounds eluting from a gas chromatograph, once both storage times of ions and sample pressures in the ion trap had been reduced below critical levels, CI mass spectra could be observed also with but minor plumbing and software changes. A detailed discussion of gas chromatography/ion trap mass spectrometry is given in Chapter 5.

F. The Informing Power of Tandem Mass Spectrometry

Fetterolf and Yost have applied the concepts of information theory to tandem mass spectrometry.[21] The range of each experimentally variable parameter and of the signal intensity is limited in time, space, and magnitude; hence these ranges may be represented by a finite numerical value which is known as the *informing power*. The informing power, P_{info}, of an analytical procedure can be expressed in terms of "binary digits" or bits and is given by

$$P_{info} = \sum_{i=1}^{n} \log_2 S_i \qquad (2.1)$$

where n is the number of quantities or parameters to be determined such as mass number, and S_i is the number of measurable steps for a given quantity. If m is the variable parameter and δm is the smallest distinguishable increment in m, then the number of steps of m is given by $m/\delta m$ which is, in mass spectrometry, the resolution, $R(m)$. Upon substitution into Eq. (2.1) and allowing for variation in the parameter m, the summation is replaced by an integral evaluated from m_1 to m_2

$$P_{info} = \int_{m_1}^{m_2} R(m)\log_2 S(m) \frac{dm}{m} \qquad (2.2)$$

When the resolution is constant and $S(m)$ is fixed by the detection system, Eq. (2.2) can be simplified to

$$P_{info} = R(m)\log_2 S(m)\ln\left(\frac{m_2}{m_1}\right) \qquad (2.3)$$

The value of P_{info} is the degree of merit of a given instrument; it can be increased by enhancing the terms in Eq. (2.3) or by the addition of other resolution elements. In tandem mass spectrometry, a second stage of mass selectivity corresponds to the addition of another resolvable parameter, n, such that

$$P_{info} = \int_{n_1}^{n_2} \int_{m_1}^{m_2} R(m)R(n)\log_2 S(m, n) \frac{dm}{m} \frac{dn}{n} \qquad (2.4)$$

Thus, the addition of a second resolvable parameter increases the informing power. Similarly, the addition of a gas chromatograph can increase further the informing power.

Let us consider then the informing power of a gas chromatograph combined with a tandem mass spectrometer, GC/MS/MS. Recalling Eq. (2.4) and assuming that the resolution for a capillary GC column with 1.0 × 10^5 theoretical plates (N) is constant during an analysis of 1 h duration, P_{info} for a GC/MS instrument is estimated[21] as 6.6 × 10^6 bits. The addition of a second mass spectrometric stage contributes a further factor of 10^3 so that P_{info} is *ca.* 7×10^9 bits. In tandem mass spectrometry, the experimental variables associated with CID, resonant frequency, amplitude and duration of resonant excitation, are other potential resolution elements. In addition to these measurable experimental parameters, other variables which are procedural in nature, such as chemical ionization, selection of positive or negative ions, or the choice of reactive collisions, can be used to increase the informing power of tandem mass spectrometry. However, the inability to scan these variables continuously makes it difficult to express these effects on informing power.

There are, however, some limitations of calculations of informing power such as in CID, where fragment ions must have a mass less than that of the parent ion; in this case, only one-half of the resolvable elements are available. In addition, real analytical systems are subject also to noise and interference, thereby reducing the informing power. On the other hand, the use of reactive collisions within the ion trap leads to the formation of adduct ions of mass greater than the parent ion, thereby increasing the informing power.

Section A
TANDEM MASS SPECTROMETRY

Raymond E. March

II. INSTRUMENTATION

A. Static and Dynamic Devices

Mass spectrometers which employ electric and/or magnetic fields, so as to produce a constant force on the ions under study, are referred to as *static* devices as the fields within the instruments remain static for transmission of a given ion species. In static devices, under these circumstances of constant acceleration, the ion trajectories are well defined and easily calculated. Quadrupole devices are associated with their respective type of mass analyzer; for example, quadrupole mass filters and quadrupole ion traps are referred to as two-dimensional and three-dimensional devices, respectively; these devices are referred to as *dynamic* as opposed to *static*. In the case of dynamic instruments, ion trajectories are influenced by a set of time-dependent forces which are derived from rapidly changing or *dynamic* fields and which render their trajectories somewhat more difficult to predict. Ions in such quadrupole fields experience strong focusing in which the restoring force, which drives the ions back toward the center of the device, increases linearly with displacement from the origin.

Instruments in which *dynamic* and *static* devices are combined in any constructive order are described as *hybrid* instruments.

The motion of ions in magnetic and electrostatic fields is described elsewhere,[10,11] as is the motion of ions in quadrupole fields.[20]

B. Instrument Dimensionality

The mass-resolving elements of magnetic and electrostatic sectors in static instruments, and of magnetic and/or electrostatic sectors and quadrupole mass filters in hybrid instruments, and of quadrupole mass filters in multistage dynamic instruments, are arranged physically in sequence so that ions pass in both space and time from the ion source, through field-free regions and mass-resolving sectors, and on to the detector. The length of a trajectory of an ion from source to detector varies from *ca.* 3/4 m in triple stage quadrupole instruments to several meters in multisector instruments such as those which are composed of four magnetic and electrostatic sectors; the corresponding flight times are *ca.* 10^{-4} and 10^{-5} s, respectively. In trapping devices, such as the quadrupole ion trap and ion cyclotron resonance mass spectrometer,[22] ion flight paths may be of the order of tens of meters or more, while their residence times are of the order of 0.1–1 s. In such trapping devices, the ions are restricted to a small region of space of the order of 3 cm^3; the various stages of *in situ* ion creation, reaction, and mass selection, though carried out sequentially in time, are not separated spatially. In ion traps, the trap itself serves as ion source, ion reactor and, with appropriate scanning of the amplitude of the RF drive potential, the trap functions mass-selectively so that ions are ejected in mass sequence and impinge upon a detector.

C. Multiple Stage Tandem Mass Spectrometry

While tandem mass spectrometry as practiced in its most common form utilizes but two mass-selective stages, that is, MS/MS or $(MS)^2$, the technique is not restricted to two mass-selective stages. The utilization of three mass-selective stages is denoted as MS/MS/MS or $(MS)^3$, while in the four sector instrument referred to above, mass-selective utilization of all four sectors is denoted as $(MS)^4$. $(MS)^4$ is the limit for mass-resolving sectors so far as sector instruments or hybrid instruments are concerned, the limit being imposed principally by the very low ion beam currents available at the final detector but also by the cost of the instrument. The first valid analytical application of $(MS)^3$ is possibly that of Burinsky et al.[23]

In the ion trap, tandem mass spectrometry can be pursued with many mass-selective stages; for example, there are four published reports of tandem mass spectrometry to the eighth, or higher degree, $(MS)^n$, $n \geq$

8.[24-27] Two examples are discussed further in Section V.C.1. Extended tandem mass spectrometry, (MS)^n, is effected when the mass-selective instability scan is preceded by an ion isolation procedure carried out on a single secondary product ion species. When this species is engaged subsequently in a bimolecular process with the formation of tertiary product ions, either one of these tertiary product ion species is isolated for yet further reaction, or all of the tertiary product ions are mass-analyzed in a mass-selective instability scan. Normally, the final mass-selective stage yields a mass spectrum of the ions remaining in the ion trap although, after several stages of tandem mass spectrometry, such a mass spectrum may consist of but a single ion species. The limiting value of n in (MS)^n is reached when only a single ion is detectable.

III. ELEMENTS OF ION TRAP THEORY

Theoretical treatments of the confinement of charged particles in a quadrupole field are to be found in several publications; reproduction of such treatments is not attempted here. These treatments are all referenced in one or more of the three monographs in which progress in the quadrupole ion trapping field has been reviewed.[12,20,28] However, it should be noted that only the theoretical treatment given in Ref. 20 takes into account the extended separation of the end-cap electrodes in commercial ion traps. For the convenience of the reader, a brief discussion of the origin and form of the stability diagram is presented here, together with amended definitions of the stability parameters and an explanation of the dependence of the fundamental secular frequencies of ion motion on the stability parameters.

A. Stability Diagram

The stability diagram shown in Fig. 2.1 is depicted in a_z, q_z space where a_z and q_z are the axial (z-direction) stability parameters; the stability diagram represents the region of overlap in which ion trajectories are stable both axially and radially (r-direction). In order to represent the two regions of stability, that is, axial and radial stability, on a single diagram, the relationship between the axial and radial stability parameters must be known. The stability parameters a_u and q_u, where $u = r,z$, are given in Eqs. (2.5) and (2.6), respectively.

$$a_r = \frac{8\,eU}{m(r_0^2 + 2z_0^2)\Omega^2}\,;\quad a_z = \frac{-16\,eU}{m(r_0^2 + 2z_0^2)\Omega^2} \qquad (2.5)$$

$$q_r = \frac{-4\,eV}{m(r_0^2 + 2z_0^2)\Omega^2}\,;\quad q_z = \frac{8\,eV}{m(r_0^2 + 2z_0^2)\Omega^2} \qquad (2.6)$$

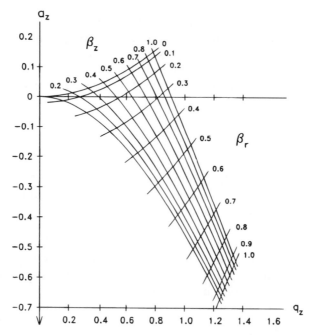

FIGURE 2.1

Stability region near the origin for the quadrupole ion trap showing the iso-β_r and iso-β_z lines. (Reproduced by permission of John Wiley & Sons)

where m is the mass of the ion, e is the electronic charge, r_0 is the radius of the ring electrode, $2z_0$ is the distance of closest approach of the end-cap electrodes, U and V are, respectively, the amplitudes of the direct (DC) and alternating (RF) components of Φ_0^R which is the potential applied to the ring electrode, and Ω is the angular frequency of the alternating component. Once the values of the trap dimensions, r_0 and z_0, and the experimental parameters Ω, U, and V are known, the values of the stability parameters a_u and q_u may be calculated for any given ion species, normally assumed to be singly charged. The working point of an ion species is defined in terms of its a_z, q_z coordinates in the stability diagram. Under all conditions of ion trap operation, the locus of the working points of all ion species is a straight line passing through the origin of the stability diagram.

The boundaries of the stability diagram are defined by $\beta_u = 0$ and 1, as shown in Fig. 2.1; β_u is a complex function of the stability parameters a_u and q_u.[12,20] The upper apex of the stability diagram is the intersection of the $\beta_r = 0$ and $\beta_z = 1$ boundaries and is located at $a_z = 0.149998$, $q_z = 0.780909$. The intersection of the q_z axis with the $\beta_z = 1$ boundary occurs at $q_z = 0.908$. When the working point of a trapped ion is moved across a stability boundary, the trajectory of the ion becomes unstable in the direction indicated by the boundary. The fundamental axial and radial secular frequencies of ion motion, ω_z and ω_r, are given by $\beta_z\Omega/2$ and $\beta_r\Omega/2$, respectively. Thus the values of ω_z and ω_r are dependent upon the working point for a given ion species. Therefore, as the working point of an ion species is changed, so are its fundamental axial and radial secular frequencies.

B. Scan Functions

Under microcomputer control, the great versatility of the ion trap can be used to advantage. The amplitudes of the RF drive and DC potentials applied to the ring electrode can be varied rapidly and precisely so as to move the working points of trapped ion species both within and beyond the stability boundaries of the stability diagram. An auxiliary RF potential applied to the end-cap electrodes can be varied, with appropriate instrumentation, in both frequency and amplitude for both resonance excitation to the point of ion ejection and for axial modulation[29] during an analytical scan of the RF drive potential. An auxiliary RF potential having several frequency components, for each of which the amplitude can be varied, can now be applied to the end-cap electrodes.[30] (See Section IV.B.6.)

The temporal variations of the potentials applied to all three electrodes, to the gated electron beam, and to the detector can be displayed in a diagram called the scan function. Thus, a scan function is a visual representation of the sequence of program segments in the software which controls ion trap operation. It is possible to discern from the scan function the movement of the working point within the stability diagram of each ion species contained within the ion trap. In some ion isolation methods, the movement of the working point within the stability diagram of the ion species undergoing isolation can be quite complex and, for this reason, the movement is sometimes referred to as the "trajectory" of the working point.

IV. ION FORMATION AND ISOLATION

A. Ion Formation

The most common ionization method used in conjunction with the ion trap is EI which is carried out normally with 70 eV electrons. The question arises here as to the effect of the RF potential on the energies of the ionizing electrons and, theretofore, on the relative abundances of primary fragment ions. Very precise trajectory calculations, which have been carried out in this laboratory on ions stored in the trap (Chapter 6 of Ref. 20), have extended to the trajectories of 70 eV electrons admitted through an end-cap electrode. The electrons pass through the trap in the order of a fraction of a nanosecond and one can see only very minor curvature of the trajectories. Thus, the transit time through the trap for the electrons is so brief that a fixed RF voltage of, say, 300 $V_{(0-p)}$ (corresponding to a low-mass cut-off of m/z 30), appears to have little influence on the trajectory; hence the energies of the electrons are minimally affected also.

If the energy of an electron is increased by the action of the RF potential, then it is expected that the extent of fragmentation would be enhanced in comparison with that obtained with 70 eV electrons. Yet, in practice, EI mass spectra obtained with the ion trap are in good agreement with library EI mass spectra, provided that the storage time is not prolonged. Thus, the energy definition of the electron beam is changed little by the usual RF levels employed during EI ionization.

When ionization is effected by electron impact, both stable and metastable ions are formed. With respect to metastable ions, we have no control over unimolecular processes, and it remains to the experimenter merely to vary the time-window for observation of metastable ion decay and analysis of product ions (Chapter 5 of Ref. 12). A wide range of possibilities exists for variation of bimolecular processes in which primary ions are reactants.

Ions can be generated also by thermospray, electrospray, and chemical ionization. Thermospray is discussed in detail in Chapter 5; electrospray is discussed in Chapters 6, 13, and 14, while chemical ionization is discussed in Chapters 7 and 13. Electrospray and the quadrupole ion trap are discussed also in Chapter 11 of Ref. 20.

B. Methods for Ion Isolation

There are two general principles upon which are based the methods by which ions may be isolated within the ion trap with a resolution approaching unit mass resolution. The two principles are (1) destabilization of ion trajectories for those ions other than the ion species to be isolated by moving the working point, or a_z, q_z coordinates in the stability diagram, of such ions across a stability boundary; and, (2) ejection of such unwanted ion species by resonance excitation of the fundamental axial secular frequency components of ion motion until ion kinetic energy can overcome the trapping potential within the ion trap. Each of the various methods discussed here is based on one or both of these general principles.

1. Upper Apex Isolation

Ions can be isolated in the ion trap when the working point of the ion species to be isolated is located within the upper apex of the stability diagram. The working point may be located there either during ionization (or introduction of externally generated ions to the ion trap), in which case RF and DC potentials are applied simultaneously, or following ionization in which case RF and DC potentials are applied sequentially.

a.   Simultaneous Application of RF and DC Potentials   RF and DC potentials can be applied simultaneously such that the ratio for $-U/V$ is maintained at a constant value, and such that a narrow range of working points falls within the upper apex of the stability diagram. At constant $-U/V$, the locus of all working points is a line passing through the origin of the stability diagram and through the $\beta_r = 0$ and $\beta_z = 1$ boundaries. Mass-selective ion isolation can be effected by variation of the amplitudes of the RF and DC potentials at constant $-U/V$ ratio. Mass resolution of the isolation process is determined by the $-U/V$ ratio.

b.   Sequential Application of RF and DC Potentials   Normally, when ions are formed *in situ* by electron impact, the DC level has been set to zero so that the locus of all working points lies along the q_z axis, and the ion trap is said to operate in the total storage mode. Let us consider three ion species of similar m/z ratio, for example, m/z 144, 146, and 148, which have been formed in the ion trap by electron impact, and examine the process by which m/z 146 can be isolated. Initially, the working points of m/z 146 lies at point **A** in Fig. 2.2(a); the working points of m/z 148 and m/z 144 lie close to point **A** with that for m/z 148 to the left and that for m/z 144 to the right. All of the working points lie on the q_z axis. The RF drive potential amplitude to which these conditions pertain is shown in Fig. 2.2(b); for an ion trap with $r_0 = 1$ cm and operated at a radial frequency (Ω) of $2\pi f$ rad s^{-1} ($f = 1$ MHz), the amplitude is 225 V$_{(0-p)}$, as shown. The working point of m/z 146 is then moved to point **B** ($q_z = 0.78$) by ramping the RF drive potential amplitude to 1409 V$_{(0-p)}$; the working points for m/z 148 and m/z 144 now lie close to point **B**. A DC potential of -135 V is then imposed on the ring potential which moves the working point of m/z 146 to point **C** in the upper apex; during the imposition of the DC potential, the working point of m/z 148 has been moved across the $\beta_r = 0$ boundary while that of m/z 144 has been moved across the $\beta_z = 1$ boundary. The working points are collinear and lie on a line through the origin of the stability diagram. However, in moving the working points of these ions across stability boundaries, the trajectories of m/z 148 and m/z 144 have been destabilized radially and axially, respectively, and these ions are lost concurrently from the ion trap. The process of trajectory destabilization can occur in microseconds. Upon removal of the DC potential and reduction of the RF drive potential amplitude, the working point of m/z 146 is restored to point **A** and this species is now isolated within the ion trap.

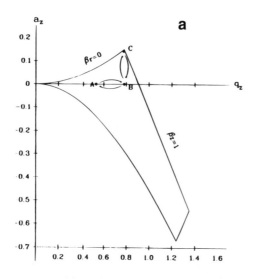

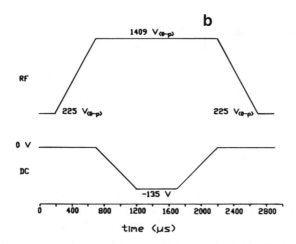

FIGURE 2.2
(a) Stability diagram for the quadrupole ion trap showing the changes in location of the working points for an ion undergoing isolation, (b) Scan function for ion isolation using RF and DC potentials. The ordinate in (b) shows the relative voltage amplitudes (not to scale), and the abscissa shows the time scale of the ion isolation process. (Reproduced by permission of Elsevier Science Publishers B.V. from the *International Journal of Mass Spectrometry and Ion Processes*, Vol. 118/119, 1992.)

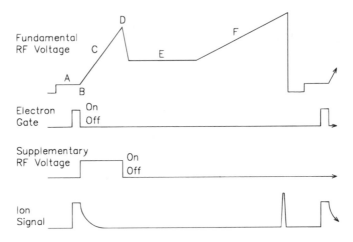

FIGURE 2.3
Scan function for a single ion species isolation mode in which the RF amplitude is varied concurrently with the activation of a frequency synthesizer in resonance with ions one mass unit heavier than the selected ion species. (Reproduced by permission of John Wiley & Sons)

Apex isolation was demonstrated by Fulford and March[31] in 1978; Strife and co-workers,[32,33] have described apex isolation for the first time in a commercial Ion Trap Mass Spectrometer (ITMS™) with respect to the tandem mass spectrometry of prostaglandins. A simulation study of apex isolation (or concurrent isolation) has been carried out for a collision-free system and shows clearly loss of m/z 144 axially and m/z 148 radially.[34]

2.   RF-Ramp/Frequency Synthesizer

Ion isolation can be effected by a combination of a ramp of the RF drive potential amplitude and a frequency synthesizer as shown in Fig. 2.3.[35] Following ionization, the RF voltage amplitude is set, **A,** to define the low mass limit for stored ions. When the RF amplitude is ramped, **C,** the trajectories of ions having lower m/z ratios than that of the selected ion species which is to be isolated are destabilized; concurrently, a frequency synthesizer is activated, **B,** in resonance with ions one mass unit heavier than the selected ion species. The duration of excitation (as determined by the ramping rate) and amplitude of the supplementary RF potential must be sufficient to remove the irradiated ions from the ion trap. The frequency synthesizer is constant during the RF ramp. Simultaneously with the destabilization of ion trajectories of lower mass ions by mass-selective instability during the RF ramp, ions having m/z ratios greater than that of the selected species come sequentially into resonance with the frequency synthesizer and are ejected from the ion trap. The upper m/z ratio limit of ions expelled from the trap, **D,** is equal to $\alpha m/z$, where

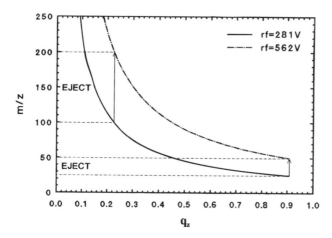

FIGURE 2.4
Diagram for selective ejection of ions at m/z 100–200 *via* the simultaneous RF-ramp/single-frequency resonance ejection mode. (Reproduced by permission of Elsevier Science Publishers B.V. from the *Journal of the American Society for Mass Spectrometry*, Vol. 2, 1991.)

α is the ratio of the maximum value of q_z for ion stability, that is, $q_z = 0.908$, to the value of q_z for the selected ion species as determined by the RF voltage amplitude during ionization. Upon completion of the RF ramp, the selected ion species remains isolated within the ion trap.

Another approach has been demonstrated[36] in which the low mass ions are ejected resonantly with a supplemental frequency of 460 kHz while the RF amplitude is ramped upwards; the scan rate is slowed as the $[M - 1]^+$ is approached. The RF amplitude is dropped slightly and a multifrequency wave form (generated by a Varian Wave-Board) is applied to the end-cap electrodes so as to eject resonantly all of the high mass ions simultaneously. Again, the selected ion species remains isolated within the ion trap.

3. Simultaneous RF-Ramp/Single-Frequency Resonance Ejection

An isolation method has been described in which all ions of m/z ratio lower than a given ion are ejected together with ions of higher m/z ratio up to such a value above which there exist no further ions. The method involves an initial RF amplitude which is ramped to a higher value simultaneously with resonance ejection at a fixed frequency.[37] The calculations shown in Fig. 2.4 were obtained with $r_0 = 1.0$ cm and $\Omega/2\pi = 1.1$ MHz. The RF voltage amplitude is set initially to a low-mass cut-off of 25 Da per charge, that is, 281 $V_{(0-p)}$; at this stage, the working point for m/z 25 lies on or close to the $\beta_z = 1$ boundary. The $\beta_z = 1$ boundary intersects the q_z axis at $q_z = 0.908$. An auxiliary frequency is applied at a frequency corresponding to resonance ejection at $q_z = 0.227$; thus, as $q_z m$ remains

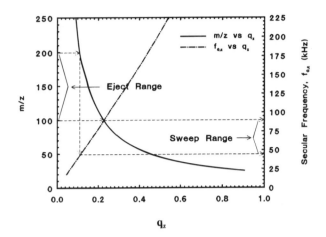

FIGURE 2.5
Diagram for selective ejection of ions at m/z 100–200 *via* the simultaneous swept-frequency resonance ejection/fixed RF mode. (Reproduced by permission of Elsevier Science Publishers B.V. from the *Journal of the American Society for Mass Spectrometry*, Vol. 2, 1991.)

constant at constant V, the ion of m/z 100 is just on the point of being or is being resonantly excited. The RF drive potential amplitude is then ramped to 562 $V_{(0-p)}$; at this juncture, m/z 50 is on the point of ejection at $q_z = 0.908$ and m/z 200 is on the point of resonant ejection at $q_z = 0.227$. During the RF ramp, all the ions in the range m/z 25 to 50 were ejected by mass-selective instability at the $\beta_z = 1$ boundary, and all ions in the range m/z 100 to 200 were ejected resonantly, so that, at the conclusion of the RF ramp only ions in the range m/z 50 to 100 remain in the ion trap, assuming the absence initially of ions with $m/z > 200$.

4. Simultaneous Swept-Frequency Resonance Ejection/Fixed RF

A wide range of masses can be ejected by sweeping the frequency of the auxiliary RF potential while maintaining the RF potential at a fixed amplitude. In this manner, the auxiliary frequency comes into resonance with the axial secular frequencies of ion species, in turn, so that the ions are ejected. The operational diagram for the ejection of ions in the range m/z 100 to 200 in this mode is shown in Fig. 2.5. In this figure, m/z is plotted as a function of q_z. On the right-hand ordinate is plotted the fundamental axial secular frequency, $f_{0,z}$, which varies inversely (to a good approximation) with mass/charge ratio. The variation of $f_{0,z}$ with q_z is plotted as a broken line in Fig. 2.5; note the slight curvature of this line. The RF trapping potential is set at an amplitude of 281 $V_{(0-p)}$ which corresponds to a high-pass $m/z_{cut-off} = 25$. The amplitude of the RF potential determines the location of the parabola shown in Fig. 2.5; note that the lower terminal point of this parabola is located at m/z 25 and $q_z =$

0.908. The m/z scale is related to the $f_{0,z}$ scale by the parabola in Fig. 2.5 in the following way. A dashed line is drawn from the secular frequency of 44.2 kHz until it intersects with the $f_{0,z}$ *versus* q_z line; the dashed line is then drawn vertically until it intersects with the RF-dependent parabola; this intersection occurs at a m/z value of 200 such that an $f_{0,z}$ value of 44.2 kHz corresponds to m/z 200. A secular frequency of 89.2 kHz corresponds to m/z 100; in this case, the corresponding dashed line from 89.2 kHz passes through the intersection of the $f_{0,z}$ *versus* q_z line with the RF-dependent parabola.

When the resonance frequency is swept from 44.2 to 89.2 kHz, ions within the mass range m/z 100 to 200 are ejected from the ion trap. With appropriate selection of the fixed RF amplitude and frequency-sweep limits for the resonance scan, other designated mass ranges can be ejected.

5. Consecutive Ion Isolation

a. Ion Ejection at Two Stability Boundaries. A further method for ion isolation has been demonstrated recently wherein ions are ejected in two consecutive operations.[38,39] In this method, ions of m/z ratio lower than that of the selected ion species are ejected at the $\beta_z = 1$ boundary of the stability diagram, while those of higher m/z ratio are ejected at the $\beta_z = 0$ boundary.

The scan function and movement of the working points within the stability diagram for this new process of ion isolation are shown in Fig. 2.6; the scan function corresponds to that given by Traldi and co-workers.[38] The removal of unwanted ion species is carried out in two consecutive stages. The order of ion ejection carried out by the Traldi group[37,38] is that of ions of low m/z followed by ions of high m/z. While in theory the order of ion ejection should be of no import, in practice, the order is critical.

In the early stages of the development of this consecutive method of ion isolation, it was observed that selected ion species could indeed be isolated when the $\beta_z = 0$ boundary was visited first: here, ions of higher m/z were ejected; upon visiting the proximity of the $\beta_z = 1$ boundary, ions of lower m/z were ejected, and the selected ion species remained isolated in the ion trap ready for photodissociation, for example. When the order of ion ejection was reversed, ions of lower and higher m/z were ejected as expected at the $\beta_z = 1$ and $\beta_z = 0$ boundaries, respectively, yet at the conclusion of the process ions of lower m/z were detected in the subsequent mass analysis scan.[40] It has been reported[41] that, while the working point of the ion undergoing isolation is in the vicinity of the $\beta_z = 0$ boundary, the selected ion species becomes kinetically excited and, after collision(s) with bath gas atoms, some ions dissociate. Fragment ions formed thus have working points which lie within the stability boundaries

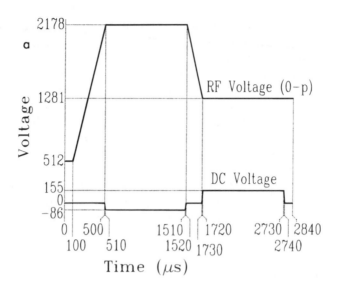

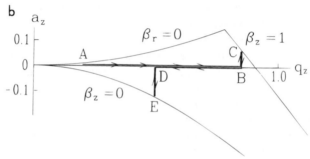

FIGURE 2.6

(a) A scan function for consecutive ion isolation using RF and DC voltages. (b) Stability diagram for the quadrupole ion trap showing the changes in location of the working point, ABCBDED, for an ion undergoing consecutive ion isolation. (Reproduced by permission of Elsevier Science Publishers B.V. from the *International Journal of Mass Spectrometry and Ion Processes*, Vol. 125, 1993.)

and are stored. This type of behavior has been described as Boundary-Activated Dissociation.[42–46]

Let us assume that the ion of m/z 207 is to be isolated from all other ion species. The "trajectory" of the working point for m/z 207 is shown schematically in Fig. 2.6, starting at point **A** for which $q_z = 0.2$. The initial position of **A** on the q_z axis is not important, although generally **A** would correspond to a value of q_z lower than 0.78 which is the threshold q_z value above which the $\beta_z = 1$ boundary can be accessed by application of a negative DC voltage to the ring electrode. The working point is moved from **A** to **B** which has a q_z value of 0.85. The working point is then moved from **B** to **C**; the position of **C** is determined by both the value chosen for **B** and the magnitude of the negative DC potential applied to the ring electrode. The coordinates of **C** are $a_z = 0.0673$, $q_z = 0.85$. The position of **C** may be defined in terms of a quantity Δm_1 which is the difference in atomic mass units (amu) between the mass of the selected ion (with its working point located at **C**) and that of a hypothetical ion of *lower* mass lying on the $\beta_z = 1$ boundary; singly-charged ions have been assumed. Δm then defines the separation, in amu, of a given working point from a boundary of the stability diagram. After a sorting time of *ca.* 1 *ms* at **C,** the working point is returned to **B** and then moved to point **D** on the q_z axis with $q_z = 0.50$. The working point is moved to point **E** in the vicinity of the $\beta_z = 1$ boundary then, after a second sorting time of *ca.* 1 ms, is returned to point **D**. The locations of **C** and **E** are critical in that they determine the proximity of the working points of the selected ion, m/z 207, to the $\beta_z = 1$ and $\beta_z = 0$ boundaries, respectively, and the degrees to which the working points for $m/z \leq 206$ and $m/z \geq 208$ lie beyond the respective boundaries.

In this study, $\Delta m_1 = 0.3$ amu such that the working point of the ion of m/z 206.7 lay on the $\beta_z = 1$ boundary. Point **E** ($a_z = -0.1213$, $q_z = 0.50$) is defined in terms of a quantity Δm_h which is the difference in amu between the mass of the selected ion and that of a hypothetical ion of *higher* mass, m/z 207.8, lying on the $\beta_z = 0$ boundary as $\Delta m_h = 0.8$ amu.

The sequence of changes in RF and DC potentials associated with consecutive ion isolation is shown schematically in the scan function of Fig. 2.6(b). A full description of the RF and DC potentials associated with a simulation study of consecutive ion isolation of m/z 207 is given elsewhere.[47]

b. Ion Ejection with Axial Modulation and at One Stability Boundary A modified form of consecutive isolation has been used in which low mass ions are ejected initially by axial modulation at $q_z = 0.86$ during the first ramp of the RF drive potential amplitude.[48] Once the low mass ions have been ejected, the remainder of the scan function is the same as that for consecutive isolation. The chief advantages compared with the method described immediately above are that an effective value for point **C** need not be determined, and the duration of the entire scan is somewhat shorter.

c. Isolation with RF-Ramp/Fixed DC In this method,[49] the RF amplitude is increased until the ion of 1 amu less than the ion to be isolated is ejected at the intersection of the q_z axis with the $\beta_z = 1$ boundary. A DC potential is than applied, and the amplitude of the RF potential is scanned down so that ions of mass/charge ratio greater than that of the ion to be isolated are ejected at the $\beta_z = 0$ boundary.

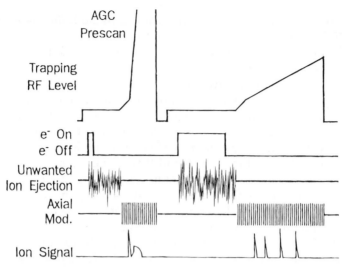

FIGURE 2.7
Scan function showing resonance ejection of unwanted ions during and immediately after the ionization period. (Reproduced by permission of S.E. Buttrill, Jr., *Proceedings of the 40th ASMS Conference on Mass Spectrometry and Allied Topics,* Washington, DC, May 31–June 5, 1992.)

6. Ion Isolation During Ionization

a. Isolation of a Single Ion Species In the above methods, the general principles of inducing trajectory instability and resonant ejection were applied to ions stored in the ion trap following ionization. Such methods clearly have implications for the number density of ions which can be isolated. While isolation is possible for ions which are present in the ion trap in low number densities amid high number densities of matrix or other ions, the number density of such ions when isolated may severely challenge the detection limits of the instrument. A novel technique has been described which seeks to overcome the problems of isolating ions present in trace quantities. The technique involves using a complex broadband waveform to eject unwanted ions during ionization.[30] The sequence of events is shown in Fig. 2.7, which depicts selected ion monitoring in the ion trap over two RF, or mass/charge, segments. As the unwanted ions are ejected throughout the ionization and cooling periods in a microscan, including the automatic gain control (AGC) pre-scan, they are removed from the reconstructed total ion current chromatogram without loss of any quantitative information. The complex waveform contains many frequencies which can be varied in amplitude so as optimize the ejection of unwanted ions and to create one or more frequency-free "windows" in which the storage of selected ions is unperturbed.[50] In this way, ions present in trace amounts in a complex matrix can be stored during ionization while all unwanted ions are ejected. The number density of trace ions which can be isolated in this manner can be enhanced by prolongation of the ionization period. An alternative procedure which has been used in ion cyclotron resonance mass spectrometry involves a stored waveform inverse Fourier transform (SWIFT)[51] with which ion trajectories may be manipulated to the point of ejection (see Chapter 4, Section V.B.2). The use of SWIFT on the ion trap for the elimination of matrix ions has been demonstrated recently.[52] A further type of multiple-frequency resonant ejection technique utilizes a noise field to excite and eject ions; the noise field can be filtered so as to isolate single ion species.[53] This technique, which is known as Filtered Noise Field (FNF), is discussed further in Chapter 4.

b. *Isolation of Two Ion Species* It should be noted that, in the above methods, only an individual or continuous range of mass/charge ratios can be isolated. However, in the case of accurate mass assignment, the isolation of two ion species differing considerably in mass/charge ratio may be required. The isolation of two such ion species can be effected by applying a complex waveform in which there are two frequency-free "windows." For example, consider the ions of m/z 131 and m/z 264 obtained by electron impact from perfluorotributylamine. When these ion species are isolated together in the ion trap, during an ionization period of duration, t, using a complex waveform with two frequency-free "windows," the ion species can be obtained in the abundance ratio of *ca.* 4:1, respectively, as shown in Fig. 2.8(a). This isolation procedure is followed with a further period of ionization but of duration $3t$ while the ions isolated initially are held within the ion trap; however, during this second ionization period, there is only one frequency-free "window" which corresponds to the ion of lower abundance, that is, m/z 264. An analytical scan following the second ionization episode shows that both ion species had been isolated in the trap and with the same abundance, as shown in Fig. 2.8(b).[50]

7. Isolation of the Ion Species Having the Highest m/z Ratio

A restricted form of ion isolation can be carried out with modified commercial ion trap detectors using the CI mode for isolation of the ion species having the highest m/z ratio. This method is particularly suitable for isolating protonated molecular species, MH^+, where M is the parent molecule. Once the normal instrumental CI procedure has been carried out, the RF drive amplitude is ramped rapidly to the low mass limit for the CI mass spectrum which is now set to just less than that of the protonated species. Provided that no cluster ions of higher m/z ratio have been formed, the protonated species remains isolated in the ion trap.

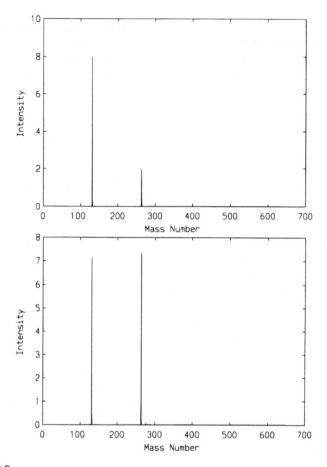

FIGURE 2.8
(a) Simultaneous isolation of m/z 131 and m/z 264 from perfluorotributylamine during ionization, (b) simultaneous isolation of m/z 131 and m/z 264 combined with isolation of m/z 264 during prolonged ionization.

Alternatively, the molecular ion $M^{+\bullet}$, if present in the normal electron impact mass spectrum, may be isolated in this manner.[54]

V. THE ION ACTIVATION AND/OR ION REACTION PERIOD

This period in the ion trap is equivalent to the transit time of an ion through a field-free region in a sector tandem mass spectrometer, in that the RF drive amplitude is held constant, or nearly so, during this period. The opportunity is presented during this period for exploration of the chemistry of mass-selected gaseous ions. However, following an ion isolation process, the ion cloud is dispersed and has a considerable range of kinetic energies; as precise control of ion trajectories is necessary for obtaining subsequently a mass spectrum with good mass resolution, it is prudent to cool the ion cloud so that it is focused close to the center of the ion trap. Once the ion cloud has been brought to the trap center, it is ready for laser irradiation, or another form of excitation, or bimolecular reactions. Thus, in this section, a discussion of collisional cooling precedes discussion of various ion activation processes and ion reactions which may occur during this period.

A. Ion Cooling

Ion cooling is that process whereby isolated ions are allowed a settling period during which they suffer collisions with helium buffer gas (Chapter 5 of Ref. 20). Normally, such a cooling period is allowed to occur when the working point of the isolated species is located on the q_z axis with 0.3 $\leq q_z \leq 0.5$. It is generally understood that during this period of collisional cooling, ions suffer a preponderance of momentum-dissipating collisions, the excursions of their trajectories from the center of the ion trap diminish in magnitude, and the ions become focused tightly near the center of the trap. In the limit, the ions form a cloud near the center in which the focusing effects of further collisions are balanced by space charge repulsion. Once the ions are in this condition, control of the trajectories of the ion ensemble is facilitated greatly in that the cloud can be dispersed readily in the central radial plane or elongated into a narrow cylinder about the z-axis by irradiation at the fundamental radial and axial secular frequencies, respectively. When ions are brought previously to a focus near the center of the trap, detection sensitivity and mass resolution are enhanced during the mass-selective instability scan. A further advantage of collisional cooling is that the entire ion cloud can be illuminated by a focused laser beam so that the degree of photodissociation is enhanced markedly compared to that obtained by irradiation of the unfocused ion ensemble within the ion trap.

It has been reported recently,[48,55] that collisional cooling can influence significantly the degree and type of fragmentation processes which occur during subsequent boundary-activated dissociation. It then appears that, while ion kinetic energies are much reduced during collisional cooling, fractional amounts of ion kinetic energy lost are deposited as internal energy in the ion, and accumulated. Upon subsequent excitation, dissociation processes having greater activation energies are favored. While this work is in the early stages, it is clear that ion cooling following ejection of low mass ions in the ion isolation process can be advantageous.

B. Ion/Molecule Reactions

Ion isolation facilitates the choice of the charged reactant in an ion/molecule encounter; however, the choice of the neutral reactant is not always as clear cut as it is for the charged reactant. There are, broadly, two types of bimolecular ion/molecule reactions, those in which the charged reactant reacts with the parent molecule and those in which a neutral reactant other than the parent molecule is involved. There are many examples of each, but let us restrict this discussion to two examples of each.

1.   Isolated-Ion/Parent-Molecule Reactions

Almost all isomers of C_5H_8 (except cis- and trans-1,3-pentadiene) could be distinguished by monitoring the products of each neutral isomer with mass-selected ions from the same isomer.[56] Such bimolecular reactions can be used to advantage in gas chromatography when ion isolation and reaction time are compatible with the gas chromatographic time scale. A second example is afforded by the endothermic ion/molecule reactions of the isolated $C_6H_4{}^{35}Cl_2{}^{+\bullet}$ molecular ion from 1,2-dichlorobenzene with the parent molecule (see Chapter 10). In this example, resonance excitation is used in order to enhance reactant ion kinetic energy.

2.   Isolated-Ion/Nonparent-Molecule Reactions

The first example is provided by the gas-phase halomethylation of several organic compounds using CH_2Cl^+ as the reactant ion;[57] the second involves methylation of the isomeric dihydroxybenzenes using the isolated dimethylfluoronium ion.[58] In each of these two examples the structures of the product ions were probed using collision-activated dissociation wrought by resonance excitation. In both of these cases, the precursor molecule of the isolated ion and the neutral reactant were present throughout the experiment and, where this arrangement can be tolerated, this arrangement is by far the more simple. An alternative approach is to separate in time the introduction of the neutral reagents using pulsed solenoid valves so as to bring about specific ion/molecule reactions,[59] or to introduce a specific collision gas.[60] The ability to change rapidly the type of neutral gas has been demonstrated for the case where one neutral CI reagent gas is introduced, utilized, and pumped away prior to introduction of a different neutral for collision-activated dissociation or as a neutral reactant.[58]

C. Ion Excitation

Ion excitation can be brought about by several methods but, as these methods are discussed in detail in Volume 1 of this series, an exhaustive review is not intended here. The five major methods of ion excitation are resonance excitation, photo-activation, boundary-activation, pulsed axial activation, and surface-activation.

1. Resonance Excitation

As discussed above, resonance excitation entails the application of a supplementary RF potential oscillating at a frequency component of motion of a mass-selected ion, usually the fundamental axial secular frequency. Energy absorbed resonantly from the supplementary RF potential appears as ion kinetic energy which causes the trajectories of the mass-selected ions to expand from the center of the ion trap. As the ion cloud expands from the center, ions experience a higher trapping field and are accelerated further. In subsequent collisions with helium buffer gas, a small fraction of ion kinetic energy is converted to internal energy which accumulates in the ion; in the limit when the excitation process is prolonged, the collisionally-activated ion either dissociates unimolecularly or escapes the trapping potential and is lost. Daughter ions produced by dissociation are trapped with efficiencies in excess of 90%.

In a study of the relevant parameters in collisional activation in the ITMS,[61] it was found that the energetics and efficiencies of the fragmentation processes were dependent on several interrelated parameters which included the frequency, amplitude and duration of excitation, the working point of the irradiated species, and the pressure and nature of the buffer gas. The extent of internal energy excitation is gauged from the activation energies, where known, of the dissociative reaction channels accessed. Maximum values for the excess internal energy have been estimated to range up to 6 eV for the tetraethylsilane radical cation, and up to 5.8 eV for dimethylphosphite molecular ions,[62] although in some systems, such as substituted aryl ketones, the maximum internal energy is limited by the presence of low-energy reaction channels.[38]

While the appearance energy of a product ion to be formed in a resonance excitation process, or cycle, may be limited to, for example, some 5 eV, the entire process can be repeated so as to access, in MS^n experiments, dissociative channels having appearance energies well in excess of the cycle limitation. A great advantage of the quadrupole ion trap is that the tandem mass spectrometric cycle of ion isolation, CID, and daughter ion analysis carried out in cycle, may be repeated but with successive generations of product ions, that is, daughter ions, granddaughter ions, etc. Thus, for example, granddaughter ions may be observed having an appearance energy of the order of 10 eV from a parent molecule, provided the appearance energies of the daughter ion from the parent molecule and the granddaughter ion from the daughter ion are some 5 eV each. The longest sequence of resonance dissociations originating from one ionization event was reported for d_{10}-pyrene, for which nine separate isolation and excitation processes, that is, $(MS)^{10}$, were carried out.[63] Using multiple activation steps, m/z 212 ($C_{16}D_{10}^{+}$) was dissociated by sequential loss of D·, D·, C_4D_2, D·, C_2D_2, D·, D·, D·, and C_3 to yield m/z 84 ($C_7^{+\cdot}$). In this sequence of 10 steps from d_{10}-pyrene, it was shown that some 29 eV of internal energy can be deposited in stages by the resonance excitation process.[62] The absorption of this large amount of energy is analogous, for example, to the step-wise increase in potential energy of a fish as it ascends a fish ladder.[15] A further example of repetitive resonance excitation is afforded by a study of the saturated sterane, cholestane;[64] mass spectra were obtained using successive stages of CID on a series of ions derived from the molecular ion of cholestane in a $(MS)^7$ experiment.

2. Photo-Activation

Photo-activation of ions confined in an ion trap has been carried out in both the multiphoton mode, using a CW CO_2 infrared laser, and the single photon mode using lasers operating in the UV and visible regions. Further discussion of photo-activation is to be found in Chapter 7 and 8.

3. Boundary-Activation

Boundary-activation[40,41] is discussed in Section IV.5.a and, at length, in Chapter 7 of Ref. 20. Boundary-activation is a recently-appreciated phenomenon wherein ions become activated, and dissociate subsequently, when the working point of the ion species is located in the proximity of, usually, the $\beta_z = 0$ stability boundary. As the boundary-activation process has been found to be both rapid and highly efficient (>90% of the mass-selected ion can be dissociated), it will probably supplant resonance excitation for CID.

4. Pulsed Axial Activation

Ion activation has been reported using fast DC pulses[65] applied to one end-cap electrode only. The amount of internal energy deposited, which can be controlled directly by variation of the amplitude of the pulse, has been estimated to exceed 10 eV in the dissociation of the n-butylbenzene molecular ion and 13 eV for dissociation of $W(CO)_6^{+\cdot}$ to W^+. Dissociation efficiencies ranged from 50% at low internal energy deposition to 5% at maximum internal energy deposition. In the high internal energy regime, collisions with an electrode surface may occur.

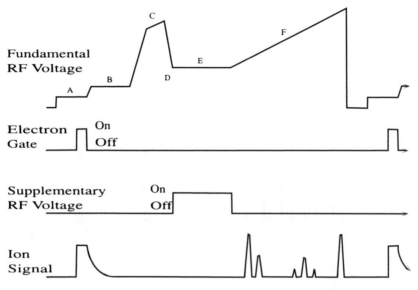

FIGURE 2.9
Operation of the ion trap in the chemical ionization MS/MS mode. Sequence of operations for obtaining a daughter ion mass spectrum of a parent ion selected from a mass spectrum produced by chemical ionization. (Reproduced by permission of John Wiley & Sons)

5. Surface Activation

Ion activation has been reported using fast DC pulses which induce ion interactions with surfaces;[66] while fragment ion yields are low at present, fragmentations requiring internal energies in excess of 15 eV were observed.

D. Chemical Ionization/Tandem Mass Spectrometry

The objective of CI is to form a pseudo-molecular ion, $[M + H]^+$, in a proton transfer reaction of the molecule of interest, M, with a CI reagent ion such as CH_5^+. A CI reagent ion is chosen usually so that the exothermicity of proton transfer is low, with the result that few fragment ions of M are formed. CI is discussed at length in Chapters 7 and 13. In this section, a typical scan function for CI/MS/MS is described, together with a single example from the author's laboratory of markedly different behavior of two similar compounds under methane CI conditions.

1. Chemical Ionization Scan Function/RF and Resonance Excitation

A typical scan function for CI/MS/MS is shown in Fig. 2.9. The amplitude of the RF drive potential is set, **A,** at a level which determines

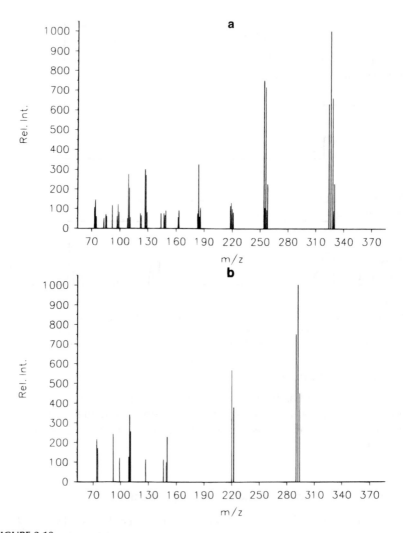

FIGURE 2.10
Mass spectra of congeners 77 and 110. (a) EI mass spectrum of congener 110, (b) EI mass spectrum of congener 77, (c) Partial CI mass spectrum of congener 110 showing the proton-ated molecular cluster, (d) Partial CI mass spectrum of congener 77 showing pronounced ion signals due to radical cations formed by charge exchange together with the protonated molecular cluster.

the low-mass cut-off; for example, in methane CI, the low-mass cut-off is set so as to confine $CH_4^{+\bullet}$ ions which are precursor ions to the CI reagent ion, CH_5^+. At **B**, CI reagent ions react with neutral sample molecules to form $[M + H]^+$ ions. It is assumed here that as the proton transfer is mildly exoergic some fragment ions may have been formed. While such

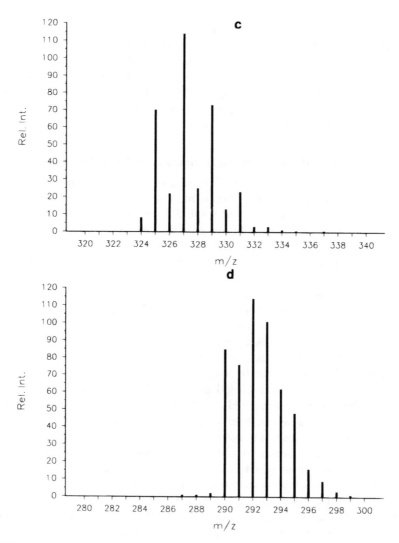

FIGURE 2.10
Continued
(b) EI mass spectrum of congener 110.

fragment ions may, in themselves, be indicative of the structure of M, we shall ignore that possibility here, and we shall proceed to remove these fragment ions in order to obtain an unambiguous daughter ion mass spectrum. During stage **C**, the RF amplitude is ramped rapidly then more slowly until the trajectories of all ions having m/z ratios less than that of $[M + H]^+$ ions are destabilized and the ions are ejected from the ion trap.

The RF amplitude is reduced then, **D**, to a value such that the q_z value for [M + H]$^+$ ions is *ca.* 0.45, whereupon the supplementary RF potential is initiated, **E**, so as to bring about CID of [M + H]$^+$ ions. Once the irradiation period with the supplementary potential is over, an analytical scan, **F**, is carried out so as to generate a daughter ion mass spectrum. Note that in this example, the removal of ions having m/z ratios greater than that of [M + H]$^+$ is not necessary.

2. An Example of Mixed Chemical Ionization/Charge Exchange

The separation of congeners 77 (3,3',4,4'-tetrachlorobiphenyl) and 110 (2,3,3',4',6-pentachlorobiphenyl) is a major analytical problem since these compounds co-elute on the DB-5 (SE-54) capillary column used most commonly for high mass resolution-gas chromatographic analysis of PCBs (Fig. 210(a) and (b)) and organochlorine insecticides.[67,68] In environmental samples, the concentration of congener 110 can so far exceed that of congener 77, that even a small proportion (for example, <1%) of 110 remaining in the non-*ortho* fraction can result in a falsely-enhanced peak area for congener 77. Yet it is congener 77 which is, by far, the more toxic.

Using a prototype Varian GC/Quadrupole Ion Storage Mass Spectrometer (QISMS) equipped with a waveform generator, EI spectra of the mass-selected molecular isotope clusters were obtained for each of congeners 77 and 110; while the reconstructed ion chromatograph showed that the two congeners co-eluted partially, the mass-selected molecular isotope clusters were separated clearly, as shown in Fig. 2.10(a) and (b). The methane CI mass spectrum of congener 110 showed a mass-selected molecular isotope cluster increased by 1 amu, in excellent agreement with the calculated abundances for chlorine and carbon isotopes (Fig. 2.10(c)). The methane CI mass spectrum of congener 77, recorded just seconds before the above CI mass spectrum, showed almost exact duplication of each chlorine isotope peak, as shown in Fig. 2.10(d). Thus 50% of the ions corresponded to pseudo-molecular ions and 50% to unchanged molecular radical cations due, presumably, to charge exchange.

This example of two similar compounds showing markedly different behavior under CI conditions is but one of several noted recently. The simultaneous observation of the products of charge exchange and proton transfer reactions is discussed in detail in Chapters 7 and 13 and, on the basis of these discussions, it appears that resolution of this problem will involve isolation of specific CI reagent ions.

E. Dynamically Programmed Scans

Let us consider a tandem mass spectrometric experiment where it is proposed to explore the effects of resonant excitation upon an isolated ion species by varying the frequency of the resonant excitation, its amplitude and duration, and the delay time prior to the onset of excitation. Such an experiment can entail the execution of many experiments each of which must be programmed separately. Todd and co-workers[69,70] have developed a technique of dynamically programmed scans wherein such experiments are strung together so that each experimental parameter is varied incrementally in sequence.

For example, the molecular ion of m/z 146 from 1,2-dichlorobenzene is to be isolated and excited resonantly. The resonant frequency is set at, for example, 150 kHz, the amplitude at 100 $mV_{(0-p)}$ and the onset delay at 2 ms; the first experiment, a microscan, is run with an excitation duration of 100 μs. The charged contents of the ion trap are then mass-analyzed. This experiment is repeated 29 times whereupon the 30 microscans are summed, averaged, and written to disk as a single macroscan. The excitation duration is incremented then by 50 to 150 μs, and the macroscan is repeated. This process, wherein the duration of resonant excitation is incremented by 50 μs each time, is repeated up to a limit of, for example, 2000 μs.

Thus far, 39 macroscans or 1170 microscans have been carried out. The excitation duration can be reset to 100 μs, and the amplitude then incremented by, for example, 50 $mV_{(0-p)}$ to an upper limit of 1600 $mV_{(0-p)}$. Following this operation, the excitation duration and amplitude are reset to their initial values and the frequency is incremented by, for example, 500 Hz to a limit of 165 kHz. Thus, at the end of this experiment some 101 macroscans (that is, 3030 microscans) have been carried out, all in a single dynamically programmed scan.

The analytical results can be portrayed on a single sheet as the variation of total ion intensity, and the intensities of up to seven ion species, or groups of ion species, with each successive macroscan. Upon examination of the results obtained in this manner, a new dynamically programmed scan function can be compiled. The application of dynamically programmed scans is described in detail in Chapter 11.

VI. ANALYTICAL SCAN RATE

In all commercial ion trap instruments, the analytical scan of the contents of the ion trap is carried out with an amplitude ramp of the RF drive potential. The rate of this ramp is fixed normally at 5555 Da s^{-1}. Recently, several advances in the development of high mass-resolution with the ion trap have been made, and the principal change which must be made in the mode of operation of the trap is to reduce the mass scanning rate. The mass scanning rate for singly-charged ions has been varied by a factor of *ca.* 8.7×10^7, that is, from 1.3×10^6 Da s^{-1} to some

0.015 Da s^{-1}. Ion ejection was brought about by resonance excitation applied at a frequency somewhat reduced from that employed in the commercial instrument ($q_z = 0.86$). The pressure of helium buffer gas was not markedly different from the normal pressure of 1 mTorr. The wide range of mass scanning rates which are accessible and which can be applied to the ion trap will permit, in the future, rapid changes in RF amplitude to the vicinity of some desired level followed by a very slow scan (with high mass resolution) over a limited mass range or ranges.

Section B
PRACTICAL MASS SPECTROMETRY

Robert J. Strife

VII. PRACTICAL ASPECTS OF TANDEM MASS SPECTROMETRY

A.   Anecdotal Introduction

Since it was pointed out recently that J.J. Thompson described the first tandem mass spectrometry (MS/MS) experiment and stated openly his optimism about the power of combining two mass spectrometers,[3] I am somewhat reticent to talk about the early days of MS/MS! My first experience with MS/MS was at Purdue in 1977, in the Department of Medicinal Chemistry. Just several hundred feet away, excitement was mounting with the analytical potential of the mass-selected ion kinetic energy, MIKES, experiment on a sector mass spectrometer of reverse geometry in the laboratory of John Beynon and Graham Cooks in the Chemistry Department.

Probably encouraged by this development, my thesis advisor at the time, Ian Jardine, suggested that we attempt to analyze a crude plasma extract for an anti-cancer drug we were working with at the time. We monitored a unimolecular first field-free region decomposition of the molecular ion on an old DuPont 490B mass spectrometer of E-B configuration. We were impressed by our ability to observe signals from just a few micrograms of material on a probe, and with a signal/noise of 10:1! With Ian's salient explanation of the experiment, a permanent interest in MS/MS was established in my mind.

After a decade and more of sector-based MS/MS experience, I was sitting at an ASMS talk by Mr. Paul Kelley (of Finnigan-MAT at the time), in my home city of Cincinnati. Over the years, I had found the low collision efficiencies observed with sector instruments to be somewhat frustrating. Paul Kelley was describing MS/MS with ion traps in which collision efficiencies of 100% were observed such that full-scan sensitivity at low-nanogram levels was possible by GC/MS/MS. I was so impressed by the data that I arranged to meet with Paul later that evening. Subsequently, I decided to write a proposal for an ion trap program at Procter & Gamble, my employer. A year later found us with the sixth commercial Ion Trap Mass Spectrometer (ITMS™) built by Finnigan and off we went. The science has been fascinating and the people in the ion trapping community have been gracious to those of us who are relative newcomers.

There are several practical concerns in conducting MS/MS experiments. Some of these include:

1. How easy is it to tune up MS/MS and how stable are particular tune-up values from day to day?

2. When sample is very limited, are there broad conditions that can give successful MS/MS results?

3. How reproducible are the spectra?

4. Are there adverse matrix effects when analyzing real-world crude extracts compared to standards?

5. How easy is it to obtain quantitative MS/MS data?

6. Can one perform selected reaction monitoring on an ion trap?

The following examples should help to clarify these issues.

B. Tuning Up for MS/MS

Whenever one is going to perform MS/MS for organic structural or analytical purposes, sufficient standard material is available usually for purposes of optimizing the MS/MS experiment. Using the ion trap, even a few micrograms of material can give a lengthy boiling profile from a heated probe. Using software menus, the time required for the optimization of parent ion isolation (RF and DC voltage values) and tickle voltage parameters (amplitude and frequency) is short compared with that required to optimize a triple stage quadrupole MS/MS experiment. Also, using a SIMS approach for ionization external to the trap (not commercially available) and injection of ions, analysis of gramicidin S at the femtomole level has been demonstrated.[19] Thus, it can be inferred that MS/MS tuning for even very small samples is practical.

C. Day-to-Day Stability

Recently, some unexpected results were obtained in one of our laboratories during the synthesis of terminally branched (i-C_3H_7) alkyl benzenes, *via* aromatic acylation and reduction. The resulting alkyl benzene was to be sulfonated and then used as a standard in environmental degradation studies of surfactants, so purity of the intermediates and final product was of utmost importance. GC/MS/MS was used over the course of several weeks to follow changes of isomer distributions as a function of synthetic conditions.

The acylating agent was pure by capillary GC analysis. In contrast, the synthesized acyl-benzene showed several peaks by capillary GC/ion trap MS. The EI mass spectra were dominated by m/z 105 (Phenyl–CO^+) and m/z 120 (McLafferty rearrangement) and showed the molecular ion at m/z 246. Capillary GC/MS/MS (m/z 246 parent ion) revealed migration of the terminal methyl group under Friedel-Craft acylation conditions.

After tuning up on a probe sample, an analysis by GC/MS/MS was performed at the 20 ng level. A typical analysis is shown in Fig. 2.11 (top trace). CID produced m/z 105 and m/z 120 as the most intense ions. The ion of m/z 228 is due to gas-phase enolization followed by rearrangement and water loss. Smaller peaks indicated the migration of the branching methyl group. The mass spectrum of peak 2 shows the first alkyl loss at m/z 157 [M $-$ H_2O $-$ C_5H_{11}]$^+$ in contrast with that of peak 3, showing the first alkyl loss from the molecular ion as $-C_3H_7$. The data suggest these compounds correspond to the 7- and 9-methyl isomers.

As samples were submitted over the next several weeks for positional-isomer analysis, it was found that by simply calling up the existing scan file for the MS/MS of the m/z 246 experiment, comparable data, day-to-day, could be obtained and more often than not, without any re-tuning of the parameters.

D. Generic Tuning

The following data demonstrate that a surrogate compound may be used for MS/MS tune-up purposes. Development of an MS/MS detection scheme was desired to allow more selective analysis of the biologically-derived inflammation mediator, LTB_4, from crude extracts of samples. LTB_4 is a 20-carbon, polyunsaturated acid (5S, 12R dihydroxy 6Z, 8E, 10E, 14Z eicosatetraenoic acid). With only 50 µg of LTB_4 on hand, a few micrograms were derivatized (methyl ester, bis-TMS ether) and an EI-GC/MS experiment was performed. An ion at m/z 383, presumed to be from α-cleavage between C_{12}–C_{13}, was proposed to lose TMSOH to give the intense ion observed at m/z 293 in the EI spectrum of the LTB_4 derivative. This ion was chosen as the parent ion.

The surrogate compound, p-coumaric acid (from our chemical stock-room), was derivatized to its bis-TMS form (MW = 308) and, as predicted, it fragmented by EI to give m/z 293 (base peak). MS/MS parameters

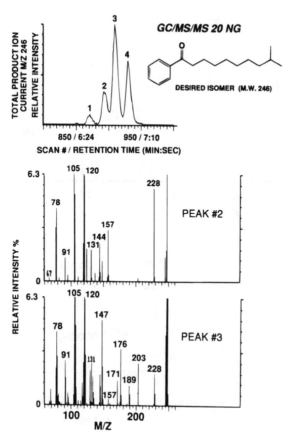

FIGURE 2.11

Top trace, GC/MS/MS analysis (20 ng, on-column injection) of an acyl-benzene isomer mixture (total product ions of the molecular ion, m/z 246). Middle trace, CID spectrum of peak 2 showing the first alkyl clip site as loss of $C_5H_{11}{}^{•} + H_2O$. Bottom trace, CID spectrum of peak 3 showing the first alkyl clip site as loss of C_3H_7.

were optimized as this derivative boiled off the solids probe. The LTB_4 derivative was re-injected (10 ng split 10:1) onto the GC column and analyzed with the same MS/MS scan function (Fig. 2.12(a)). Better than 50% observed CAD efficiency was obtained in acquiring the product ion spectrum on 1 ng of this analyte (Fig. 2.12(b)). Extensive rearrangement of the double bond structure of the parent ion (Fig. 2.12(c)), observed in the product ion spectra of other eicosanoid oxonium ion fragments, was proposed to account for the abundant and highly conjugated carbonium ion of m/z 203 ($-TMSOH$). Other fragment ions are discussed in the figure legend.

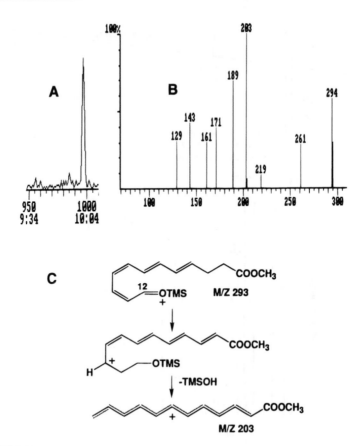

FIGURE 2.12

Capillary GC/EI-MS/MS analysis of 1 ng of derivatized LTB_4; (A) Relative intensity of the total product ion current of m/z 293 *versus* time (min). (B) Background-subtracted product ion spectrum, relative intensity *versus* m/z. (C) Structure of m/z 293 and genesis of highly conjugated m/z 203 *via* 2-H rearrangement. Other ions: m/z 189 $[-CH_2OTMS]^+$, m/z 171 $[203-CH_3OH]^+$, m/z 161 $[189-C_2H_4]^+$, m/z 143 $[171-C_2H_4]^+$, m/z 129 $[161-CH_3OH]^+$.

E.   Spectral Reproducibility

During the six years that we have been operating our ITMS, we have had occasion to analyze eicosanoids such as prostaglandins and leukotrienes many times. One of these eicosanoids is prostaglandin E_2 (PGE_2). EI analysis of PGE_2 as the methyl ester, methoxime, bis-TMS-ether derivative produces an intense ion at m/z 225, derived from C_{11}–C_{20} of the molecule. A comparison of two CID spectra of m/z 225, obtained about 17 months apart, is shown in Fig. 2.13, along with the parent ion structure. The He collision gas pressure was regulated only by flow from the capillary GC. A q_z value of 0.2 was chosen for the parent ion in each case; tickle voltage and times were optimized independently. Even the

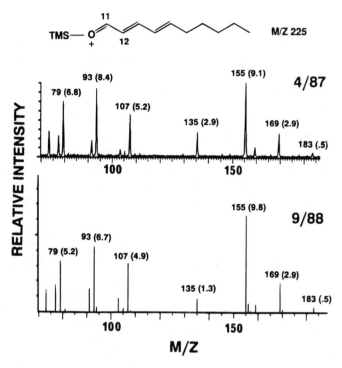

FIGURE 2.13

Comparison of CID spectra of m/z 225 parent ion produced by EI of PGE$_2$ methyl ester, methoxime, bis-TMS ether derivative. The spectra were taken about 1.5 years apart.

very low intensity ions are reproducible. Our qualitative experience to date is favorable toward spectral reproducibility of various analytes. The spectral interpretation is discussed below.

F. Matrix Effects

A principal advantage of MS/MS is the selectivity of the detection scheme. Therefore, in the analysis of "real-world" samples by MS/MS, it is not unusual for the sample clean-up to be minimized. The fact that the analyte is present as a minor component in a relatively large amount of matrix presents problems unique to the ion trap.

It is essential to use an ionization time long enough to create analyte ions in sufficient quantities for an adequate S/N to be achieved at the detection stage. At the same time, such an ionization pulse (10 ms for 100 fmol of analyte, for example) will create a large excess of ions from the more-abundant matrix. The main complications are (1) creation of a space-charged condition in the trap, thereby altering some of the fundamental characteristics of the parent ions (for example, axial frequency); and (2) potential ion/molecule reactions stripping away parent ion populations.

In practice, we have found that by tuning up parent ion isolation parameters *under space-charged conditions* for the standard, the first problem can be overcome routinely using a post-ionization isolation scheme with RF and DC voltages, to effect mass-selective storage (apex isolation). Under the same ionization conditions, the approach using an RF voltage, to effect only a high-pass mass filtering before carrying out MS/MS, was unsuccessful for crude tissue and plasma extracts.[71,72] It was suspected that the space-charge condition is not corrected prior to MS/MS, whereas the RF/DC voltage sequence does correct the condition by more efficient ejection of a wider range of mass/charge ratios of matrix ions.

It has been stated also that the tickle frequency of the parent ion shifts during elution of a capillary GC peak because of the changing concentration of the analyte and the occurrence of a space-charged condition at the high-concentration point on the peak, and that this is problematic. However, linear calibrations of a drug from crude plasma extracts, down to the parts per billion range, have been obtained by the above methods (see below). Reproducible isolation and decomposition of the parent ion species would be necessary to obtain a linear relationship. On the other hand, when the analyte concentration becomes *very* high, even an RF/DC voltage sequence may not correct the space-charge condition. Thus, dynamic range as determined by this method may be limited at the high concentration end.

The second drawback mentioned above is governed by physico-chemical laws and, depending on the type of parent ion isolated (for example, odd-radical cation *versus* even-electron, protonated species) and the properties of other matrix species (for example, ionization potentials, gas-phase proton affinity, etc.), may not be problematic. It is potentially more of a problem for ion traps than conventional mass spectrometers because of the long residence times of the ions in the mass analyzer, which also serves as the source.

VIII. TANDEM MASS SPECTROMETRY CASE HISTORIES

A. Quantitation

In GC/MS work using an ion trap, the ionization time is adjusted at the beginning of each scan by a software feature called automatic gain control (AGC). AGC takes a fast, short *pre-scan* before each full scan, to estimate the sample concentration based on the *total* ion current over a small mass range. Then, the software picks the appropriate ionization time. AGC allows also quantitation of samples containing unknown analyte concentrations to be carried out.

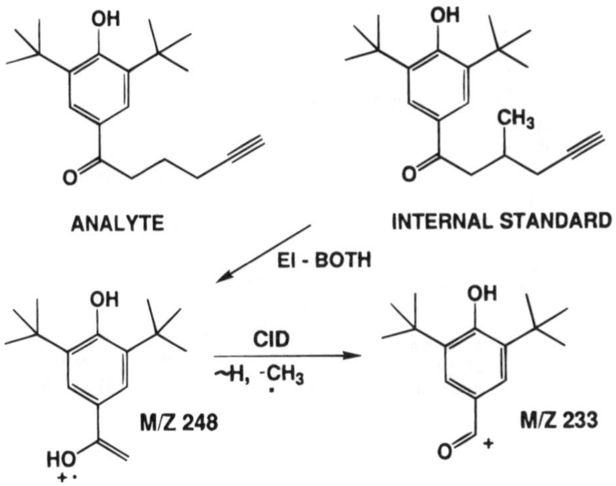

FIGURE 2.14
EI and CID fragmentations of a di*tert*butylphenol drug and a methylene analog internal standard.

In GC/MS/MS work, a similar feature designed to optimize *parent* ion intensity is needed. Parent-ion-based AGC is not available yet commercially for GC/MS/MS, although a working software system has been written and demonstrated in one laboratory (see below).[73] With sufficient patience, quantitative MS/MS can be performed without AGC.

A di*tert*butylphenol drug which was under development and a methylene analog (as internal standard) were quantitated by GC/MS/MS using the ion trap. Both compounds produced m/z 248 (base peak) by EI, *via* a McLafferty rearrangement. This ion decomposed under CID conditions by loss of a methyl radical to m/z 233 with 100% observed collision efficiency (Fig. 2.14).

For the calibration line derived from plasma extracts, an appropriate fixed ionization time was chosen for each of these two components, eluting during the GC/MS/MS run, based upon their known concentrations. High concentrations of analyte received low ionization times and *vice versa*. At a spike level of 1 ppb, the drug was barely detectable in the total product ion trace of m/z 248 (Fig. 2.15, upper trace). The lack of specificity is somewhat equivalent to that found in selected ion monitoring of m/z 248. By reconstructing the specific reaction for the MS/MS transition, the drug is clearly visible (Fig. 2.15, lower trace). The 1 ppb spike level corresponds to only 40 pg injected. Note that the observed (raw) chromatographic peak areas are about the same for the analyte and internal standard, because the ionization time is chosen in an inverse relationship to the expected concentration. The raw peak areas were then scaled according to the chosen ionization time and a linear plot was obtained for the ratio

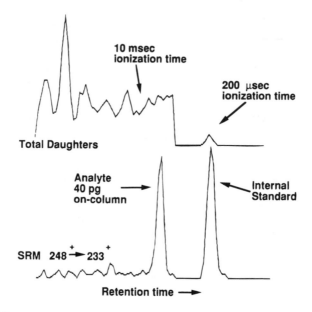

FIGURE 2.15

GC/MS/MS analysis of a single extract of a plasma sample spiked at 1 ppb with the drug from Fig. 2.14. Upper trace, relative intensity of total product ions of m/z 248 as a function of time with an ionization time of 10 ms. A software program changes the ionization time to 200 μs before the internal standard (spiked at 50 ppb) elutes. Lower trace, reaction chromatogram for the specific decomposition of m/z 248 to m/z 233. (Reprinted with permission of Elsevier Science Publishing Co., Inc. from the *Journal of the American Society for Mass Spectrometry*, Vol. 3, 1992.)

of analyte peak area/internal standard peak area *versus* the mole ratio of the two compounds (Fig. 2.16).

Plasma samples taken from a human clinical trial and containing unknown drug concentrations were extracted and analyzed twice. The first run was used to estimate an approximate drug concentration and to select an optimized fixed ionization time. The second run, using the optimized time, gave a more accurate determination. In GC/MS analyses, automatic gain control accomplishes this same two-step task, but continuously in real time, while the GC peak is eluting. Thus, the development of a GC/MS/MS AGC function, optimized for parent ion isolation, will be essential to implementing practical quantitative GC/MS/MS.

A comparison of the ITMS results with those obtained on a triple stage quadrupole, TSQ, (courtesy of R.L. Dobson and S. Ward) for the same plasma samples is shown in Fig. 2.17. The correlation is somewhat remarkable in that the TSQ results relied on a different MS/MS detection scheme (molecular ion, m/z 300, decomposing to m/z 248) and utilized a [^{13}C], [^{18}O]-labeled internal standard. The TSQ assay is much easier to

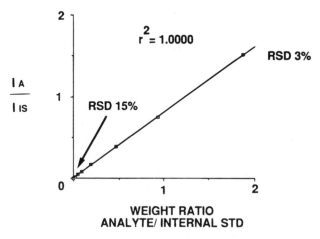

FIGURE 2.16

Calibration line for the drug spiked in plasma from 1–100 ppb. I_A is the chromatographic peak area for the analyte scaled by a factor equal to the ratio of the ionization time for the internal standard:ionization time for the analyte. I_{IS} is the peak area for the internal standard.

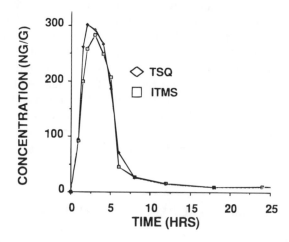

FIGURE 2.17

Comparison of quantitation results on split plasma samples for the TSQ-based assay *versus* the ITMS-based assay. In the ITMS assay, a methylene analog was used as the internal standard (IS) and the decomposition m/z 248 → m/z 233 was monitored. In the TSQ assay, a stable-isotope-labeled IS was used and the decomposition of m/z 300 → 248 was monitored.

carry out and is fully automated for data reduction as well. However, the ITMS results were an important milestone, showing that comparable quantitative results could be obtained using an ion trap MS/MS approach and that development of an AGC-MS/MS function was the next logical step.

B. The Co-Elution Problem

Stable isotope-labeled internal standards are preferred in quantitative mass spectrometry. More often than not, they co-elute with the analyte. The determination of low isotope ratios is challenging because it presents a problem similar to that described for trace analysis in general, namely that of determining a low concentration of analyte in the presence of a much more abundant species (for example, spurious matrix or, in this case, internal standard). The determination of low isotope ratios is important in trace analysis, where an excess of labeled internal standard can lessen adsorptive losses of the analyte during sample preparation. The reverse situation can occur in biological studies such as "pool-size" determinations where large dilutions of dosed, labeled material, by the endogenous compound, may occur.

Recently, a parent-ion based AGC-optimization has been developed using a two-step, RF/DC voltage isolation sequence.[21] Fig. 2.18 shows the analysis of an 80:1 mixture of a compound and its stable isotope-labeled congener. The spectrum obtained under total ion current-based AGC shows only the major component because an ionization time of 88 ms was selected by the software (upper trace, Fig. 2.18). When the same mixture was analyzed using the new parent-ion-based AGC software, with m/z 167 selected for the deuterated internal standard, an ionization time of over 9600 ms was chosen by the software. This situation creates a space-charged trap initially, but the RF/DC voltage sequence corrects the condition and m/z 167 is observed. Based on the absolute intensity of the two spectra shown, the mole ratio of the two compounds (81:1) and the ratio of the ionization times (1:109), better than 50% of the predicted minor-component intensity is obtained using the long ionization time.

C. Capillary GC/Full-Scan MS/MS *versus* Selected Reaction Monitoring

Once an isolated parent ion species has been dissociated during the MS/MS experiment using the ion trap, all product ions, down to about 10% of the parent ion mass, are trapped efficiently. These ions may be scanned out fairly rapidly (5500 Da/s) during the analytical scan at unit mass resolution. Several scans per second may be achieved. Selected reactions are then "monitored" by reconstructing particular product-ion chromatograms, post-run, *via* the data system. Since the ion trap scan rate is fast, there is no need to develop a reaction-monitoring scheme. In fact, the ability to reconstruct many reactions, without sacrificing sensitivity, from a single GC/MS/MS run can be advantageous.

Development of *in vitro* models of skin irritation is important to many consumer product companies. The response of cultured skin cells (*in vitro*)

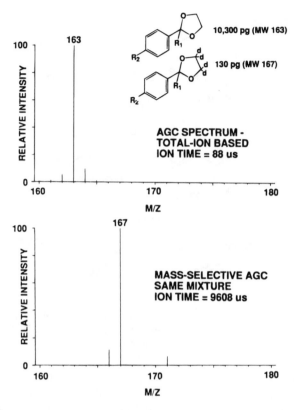

FIGURE 2.18
Analysis of an approximately 80:1 mix of an unlabeled (MW = 163) and labeled (d_4-MW = 167) substance. Upper trace, spectrum from a total-ion-current-based AGC EI/MS analysis of the mixture. Lower trace, spectrum from a parent-ion-based AGC analysis of the same mixture with m/z 167 specified as the desired parent mass. (Data used by permission of Prof. R. A. Yost, Univ. Florida, Gainesville; see Ref. 21.)

to various irritants is sometimes gauged by changes in various biomarkers. The 20-carbon, arachidonic acid metabolite, Prostaglandin E_2, was being examined as a possible biomarker to gauge irritation potential in one particular cell line, with quantitation of a single organic extract by GC/MS/MS. The methyl ester, methoxime, bis-TMS ether of PGE_2 produces an intense oxonium ion of m/z 225 under EI. The ion is derived from C-11 through C-20 of the molecule and it shows many structurally-specific fragments under CID conditions (Fig. 2.19).

Analysis of skin cell extracts (product ions of m/z 225) revealed several capillary GC peaks separated by retention time in the total product ion chromatogram (Fig. 2.20, upper trace). But the surprise was that each one of these components displayed the five specific decomposition reactions shown in Fig. 2.19. Thus, these GC peaks represent five individual com-

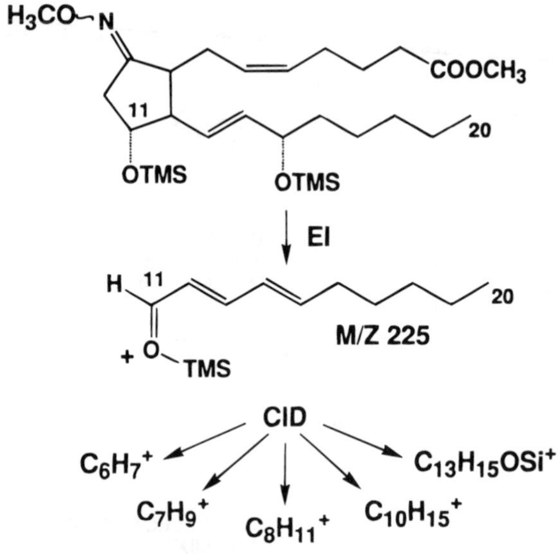

FIGURE 2.19

EI genesis of m/z 225 from PGE$_2$-methyl ester, methoxime, bis-TMS ether derivative and subsequent CID of m/z 225. The structure rearranges as in the case of LTB$_4$ (Fig. 7.2) to lose TMSOH and produces a series of highly unsaturated ions.

pounds, all related to PGE$_2$ and all possessing the common structural subunit, C_{11}–C_{20} (reaction chromatograms as labeled, Fig. 2.20). Since their overall structures are different, the peaks must represent upper side chain metabolites (C_1–C_8). The next logical step is to determine which metabolite represents a rate limiting or end point step in the skin cells and then to use it potentially as a biomarker.

It is important to note that this discovery was totally dependent on the fact that full-scan MS/MS data were being collected. Selected reaction monitoring of one reaction would have revealed several capillary GC peaks. It is quite possible that the peaks at retention times outside of that for the analytical standard could have been attributed to nonspecific interferences, instead of other PGE$_2$ metabolites.

D.   GC/MS^n for Metabolite Structures

Many times, an analyte is derivatized prior to GC analysis. It was pointed out several years ago[74] that, in subsequent MS/MS, the product ion spectrum is often dominated by a derivative-specific fragmentation. This observation impacts selectivity of analysis in a negative fashion. Also, it was shown[75] recently how this problem can lead to ambiguity in

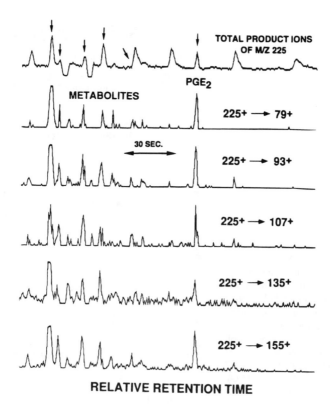

FIGURE 2.20

GC/MS/MS traces for the analysis of PGE$_2$ (methyl ester, methoxime, bis-TMS ether derivative) from skin-cell extracts, relative intensity *versus* relative retention time. Upper trace,
total product ions of m/z 225 (see Fig. 7.9). Other traces are reconstructed product ions for
particular CID reactions, as labeled.

identifying positional isomers of hydroxylated metabolites of methylene-
dioxy amphetamine (trifluoroacetate/trifluoroacetamide derivative).
While the structural proof of the metabolites was solved ultimately by
synthesis of the respective isomeric standards, an interesting demonstration was given.

Under CI conditions, the [M + H]$^+$ ion of all of the isomers fragmented
via CID by loss of the derivative group CH$_3$N=C (OH) CF$_3$ to give an
intense product ion of m/z 275, with little other spectral information
(Fig. 2.21, upper scheme). Re-analysis by GC/MS/MS/MS (protonated
molecule → m/z 275 → new product ions) gave distinct spectra for each
isomer. Obviously, having the standards on hand to study made this
problem easier. But as experience is gained in interpretation of MS/MS
data, it is not difficult to imagine that the GC/MS/MS/MS approach will
be used eventually with *a priori* interpretation to assign structures.

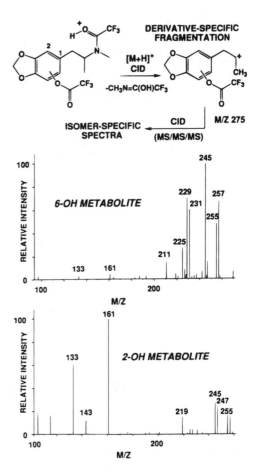

FIGURE 2.21
GC/MS/MS/MS analysis of methylenedioxy-methamphetamine hydroxylated metabolites (bis-TFA ester derivative). Only MS/MS/MS gives isomer-specific spectra. The second-generation product ions are: m/z 257 ($-H_2O$), m/z 255 ($-HF$), m/z 247 ($-CH_2CH_2$), m/z 245 ($-CH_2O$), m/z 231 ($-CO_2$), m/z 229 ($-HCOOH$), m/z 161 ($-CF_3COOH$), m/z 133 (161$-C_2H_4$). (Data used by permission of Prof. R. Foltz, Univ. Utah, Salt Lake City, UT.)

So far, this section has been biased toward analytical applications of tandem mass spectrometry using the ion trap. But there are a wealth of interesting experiments from the physico-chemistry that have been demonstrated on the commercial ion trap instrumentation.

E.   Filtered Noise Fields for High-Mass MS/MS

The use of a digitally-synthesized, broad frequency spectrum for MS/MS applications in the ion trap is becoming increasingly popular and important. Commercially-available equipment interfaces with the Finnigan ITMS. The power distribution across a digitally-synthesized waveform is very flat, from 10 kHz to 600 kHz. Holes or notches may be placed in the frequency band (that is, deleting certain frequency components) to effect the mass-selective isolation of an ion prior to MS/MS.

Consider the case of trisperfluorononyl triazine (TPFNT), which under EI conditions produces a relatively abundant fragment at m/z 1466 (loss of F·). This ion can not be apex-isolated (mass-selectively) by an RF/DC voltage sequence because the required voltages cannot be generated by the commercial equipment. Neither can this ion be mass-analyzed using a standard RF-voltage ramp, because at maximum RF voltage the ion is still trapped.

The frequency synthesizer can be used for three purposes in this one experiment: (1) The mass range of the trap may be tripled by application of an AC potential oscillating at a single frequency (*ca.* 120 kHz, 20 $V_{(p-p)}$) across the end-cap electrodes during the mass scan. This frequency corresponds to a q_z value of about 0.3, or one-third of the value ions normally need to achieve for mass analysis as they leave the trap. Thus if m/z 650 leaves at the maximum voltage, in this application up to m/z 1950 may be forced to leave the trap during the mass scan. (2) After EI, a notched power band can be applied to the end-cap electrodes to isolate m/z 1466 only, to prepare it for MS/MS. (3) Finally, a narrow power band from 10.5 to 12.5 kHz (the axial frequency of m/z 1466 lies within this band) can be applied to form product ions of m/z 1466.

In Fig. 2.22, the top trace shows the EI mass spectrum obtained for a TPFNT sample boiled off the probe. The mass axis is listed as *observed mass* because the system can only be calibrated under normal ion ejection conditions ($q_z = 0.868$). In this case, the higher-mass ions (above m/z 150) are forced out of the trap at one-third of this q_z value and the observed mass must be tripled to obtain the actual value. Not only is m/z 1466 observed, but also an ion/molecule reaction artifact is seen, producing a broadened mass peak above m/z 1500. While the molecular ions are formed and then cooled continuously *via* collisions following ionization, the ion/molecule reaction product ions may not be cooled adequately and focused at the time of the analytical scan, thus leading to broadness in the peak.

When the experiment is modified by application of a notched frequency band, mass-selective storage of m/z 1466 is effected (middle trace, Fig. 2.22). Finally, a third sequential modification is made by adding the reverse-notch excitation band, causing CID of the parent ion (bottom trace, Fig. 2.22). Both the time of application of the field and the amplitude may be varied to effect the disappearance of the parent ion. In this case, Ar was added to the manifold at a partial pressure which was just below the threshold for causing scattering losses (decreasing intensity) for m/z 1466 in the mass-selective isolation experiment (middle trace, Fig. 2.22).

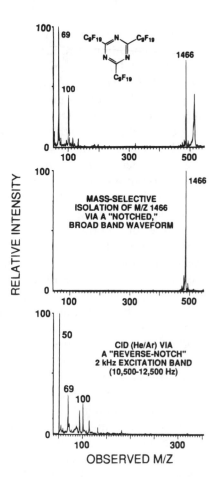

FIGURE 2.22

High-mass MS/MS. Top trace, EI mass spectrum of trisperfluorononyl triazine; m/z 1466 [M–F$^{\cdot}$]$^+$, m/z 100 [C$_2$F$_4$]$^+$, m/z 69 [CF$_3$]$^+$; also, an ion/molecule reaction product adduct (probably [M + CF$_2$]$^+$ is observed. Note, the mass range of the trap is tripled above m/z 150. Middle trace, mass-selective isolation of m/z 1466 with a notched waveform. Bottom trace, CID of the isolated m/z 1466 using a reverse notch 2 kHz wide.

F. Final Precautions

Ion "trappists" have been warned that they may fall into a black hole or worse yet, a black canyon, if they are not careful.[76–79] While use of these terms has no doubt caused confusion for more than one astrophysicist performing a literature search, only to find himself engrossed in mass spectrometry journals, the terms do have practical significance for ion trap operation. The terms *black hole* and *black canyons* have been used to describe the *selective* and unintentional loss of ions in the trap.

The commercial ion trap has a distorted geometry,[80] causing a non-ideal electric field. In fact, nonlinear resonances from octapole and hexapole components of the field occur at iso-β_z lines = 1/2 and 2/3 on the stability diagram. These nonlinear resonances will oscillate ions of the correct a_z and q_z values, corresponding to positions on the iso-β_z lines, and cause them to be lost from the trap. Longer residence times prior to scanning (CI and MS/MS experiments in particular) favor observation of the phenomenon. Thus, short EI-MS scans do not appear to be affected, but longer MS/MS and chemical ionization experiments may be, which is more of a minor annoyance than anything. Thus the experimenter needs to be aware and to be careful of these operating points when setting up experiments or interpreting data.

IX. CONCLUSION

Many of the concepts utilized in today's computerized-hardware and software driven MS/MS experiments (for example, mass-selective storage and using ion axial frequencies to one's advantage) were demonstrated by the early workers in the field. Though some of the early demonstrations were rudimentary, the advances in instrument control through computerization have allowed newcomers in the field to re-apply these early fundamental learnings in inventive, new ways.

The commercial ion trap mass spectrometer can be used for a wide variety of analytical, organic or physico-chemical studies. While custom software has been developed in some laboratories for special applications (automated MS/MS optimization,[74] parent-ion AGC controlled MS/MS,[71]), the second generation commercial ITMS is awaited by many. It should prove to be more automated and even simpler to operate than the current ITMS, and its application to problem solving will be highly advanced from the prototype instrument marketed several years ago in the late 1980s. Moreover, new methods of trap operation and the arrival of other commercial instruments will drive prices downward, so that the use of benchtop MS/MS may become routine in the near future.

The instrumental and operational simplicity by which single-stage ion trap mass spectrometry is extended to tandem mass spectrometry, by the addition of a second mass-selective stage, will facilitate greatly the increased use and wider application of tandem mass spectrometry. The speed with which the elements of tandem mass spectrometry may be performed, and the high value of the informing power, particularly when gas chromatography is combined with tandem mass spectrometry, makes possible the analysis of complex samples at the sub-ppb level. The combination of instrumental and operational simplicity with the high value of the informing power can translate into a considerably reduced cost per bit of informing power, hence to enhanced accessibility to this technique.

Section C
ION TRAP MASS SPECTROMETRY AND LASERS

Colin S. Creaser

X. PRACTICAL ASPECTS OF LASER DESORPTION AND PHOTODISSOCIATION

A.   Introduction

Quadrupole ion trap mass spectrometers are particularly well-suited to laser based applications and there have been several reported in recent years. Most of these developments have been in the area of laser desorption/ionization (LD) or the application of photodissociation in tandem mass spectrometry. In the former case, the capacity of the trap to store ions formed during the rapid laser desorption process is important, while in the latter case, the long irradiation times for a selected precursor ion held in the trap lead to high fragmentation efficiencies. While these techniques have been investigated using research ion traps, recent reports of modifications to commercial benchtop instruments point to the potential for the wider use of lasers with ion traps in analytical applications.

B.   Laser Desorption

Several instrumental arrangements have been reported for laser desorption in ion traps using direct laser introduction or fiber optic interfaces. Fig. 2.23 illustrates the three most common interfacing configurations. These instrumental arrangements have been used with a range of laser wavelengths, notably the 1064, 532, 355, and 255 nm lines of the Nd-YAG laser, the UV output of the nitrogen laser at 337 nm and the 10.6 μm line of the CO_2 laser.

The earliest report of LD in ion trap mass spectrometry[81] described the injection of externally-desorbed metal ions into the trap through holes in one of the end-cap electrodes. The instrumental arrangement is shown schematically in Fig. 2.23(a). In this configuration, the ions are generated by pulsed laser irradiation in an external ion source and injected through a lens assembly. This approach has been employed by other groups for LD studies using different lens systems.[82,83] However, axial introduction of laser-desorbed ions into the trap has also been demonstrated using a fiber optic probe assembly without auxiliary injection optics.[84] An alternative and widely-used configuration in which ions are desorbed within the trap is shown in Fig. 2.23(b). In this arrangement, the sample is mounted on a probe, introduced through a small hole in the ring electrode, and

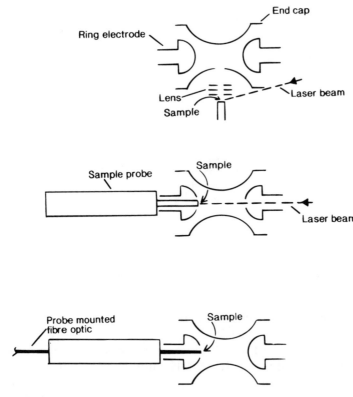

FIGURE 2.23
Ion trap configurations for laser desorption.

desorbed by a pulse from a focused laser beam which enters the trap through a second hole on the opposite side of the ring.[85–90] This arrangement has the advantage of leaving the electron filament assembly in place behind one end-cap electrode. A third type of interface (Fig. 2.23(c)) is even simpler in design and the sample, either on its own or in a matrix, is placed directly on to the tip of a 0.6 mm probe-mounted silica fiber optic, which is inserted into the vacuum system *via* the probe lock and through a hole in the ring electrode. The other end of the fibre is coupled to the laser by a three-axis delivery system outside the vacuum housing, and the sample/matrix is desorbed into the trap by a pulse of photons passing along the fiber.[91,92] This probe-mounted interface implemented on a research ion trap is shown in Fig. 2.24.

Ions formed by laser desorption in the trap cavity or injected from an external source have high kinetic energies which can lead to poor trapping efficiencies. The efficiency is determined by the amplitude of the trapping RF voltage applied to the ring electrode, and the pressure and nature of the buffer gas in the trap. At low RF amplitudes (low-mass

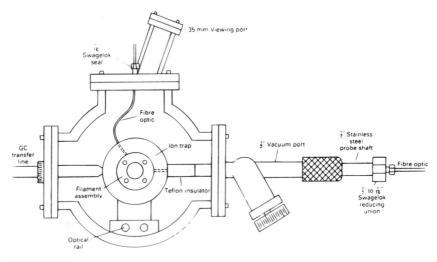

FIGURE 2.24
Fibre optic interfaces for ion trap laser desorption and photodissociation.

cut-off levels), trapping efficiency may be so poor that no ions are retained in stable trajectory following the laser pulse.[81] However, efficiency increases rapidly with RF amplitude to reach a maximum which is dependent on the mass of the analyte.[86,89] Helium buffer gas pressures in the range 1 to 4 mTorr are usually required to retain the desorbed ions, although higher helium pressures, or the use of alternative bath gases such as argon, neon and xenon, may be necessary to dampen the trajectories of high mass ions.[81,84,86,89] The trapping efficiency is also improved when the laser pulse timing is synchronized to the phase of the RF voltage amplitude[88] since the laser pulse length, typically 10 ns, is much shorter than the period of the RF applied to the ring electrode (1 μs). Spectral quality may be enhanced further by introducing a delay period of up to several tens of milliseconds into the scan routine following desorption and prior to the mass selective ejection scan or other procedure. This delay period is necessary to permit collisional cooling of the ions to the center of the ion trap. Under these conditions, the trap can usually be filled to the space-charge limit within a few laser pulses.

All of the interface configurations described above have been employed for analytical applications using both thermal and matrix-assisted (MALDI)[93] laser desorption/ionization. An example of the former is the direct desorption of the ionic dye rhodamine 6G using the probe-mounted fiber optic arrangement.[92] The positive ion spectrum, shown in Fig. 2.25(a), was obtained by 532 nm laser pulses from a Nd-YAG laser. At this wavelength, no fragments are formed and the spectrum is characterized by gas-phase ions corresponding to the cationic part of the mole-

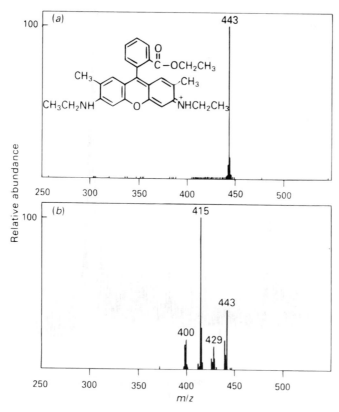

FIGURE 2.25
Laser desorption ion trap mass spectra of rhodamine 6G, (a) LD/MS and (b) LD/MS-MS
of m/z 443. (Reproduced from Ref. 92)

cule at m/z 443. The use of shorter laser wavelengths, such as the tripled
and quadrupled Nd-YAG lines at 355 nm and 266 nm, results in fragment
formation because of the additional internal energy deposited in the
desorbed ions.

An advantage of the ion trap for LD studies is that once the ion cloud
is retained in stable trajectory in the trap, a wide variety of procedures
can be employed, such as tandem mass spectrometry (MS^n) of precursor
ions[83,84,86,89] and ion/molecule reactions.[81,84] Hence, fragment ion information may be generated for rhodamine 6G by the LD/MS/MS analysis of
the isolated m/z 443 ion. The resulting product ion spectrum is shown
in Fig. 2.25(b).[92] The use of laser desorption to initiate ion/molecule reactions is illustrated by the interaction of desorbed Au^+ ions with benzene
vapor[81] leading to the formation of a variety of ionic products including
the charge exchange product $C_6H_6^+$ and the cluster ions $[Au-C_6H_6]^+$,
$[Au(C_6H_6)_2]^+$ and $[Au(C_6H_6)(H_2O)]^+$. Alkali-metal crown ether adducts

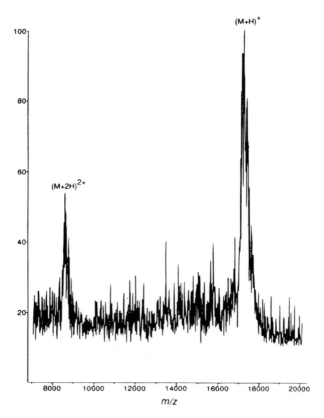

FIGURE 2.26

UV-LD (337 nm) mass spectrum of 10 fmol whale myoglobin ($[M + H]^+$ = 17,294 (average)) in sinapinic acid at a 1:11 × 10^6 ratio. (Reproduced from Ref. 83)

and transition metal (Cr^+ and Fe^+) bound dimers with naphthalene have also been observed following laser desorption of metal ions.[84]

Many molecules of biological interest have been analyzed using laser desorption combined with ion trap mass spectrometry. For example, the spectra of leucine-enkephalin, α-endorphin, bovine adrenal medulla, methionine enkephalinamide, parathyroid, angiotensin, digitoxigenin and several disaccharides have been reported.[84,85,88] A variety of high mass peptides and proteins have also been detected by matrix-assisted LD/ion trap mass spectrometry using resonance ejection to extend the mass range.[82,83,88,89,94] The high mass and high sensitivity of the ion trap have been illustrated clearly by the detection of the $[M + H]^+$ ion of chicken egg albumin (MW = 43,300),[82] and analysis of as little as 10 fmol of whale myoglobin ($[M + H]^+$ = 17,294)[83] by matrix-assisted (sinapinic acid) LD/ion trap mass spectrometry. The partial mass spectrum of the whale myoglobin is shown in Fig. 2.26.

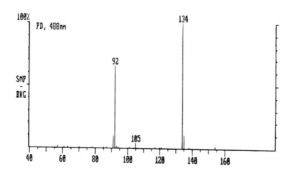

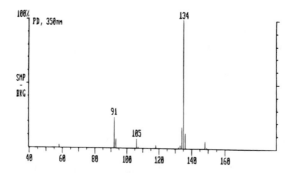

FIGURE 2.27
Effect of laser wavelength on the m/z 91/92 branding ratio for photodissociation of the m/z 134 ion of n-butylbenzene at 488 and 350 nm.

C. Photodissociation

In tandem mass spectrometry, it is possible also to use photodissociation (PD) for structural studies of ions with suitable absorption characteristics. PD has been investigated extensively in ion cyclotron resonance trapping devices[95] and the analytical potential of PD in ion traps has also received attention.

The earliest reports of PD in an ion trap used direct introduction of the laser to study the multiphoton infrared dissociation of small ions such as H_2^+ and propan-2-ol dimers.[96–98] An alternative to direct introduction of the laser beam is the use of a fiber optic interface, first reported in 1987, where a 1 mm silica fiber was used to introduce light from a Nd-YAG laser.[99] Fig. 2.24 shows two fiber optic interface configurations suitable for PD studies on a research ion trap mass spectrometer.[92] These interfaces are easy to install and require little modification of the trap. In both cases, the end of the 0.6 mm or 1 mm silica fiber is inserted into the trap through a small hole drilled in the ring electrode. The other end of the fiber is coupled to a

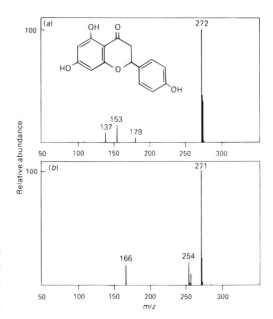

FIGURE 2.28
MS/MS product ion spectra of
the molecular ion (m/z 272) of
4′,5,7-trihydroxyflavanone
using (a) PD (350 nm) and (b)
CAD (Reproduced from Ref. 92)

continuous wave (e.g. Argon ion) or pulsed (e.g. Nd-YAG) laser located outside the vacuum housing.

Quadrupole ion traps are particularly well-suited to PD studies because of the long irradiation times which are possible when selected precursor ions are held in stable trajectories in the trap. These long irradiation times result in high dissociation efficiences in ion trap photodissociation (10 to 100%). The sensitivity of the ion trap, together with this high efficiency makes it possible for PD to be used routinely for trace analytes. For example, ion trap PD/MS/MS has been shown to be compatible with chromatographic introduction of samples at the low nanogram level.[100]

The extent of fragmentation is determined by the wavelength of the laser, the laser power, and the irradiation time. The energy deposited in the ion during the dissociation step may be controlled, in principle, by the energy of the incident photons. Thus, a change in the photon energy from 2.6 eV (488 nm) to 3.7 eV (350 nm) on the dissociation of the isolated molecular ion of n-butylbenzene at m/z 134, is reflected in the intensities of the simple cleavage and rearrangement product ions at m/z 91 ($C_7H_7^+$) and m/z 92 ($C_7H_8^{+\bullet}$). The intensity of m/z 91, the high energy dissociation product, is enhanced with respect to the ion at m/z 92 when the 350 nm line is used for PD (Fig. 2.27).[101]

The product ion spectra obtained using PD and CAD may also provide complementary structural information. An example of such complementarity is the tandem mass spectrometry of the molecular ion (m/z 272) of naringenin (4',5,7-trihydroxyflananone). The product ion spectra for the PD (350 nm, Ar$^+$ ion laser) and CAD MS/MS analysis are shown in Fig. 2.28. The PD spectrum shows products at m/z 179, [M–C$_6$H$_4$OH]$^+$, m/z 153, [(HO)$_3$C$_6$H$_2$CO]$^+$, and m/z 137, [(HO)$_2$C$_6$H$_3$CO]$^+$, while the CAD product spectrum has none of these ions, but exhibits peaks at m/z 254, [M–CO]$^+$ and m/z 166, [M–CHC$_6$H$_4$OH]$^+$. These differences may be attributed to the activation of entirely separate fragmentation pathways by the two dissociation processes. In this case, the collisional process is clearly different to the specific deposition of the photon energy into the chromaphore in PD. For this reason, PD has shown promise for differentiation of isomeric compounds[100] which yield identical spectra under CAD. The potential of PD in the ion trap is likely to be realized further by the excitation of high mass precursor ions (>3000 u) derived from laser desorption experiments, since these ions are difficult to fragment by collisional processes.

REFERENCES

1. Thomson, J. J. *Philos. Mag. V.* 1897, *44*, 293.
2. Thomson, J. J. *Proc. Camb. Philos. Soc.* 1897, *9*, 243.
3. Thomson, J. J. *Rays of Positive Electricity and the Application to Chemical Analysis.* Longmans Green, London, 1913, p. 56.
4. ASMS Report on Nomenclature, *Proc. 32nd ASMS Conf. Mass Spectrometry and Allied Topics.* San Antonio, 1984, p. 921.
5. Beynon, J. H.; Morgan, R. P. *Int. J. Mass Spectrom. Ion Phys.* 1978, *27*, 1.
6. Kruger, T. L.; Litton, J. F.; Kondrat, R. W.; Cooks, R. G. *Anal. Chem.* 1976, *48*, 2113.
7. Levsen, K.; Schwarz, H. *Angew. Chem. Int. Ed. Engl.* 1976, *15*, 509.
8. McLafferty, F. W.; Bockoff, F. M. *Anal. Chem.* 1978, *50*, 69.
9. Haddon, W. E. In: *High Performance Mass Spectrometry.* M. L. Gross, (ed.), American Chemical Society, Washington, DC, 1978, pp. 97–119.
10. McLafferty, F. W. (ed.), *Tandem Mass Spectrometry.* John Wiley & Sons, New York, 1983.
11. Busch, K. L.; Glish, G. L.; McLuckey, S. A. *Mass Spectrometry/Mass Spectrometry; Techniques and Applications of Tandem Mass Spectrometry.* VCH Publishers, New York, 1988.
12. March, R. E.; Hughes R. J. *Quadrupole Storage Mass Spectrometry.* Chemical Analysis Series, Vol. 102, John Wiley & Sons, New York, 1989.
13. Todd, J. F. J. *Mass Spec. Rev.* 1991, *10*, 3.
14. Cooks, R. G.; Glish, G. L.; McLuckey, S. A.; Kaiser, Jr., R. E. *Chem. Eng. News.* 1991, *69(12)*, 26.
15. March, R. E. *Int. J. Mass Spectrom. Ion Processes.* 1992, 118/119, 71.
16. Todd, J. F. J.; Penman, A. D. *Int. J. Mass Spectrom. Ion Processes.* 1991, *106*, 1.
17. Griffiths, I. W. *Rapid Commun. Mass Spectrom.* 1990, *4(3)*, 69.
18. Nourse, B. D.; Cooks, R. G. *Anal. Chim. Acta.* 1990, *228*, 1.

19. Cooks, R. G.; Kaiser, Jr., R. E. *Acc. Chem. Res.* 1990, *23(7)*, 213.

20. *Practical Aspects of Ion Trap Mass Spectrometry*, R. E. March and J. F. J. Todd (eds.), Vol. 1, "Fundamentals and Instrumentation", Modern Mass Spectrometry series, CRC Press, Roca Baton, FL, 1995.

21. Fetterolf, D. D.; Yost, R. A. *Int. J. Mass Spectrom. Ion Processes.* 1984, *62*, 33.

22. Lehman, T. A.; Bursey, M. M. *Ion Cyclotron Resonance Spectrometry.* John Wiley & Sons, New York, 1976.

23. Burinsky, D. J.; Cooks, R. G.; Chess, E. K.; Gross, M. L. *Anal. Chem.* 1982, *54*, 295.

24. Louris, J. N.; Brodbelt-Lustig, J. S.; Cooks, R. G.; Glish, G. L.; Van Berkel, G. J.; McLuckey, S. A. *Int. J. Mass Spectrom. Ion Processes.* 1990, *96*, 117.

25. Nourse, B. D.; Cox, K. A.; Cooks, R. G. *Proc. 39th ASMS Conf. Mass Spectrometry and Allied Topics.* Nashville, 1991, p. 1479.

26. McLuckey, S. A.; Glish, G. L.; Van Berkel, G. J. *Int. J. Mass Spectrom. Ion Processes.* 1991, *106*, 213.

27. Todd, J. F. J.; March, R. E.; Franklin, A. M.; Penman, A. D. *Gas Phase Ion Chemistry Conf.* Padova, Italy, April 22–24, 1991.

28. *Quadrupole Mass Spectrometry and Its Applications.* Dawson, P.H. (ed.), Elsevier, Amsterdam, 1976.

29. Tucker, D. B.; Hameister, C. H.; Bradshaw, S. C.; Hoekman, D. J.; Weber-Grabau, M. *Proc. 36th ASMS Conf. Mass Spectrometry and Allied Topics.* San Francisco, 1988, p. 628.

30. Buttrill, Jr., S. E.; Shaffer, B.; Karnicky, J.; Arnold, J. T. *Proc. 40th ASMS Conf. Mass Spectrometry and Allied Topics.* Washington, DC, 1992, p. 1015.

31. Fulford, J. E.; March, R. E. *Int. J. Mass Spectrom. Ion Phys.* 1978, *26*, 155.

32. Strife, R. J.; Kelley, P. E.; Weber-Grabau, M. *Rapid Commun. Mass Spectrom.* 1988, *2(6)*, 105.

33. Strife, R. J.; Keller, P. R. *Org. Mass Spectrom.* 1989, *24*, 201–201.

34. March, R. E.; Londry, F. A.; Alfred, R. L.; Franklin, A. M.; Todd, J. F. J. *Int. J. Mass Spectrom. Ion Processes.* 1992, *112*, 247.

35. Kelley, P. E.; Syka, J. E. P.; Ceja, P. C.; Stafford, G. C., Louris, J. N.; Grutzmacher, H. F.; Kuck, D.; Todd, J. F. J. *Proc. 34th ASMS Conf. Mass Spectrometry and Allied Topics.* Cincinnati, 1986, p. 963.

36. Bolton, B.; Wells, G.; Wang, M. *Proc. 41st ASMS Conf. Mass Spectrometry and Allied Topics.* San Francisco, 1993, p. 474a.

37. McLuckey, S. A.; Goeringer, D. E.; Glish, G. L. *J. Am. Soc. Mass Spectrom.* 1991, *2*, 11.

38. Gronowska, J.; Paradisi, C.; Traldi, P.; Vettori, U. *Rapid Commun. Mass Spectrom.* 1990, *4*, 306.

39. Ardanaz, C. E.; Traldi, P.; Vettori, U.; Kavka, J.; Guidugli, F. *Rapid Commun. Mass Spectrom.* 1991, *5*, 5.

40. Todd, J. F. J. personal communication.

41. Paradisi, C.; Todd, J. F. J.; Traldi, P.; Vettori, U. *Org. Mass Spectrom.* 1992, *27*, 251.

42. Paradisi, C.; Todd, J. F. J.; Vettori, U. *Org. Mass Spectrom.* 1992, *27*, 1210.

43. Creaser, C. S.; O'Neill, K. E. *Second European Meeting on Tandem Mass Spectrometry.* Warwick, UK, 1992.

44. Curcuroto, O.; Fontana, S.; Traldi, P.; Celon, E. *Rapid Commun. Mass Spectrom.* 1992, *6*, 322.

45. Matthews, L. S. M.Sc. Thesis, University of East Anglia, UK, 1992.

46. Creaser, C. S.; O'Neill, K. E. *Org. Mass Spectrom.* 1993, *28(5)*, 564.

47. March, R. E.; Tkaczyk, M.; Londry, F. A.; Alfred, R. L. *Int. J. Mass Spectrom. Ion Processes,* 1993, *125*, 9.

48. March, R. E.; Weir, M. R.; Londry, F. A.; Catinella, S.; Traldi, P.; Stone, J. A.; Jacobs, B. *Proc. 41st ASMS Conf. Mass Spectrometry and Allied Topics.* San Francisco, 1993, 446a.

49. Todd, J. F. J.; Bexon, J. J.; Smith, R. D.; Weber-Grabau, M.; Kelley, P. E.; Syka, J. E. P.; Stafford, G. C. *Proceedings 16th Meeting British Mass Spectrometry Society.* York, UK, 1987, 206.

50. Londry, F. A.; Wells, G. J.; March, R. E. *Proc. 41st ASMS Conf. Mass Spectrometry and Allied Topics.* San Francisco, 1993, 790a.

51. Chen, L.; Wang, T-C. L.; Ricca, T. L.; Marshall, A. G. *Anal. Chem.* 1987, *59*, 449–454.

52. Julian, R. K.; Cooks, R. G. *Anal. Chem.* 1993, *65*, 1827–1833.

53. Kelley, P. E.; Hoekman, D. J.; Bradshaw, S. C.; Stiller, S. W. *Proc. Pittsburgh Conf. Analytical Chemistry and Applied Spectroscopy.* Atlanta, March 8, 1993, p. 1081.

54. Wang, X.; Bohme, D. K.; March, R. E. Unpublished results.

55. March, R. E.; Londry, F. A.; Fontana, S.; Catinella, S.; Traldi, P. *Rapid Commun. Mass Spectrom.* 1993, *7*, 929.

56. Kascheres, C.; Cooks, R. G. *Anal. Chim. Acta.* 1988, *215*, 223.

57. Brodbelt, J. S.; Cooks, R. G. *Anal. Chim. Acta.* 1988, *206*, 239.

58. Nourse, B. D.; Brodbelt, J. S.; Cooks, R. G. *Org. Mass Spectrom.* 1991, *26*, 575.

59. Emary, W. B.; Kaiser, R. E.; Kenttamaa, H. I.; Cooks, R. G. *J. Amer. Soc. Mass Spectrom.* 1990, *1(4)*, 308.

60. Curtis, J. E.; Kamar, A.; March, R. E.; Schlunegger, U. P. *Proc. 35th ASMS Conf. Mass Spectrometry and Allied Topics.* Denver, 1987, p. 237.

61. Johnson, J. V.; Pedder, R. E.; Kleintop, B.; Yost, R. A. *Proc. 38th ASMS Conf. Mass Spectrometry and Allied Topics.* Tucson, 1990, p. 1130.

62. Louris, J. N.; Cooks, R. G.; Syka, J. E. P.; Kelley, P. E.; Stafford, Jr., G. C.; Todd, J. F. J. *Anal. Chem.* 1987, *59*, 1677.

63. Nourse, B. D.; Cox, K. A.; Morand, K. L.; Cooks, R. G. *J. Am. Chem. Soc.* 1992, *114*, 2010.

64. Louris, J. N.; Brodbelt-Lustig, J. S.; Cooks, R. G.; Glish, G. L.; Van Berkel, G. J.; McLuckey, S. A. *Int. J. Mass Spectrom. Ion Processes.* 1990, *96*, 117.

65. Lammert, S. A.; Cooks, R. G. *Rapid Commun. Mass Spectrom.* 1992, *6*, 528.

66. Lammert, S. A.; Cooks, R. G. *J. Am. Soc. Mass Spectrom.* 1991, *2*, 487.

67. Duinker, J. C.; Hillebrand, M. T. J.; Palmork, K. H.; Wilhelmsen, S. *Bull. Environm. Contam. Toxicol.* 1980, *25*, 956.

68. Mullin, M. D. *Congener Specific PCB Analysis Techniques Workshop.* Environmental Protection Agency, Large Lakes Research Station, Gross Iles, MI, June 11–14, 1985.

69. Todd, J. F. J.; Penman, A. D.; Thorner, D. A.; Smith, R. D. *Rapid Commun. Mass Spectrom.* 1990, *4*, 108.

70. Todd, J. F. J.; Penman, A. D.; Thorner, D. A.; Smith, R. D. *Proc. 38th ASMS Conf. Mass Spectrometry and Allied Topics.* Tucson, 1990, 532–533.

71. Strife, R. J.; Simms, J. R. *Anal. Chem.* 1989, *61*, 2316.

72. Strife, R. J.; Simms, J. R. *J. Am. Soc. Mass Spectrom.* 1992, *3*, 372.

73. Griffin, T. P.; Yates, N. A.; Yost, R. A. *Proc. 41st ASMS Conf. Mass Spectrometry and Allied Topics.* San Francisco, 1993, p. 715a.

74. Strife, R. J.; Simms, J. R. *Anal. Chem.* 1988, *60*, 1800.

75. Lim, H. K.; Foltz, R. L. *Biol. Mass Spectrom.* 1991, *20*, 677.

76. Guidulgi, F.; Traldi, P.; *Rapid Commun. Mass Spectrom.* 1991, *5*, 343.

77. Morand, K. L.; Lammert, S. A.; Cooks, R. G. *Rapid Commun. Mass Spectrom.* 1991, *5*, 491.

78. Guidulgi, F.; Traldi, P.; Franklin, A. M.; Langford, M. L.; Murrell, J.; Todd, J. F. J. *Rapid Commun. Mass Spectrom.* 1992, *6*, 229.

79. Eades, D. M.; Yost, R. A. *Rapid Commun. Mass Spectrom.* 1992, *6*, 573.

80. Louris, J; Schwartz, J.; Stafford, G.; Syka, J.; Taylor, D. *Proc. 40th ASMS Conf. Mass Spectrometry and Allied Topics.* Washington, 1992, p. 1003.

81. Louris, J. N.; Amy, J. W.; Ridley, T. Y.; Cooks, R. G. *Int. J. Mass Spectrom. Ion Processes.* 1989, *88*, 97–111.

82. Schwartz, J. C.; Bier, M. E. *Rapid Commun. Mass Spectrom.* 1993, *7*, 27–32.

83. Jonscher, K.; Currie, G; McCormack, A. L.; Yates, J. R. *Rapid Commun. Mass Spectrom.* 1993, *7*, 20–26.

84. McIntosh, A.; Donovan, T.; Brodbelt, J. *Anal. Chem.* 1992, *64*, 2079–2083.

85. Heller, D. N.; Lys, I.; Cotter, R. J. *Anal. Chem.* 1989, *61*, 1083–1086.

86. Glish, G. L.; Goeringer, D. E.; Asano, K. G.; McLuckey, S. A. *Int. J. Mass Spectrom. Ion Processes.* 1989, *94*, 15–24.

87. Gill, C. G.; Daigle, B.; Blades, M. W. *Spectrochim. Acta.* 1991, *46B*, 1227–1235.

88. Doroshensko, V. M.; Cornish, T. J.; Cotter, R. J. *Rapid Commun. Mass Spectrom.* 1992, *6*, 753–757.

89. Chambers, D. M.; Goeringer, D. E.; McLuckey, S. A.; Glish, G. L. *Anal. Chem.* 1993, *65*, 14–21.

90. Vargas, R. R.; Yost, R. A.; Lee, M. S.; Moon, S. L.; Rosenberg, I. E. *Proc. 40th ASMS Conf. Mass Spectrometry and Allied Topics.* Washington, DC, 1992, p. 1751.

91. Creaser, C. S.; McCoustra, M. R. S.; O'Neill, K. E. *Proc. XIX Conf. British Mass Spectrometry Society.* 1992, p. 110–111.

92. Creaser, C. S.; O'Neill, K. E. *Anal. Proc.* 1993, *30*, 246–248.

93. Karas, M.; Bachmann, D.; Bahr, H.; Hillenkamp, F. *Int. J. Mass Spectrom. Ion Processes.* 1987, *78*, 53.

94. Cox, K. A.; Williams, J. D.; Cooks, R. G.; Kaiser, R. E. *Biol. Mass Spectrom.* 1992, *21*, 226.

95. *Lasers and mass spectrometry,* Lubman, D. M. (ed.), Oxford University Press, New York, 1990.

96. Hughes, R. J.; March, R. E.; Young, A. B. *Int. J. Mass Spectrom. Ion Phys.* 1982, *42*, 255.

97. Hughes, R. J.; March, R. E. *Can. J. Chem.* 1983, *61*, 824–833.

98. Hughes, R. J.; March, R. E. *Can. J. Chem.* 1983, *61*, 834.

99. Louris, J. N.; Broadbelt, J. S.; Cooks, R. G. *Int. J. Mass Spectrom. Ion Processes.* 1987, *75*, 345.

100. Creaser, C. S.; McCoustra, M. R. S.; O'Neill, K. E. *Org. Mass Spectrom.* 1991, *26*, 335–338.

101. O'Neill, K. E. Ph.D. Thesis, University of East Anglia, U.K. 1992.

Chapter 3

REVIEW OF MODERN ION TRAP RESEARCH

M. Judith Charles and Gary L. Glish

CONTENTS

0-8493-8251-3/95/$0.00+$.50
© 1995 by CRC Press, Inc.

I. INTRODUCTION

The advent of the mass-selective instability mode of operation by Stafford et al.[1] marks the beginning of the modern era of ion trap technology. Following this innovation, advances in the ion trapping field have occurred at an unprecedented rate, compared to the first 25 years after Paul and Steinwedel[2] invented the ion trap. The result is a new generation of ion trap that has the capacity to perform more different types of mass spectrometry experiments than any other mass spectrometer. These capabilities will allow the ion trap to be used not only as a powerful analytical instrument, but also as a valuable tool for fundamental studies in chemistry and related areas.

This chapter will provide an overview of the advances in ion trap technology and the capabilities of the ion trap by reviewing recent developments in the field. The chapter is divided into three major areas of ion trap research and development: combining various ionization techniques with the ion trap; the operation and performance of the ion trap as a mass analyzer; and performing tandem mass spectrometry (MS/MS) experiments with the ion trap.

II. IONIZATION TECHNIQUES

Recent work has shown the ion trap to be amenable to common ionization techniques used with sector and linear quadrupole instruments (for example, electron and chemical ionization, fast atom bombardment, electrospray, etc.). In addition, the ion trap is well-suited to pulsed ionization techniques (for example, laser ionization) normally implemented on time-of-flight (TOF) mass spectrometers or ion cyclotron resonance (ICR) instruments.

Sector, linear quadrupole, and TOF instruments are so-called "beam" instruments with the ions being formed in an ion source and then transported through an electric and/or magnetic field for mass-to-charge analysis. Conversely, the ion trap and the ICR are trapping instruments and the ions can be formed *in situ*, that is, within the trapping volume, or injected from an external ion source. While the ion trap and ICR have many similarities, an advantage of the ion trap is that it is capable of operating at much higher pressures. This capability allows atmospheric pressure ionization (API) techniques to be implemented more easily on the ion trap. In addition, the higher operating pressure of the ion trap as compared with the ICR instrument leads to better trapping efficiency of ions injected from external ionization sources. While API and other external ionization methods are probably going to be the more important ionization techniques used with the ion trap in the future, it is useful to discuss first the more conventional ionization techniques used with the quadrupole ion trap, in which the ions are formed in the trapping volume of the instrument.

A. Ion Formation Within the Trapping Volume

Ion formation within the trapping volume by electron ionization (EI) was the dominant mode of operation of the ion trap before development of the mass-selective instability mode of operation. Formation of the ions within the analyzing field presumably leads to more efficient collection of the ions, although this assumption is based mainly on the observation of higher sensitivities rather than quantitative measurements. While internal ionization is very effective, a major limitation of internal ion formation is that the sample must be volatile.

1. Electron Ionization

Electron ionization is performed in the trapping volume of the ion trap by gating electrons into the ion trap from an external filament. Since the ion trap operates with a time-varying voltage, a distribution of energies of the electrons exists that depends on the phase of the radio frequency (RF) when the electrons enter the ion trap, and on the amplitude of the RF. Calculations suggest an average electron energy of about 50 eV when operating at a typical low mass cutoff of 20 to 30 Da.[3] This average energy is consistent with the observation that under appropriate conditions (for example, when the ion trap is not overloaded with sample (*vide infra*)), the ion trap provides EI mass spectra comparable to standard 70 eV EI mass spectra obtained on sector or quadrupole instruments, allowing the use of mass spectral libraries. The trapping voltages (RF and DC) also influence the trajectories of electrons and the volume of the trapping field in which the ions are stored.[4] Such influences have implications for ionization strategies in which the analysis is focused on certain mass/charge ratios. For a given analysis the preferred combination of trapping voltages during ionization may depend upon the nature of the sample such as concentration of the analyte and the nature of the matrix.

Because the physics of EI in an ion trap is the same as other EI sources, the number of analyte ions formed is dependent on the ionization cross section, electron path length, analyte number density, and electron flux. Assuming a fixed analyte number density, the only variable term is the electron flux. This parameter can be controlled crudely for all EI sources by altering the emission current of the filament. In the ion trap, however, the electron "flux" (that is, the number of ionizing electrons per ionization pulse) can also be varied by changing the time that the electrons are gated into the ion trap. The time can be varied over several orders of magnitude and can lead to substantial changes in sensitivity (ion signal/quantity of analyte) and an increase in the dynamic range.[5] When a small quantity of sample is present in the ion trap, a long ionization time can be used to form a sufficient number of ions for efficient detection. Conversely, when a large quantity of sample is present, shorter ionization times can be used to prevent the accumulation of too many ions in the ion trap; too many ions result in space-charge effects which have a deleterious effect on the quality of the mass spectrum. For example, in the analysis of benzophenone, long ionization times can be used to increase the sensitivity at analyte pressures of 8.1×10^{-9} Torr without observing space-charge effects (Fig. 3.1(a)). At higher analyte pressures (7.0×10^{-8} Torr, Fig. 3.1(b)), however, ionization times greater than 8 ms result in a loss of mass resolution and a nonlinear response of the analyte *versus* ionization time, which indicates space-charge and/or competitive reactions. The optimum ionization time that avoids space-charge effects can be determined automatically on commercial instruments using automatic gain control (AGC). When AGC is implemented, a short ionization/detection cycle is used to determine the relative analyte concentration. From this determination of analyte concentration, the ionization time for the analytical scan is set so that an optimal number of ions are formed (i.e., a number just below the level at which space-charge effects begin to be observed). AGC allows the dynamic range of the ion trap to be enhanced by maximizing the sensitivity (long ionization times) when sample concentration is low, and reducing the sensitivity (shorter ionization time) when the analyte concentration is larger.

Other strategies have been implemented to enhance the dynamic range and sensitivity of the ion trap, particularly in cases where a small quantity of analyte exists in the presence of a larger quantity of interferant(s). One approach, which was implemented prior to the current era of ion trap development, is the use of both RF and DC voltages on the ion

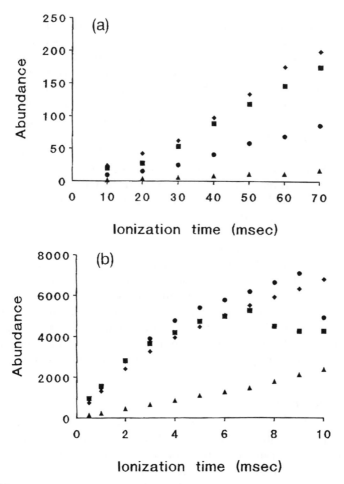

FIGURE 3.1

Plot of the abundance of m/z 182 (▲), m/z 105 (●), m/z 77 (■) and m/z 51 (♦) formed by electron ionization of benzophenone as a function of time at analyte pressures of (a) 8.1×10^{-9} Torr, and (b) 7.0×10^{-8} Torr. (Reprinted from Ref. 5 with permission.)

trap.[6] However, the previously mentioned work indicates that operating the ion trap in this mode during ionization drastically reduces the volume in which ions can be trapped when they are formed.[4] A better alternative to improve the sensitivity for a small component in the presence of a larger one is to perform multiple, short ionization pulses, rather than one long ionization pulse, while using RF/DC mass-selective stability between ionization pulses to eject ions other than those of interest.[7] More recently, use of resonant ejection (*vide infra*) has been shown to be a very useful means to increase dynamic range during EI in an ion trap.[8] Unwanted ions can be ejected resonantly during the electron pulse by a broadband frequency applied to the end-cap electrodes, with the resonant frequency of the ions of interest excluded from this broadband resonant excitation signal. There are a number of ways in which such a broadband signal can be generated such as with an arbitrary waveform generator[9] or an inverse Fourier transform of the desired frequency spectrum.[10]

It is generally accepted that ions are extracted from a conventional EI source a few microseconds after formation. In the ion trap they may be stored for up to tens of milliseconds prior to analysis. The major consequence of this time difference is a greater likelihood for the ions to undergo an ion/molecule reaction. While ion/molecule reactions can be useful, unwanted reactions can be deleterious to a mass spectral analysis.

The loss of analyte ions by ion/molecule reactions depends upon the reaction rate constant, the reaction time, and the number density of the *neutral* reactant. For a given reaction rate constant, compensation for the longer potential reaction time in the ion trap can be achieved only by lowering the concentration of the neutral reactant, which is often the analyte itself. Thus, for reactions with relatively large reaction rate constants, smaller amounts of analyte must be used to prevent a commonly-observed ion/molecule reaction, self-chemical ionization.[11] Fortunately, because of the sensitivity of the ion trap, it is usually possible to obtain EI spectra with lower sample amounts than are typically used in beam instruments, reducing the likelihood of ion/molecule reactions.

2. Chemical Ionization

Chemical ionization (CI) in an ion trap makes use of the long reaction times accessible with the ion trap, rather than high reagent gas number densities of conventional high pressure CI sources. Low pressure CI in an ion trap was first demonstrated when an ion trap was used as a "source" for a quadrupole mass filter.[12] It was more than a decade later when it was demonstrated that chemical ionization and subsequent mass analysis, could be done with just an ion trap.[13] In this study, in addition to conventional CI, low mass reagent ions (for example CH_5^+ from methane) were ejected using mass-selective instability, allowing higher mass reagent ions (e.g., $C_2H_5^+$) to be the primary proton transfer agents.

Mass-selective stability was incorporated subsequently into the CI experiment to allow any mass/charge to be isolated for use as the reagent ion.[14,15] Selection of the reagent ion with which to perform chemical ionization can be useful for determining the genesis of unusual ions in CI spectra[16,17] and also provides more control over the ionization process, compared to the CI experiment on a beam instrument. For example, $C_2H_5^+$ imparts less internal energy to the molecule upon ionization by proton transfer than does CH_5^+ (1.43 eV as compared with 2.78 eV, for phenylacetonitrile; and 1.26 eV as compared with 2.61 eV, for *n*-butylbenzene). Thus,

TABLE 3.1

Comparison of chemical ionization mass spectra* obtained with $(CH_5)^+$ and $(C_2H_5)^+$

Compound	m/z	Reactant Ion	
		$(CH_5)^+$	$(C_2H_5)^+$
		(% Relative Abundance)	
Phenylacetonitrile (MW = 117)	118	37	100
	116	2	<1
	91	100	45
n-Butylbenzene (MW = 134)	135	70	100
	133	18	8
	91	27	8
	57	100	27

* Table derived from Ref. 15.

proton transfer using $C_2H_5^+$ results in mass spectra characterized by higher relative abundances of the $(M + H)^+$ ion and lower relative abundances of fragment ions compared to the CH_5^+ CI mass spectra (Table 3.1). Selecting a specific reagent ion is also useful for charge exchange reactions. Using N_2O as the reagent gas, both NO^+ and N_2O^+ are produced *via* EI, with the former having a recombination energy more than 3.5 eV lower than the latter. Thus, in the charge exchange ionization of etioporphyrinogen III, the selection of NO^+ as the reagent ion produces only a molecular ion, whereas $N_2O^{+\cdot}$ charge exchange generates structurally significant fragments in addition to the molecular ion.[18]

Due to the low reagent gas pressure of CI in the ion trap, a vast array of ions, which cannot be formed in sufficient abundance in a high pressure CI source, can be generated to serve as CI reagent ions. Thus, investigation of new, "unconventional" CI reagent ions which undergo reactions other than simple proton or charge transfer, is a promising future area of research with ion traps. One example of such work is that of using reagent ions from dimethylether to form $(M + 13)^+$ ions by reaction with certain organic functional groups.[19]

The vast majority of CI work using ion traps involves creating and detecting positive ions. Negative ion CI experiments generally require that electrons be captured as the initial step in the reaction sequence, that is, for the formation of negative reagent ions. Electrons are not trapped under typical ion trap operating conditions and the probability of electron capture is small for electrons injected into the ion trap. Additionally, the injected electrons can (and do) form positive ions, which can then neutralize the negative ions *via* ion/ion recombination reactions. Chemical ionization using H_2O as the reagent gas and detection of the negative ions formed has been demonstrated in the ion trap, although as expected, the sensitivity was not very good.[20] However, when negative reagent ions are formed externally and injected subsequently into the ion trap, then negative ion CI can be as sensitive as positive ion CI.[21]

3. Photo-Ionization

Few studies have been performed coupling photo-ionization with an ion trap, even though this technique is compatible with the ion trap. Prior to the current generation of ion traps, a cylindrical ion trap with photo-ionization was used to study kinetic shifts.[22] Only two studies have described photo-ionization coupled to the current generation of ion traps.[23,24] Both of these studies used resonance enhanced multiphoton ionization (REMPI).

In one case, the 2 + 2 ionization of NO was studied to ascertain the potential of combining the ion trap with REMPI.[23] The NO system was chosen since its spectroscopy is fairly well understood. The ion trap gave a concentration detection limit 30 times better than a supersonic beam TOF apparatus using a 2 + 1 ionization scheme, while the number of molecules detected in the ion trap was 150 times lower.

The other study using photo-ionization demonstrated the utility of REMPI combined with MS/MS for improved selectivity and sensitivity in mixture analysis.[24] Such was the case even though fragmentation during ionization could be controlled *via* the laser power, and that the same fragment ions were observed during ionization as *via* MS/MS.

B. Injection of Externally Formed Ions

The capabilities of the ion trap can be extended to the analysis of involatile compounds by injecting externally formed ions into the ion trap. Early work exploring ion injection into an ion trap, much of it theoretical, has been reviewed.[25] The current era of ion injection into ion traps began with work performed at Purdue University[26] and Oak Ridge National Laboratory.[27] Since this initial work, injection of externally formed ions from a wide variety of ionization sources has been accomplished.

1. Electron and Chemical Ionization

As discussed previously, electron and chemical ionization can readily be accomplished within the trapping volume of the ion trap. However, it is possible to use external EI and CI sources with an ion trap. Forming the ions in an external source has the advantage of separating the analyte and the CI reagent gas from the mass analysis region, which greatly reduces the potential for ion/molecule reactions. It was noted previously that self-CI can occur under EI conditions in the ion trap. It has also been shown in chemical ionization studies that analyte ions can react with the neutral reagent gas, resulting in the formation of ions that are not characteristic of the analyte.[28] Avoiding such potential reactions was cited as one reason for using an external EI source in the initial ion injection experiments.[26] An external EC/CI source was combined with the ion trap in another study, although the main focus of this work was the investigation of injection of ions from a source not in line-of-sight with the ion trap.[29] While neither of these efforts was very intensive, they demonstrated the ability to perform such experiments and it seems likely that external EI and CI sources will be used more frequently in the future.

2. Atmospheric Sampling Glow Discharge Ionization

The atmospheric sampling glow discharge ionization (ASGDI) source was the first high-pressure source interfaced to an ion trap.[27,30] The ability of the ion trap to operate at relatively high pressures facilitates coupling of high-pressure sources such as the ASGDI source. Initial work using the ASGDI/ion trap combination reported mass spectra of TNT sampling an estimated 100 fg in air (based on head space vapor pressure) in 1 ms; MS/MS spectra of five times this amount of material; and ion dissociation during injection. Important parameters for dissociation during injection were the RF level, bath gas pressure, and nature of the bath gas. A linear relationship between the RF level for maximum injection efficiency and mass was also noted. Others have also noted the relationship between RF level and injection efficiency as reported for ASGDI but, in addition, they observed local maxima and minima when observing ion intensity *versus* RF level during injection.[29,31]

The glow discharge source has also been used to generate negative CI reagent ions for injection into the ion trap. Much higher abundances of negative CI reagent ions are formed in this manner, leading to greatly improved negative CI sensitivities in the ion trap.[21]

3. Laser Desorption

Laser desorption was the first ionization technique that was applied for analyzing involatile compounds using an ion trap. Initial work evaluated two approaches for performing the experiment. In one approach, externally-ionized metal ions were formed by laser ablation and a lens system was used to inject metal ions into the ion trap.[26] In the other approach,[32,33] organic ions were laser-desorbed directly into the ion trap by bringing the laser beam through a hole in the ring electrode, that is, opposite a second hole into which the sample probe was inserted. This approach is somewhat intermediate between ion injection and internal

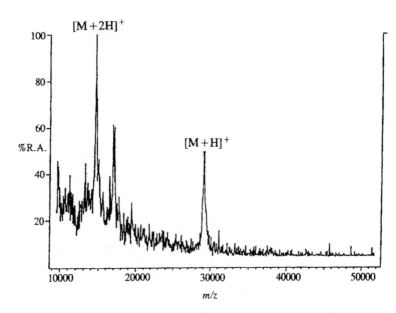

FIGURE 3.2
The MALDI ion trap mass spectrum of bovine carbonic anhydrase (MW 29,025) using a reduced fundamental frequency. (Reprinted from Ref. 35 with permission.)

ion formation since the ions are formed right at the edge of the trapping field. For the laser-desorped organic ions it was found that increasing the helium bath gas pressure increased the trapping efficiency of the desorbed ions.[32] Detections limits in the 10 pmol range were reported and MS/MS of laser-desorbed ions was demonstrated in this study.

More recently, matrix-assisted laser desorption ionization (MALDI) has been implemented on the ion trap. Detection limits in the low femtomole range for peptides under mass 1500 were reported by desorbing the sample directly into the ion trap.[34] A relationship was noted between the optimum RF potential for trapping the desorbed ions and the mass of the ion trapped. The sensitivity was improved for higher mass species (bovine insulin MW = 5734, and cytochrome c MW = 12,364) by using higher pressures of the bath gas and using argon instead of helium. MS/MS and MS^n was demonstrated also for MALDI-generated peptide ions.

MALDI experiments have been carried out also by performing laser desorption external to the ion trap.[35,36] In one of these studies,[35] the fundamental RF frequency was reduced to provide a deeper trapping well, so as to allow analysis of higher mass ions. Fig. 3.2 shows the MALDI mass spectrum of bovine carbonic anhydrase (MW 29,025); other data were presented showing the detection of ions with mass/charge ratios in excess of 40,000. MS/MS of bovine insulin was reported also, demonstrating the capability of the ion trap to dissociate singly-charged ions of much greater mass than can be done with other types of mass spectrometers. The other MALDI study focused on detection limits and the relationship between optimum RF trapping level and mass.[36] A linear relationship between the square root of the mass and the optimal RF level was found, and a detection limit of 10 fmol for sperm-whale myoglobin (avg. MW = 17,294) was reported. This study also compared MS/MS spectra for MALDI- and liquid SIMS-derived ions and found them to be quite similar.

4. Electrospray

Electrospray ionization was first coupled with the ion trap via a modified ASGDI source. The modification consisted of a lens assembly added in the intermediate pressure region and an atmosphere aperture half the diameter of that used for ASGDI.[37] This electrospray arrangement is noteworthy for its simplicity. While most electrospray sources use either a countercurrent gas flow or heat to desolvate the electrosprayed ions, neither is required when using an ion trap. A unique feature of the ion trap is that the ions can be desolvated after they are trapped. Desolvation occurs under normal operating conditions for weakly solvated ions[37] or it can be induced for more strongly solvated ions by resonant excitation.[38] Desolvation by resonant excitation is shown in the myoglobin mass spectra in Fig. 3.3. The mass spectrum in Fig. 3.3(a) is that obtained without any resonant excitation over the mass range shown. The ion current is spread out over a wide range of mass/charge ratios with a large distribution of solvent molecules attached to the ions. The spectrum in Fig. 3.3(b) was obtained by exciting the ions resonantly and causing them to desolvate prior to detection. As can be seen, this desolvation procedure leads to a typical electrospray mass spectrum with predominately one mass for each charge state.

Since electrospray is a soft ionization technique, protonated molecules and other molecular adducts are typically formed. The ions must be energized after ionization to induce fragmentation in order to yield structural information. Ion activation is often accomplished on beam instruments by having a voltage gradient in the interface region to effect collision-induced dissociation.[39] The interface CID method can be employed with the ion trap or, alternatively, injection-induced dissociation can be effected by varying the RF amplitude during ion injection.[40] For porphyrins, similar fragmentation can be induced by either approach as shown in Fig. 3.4. While it is believed that the injection-induced dissociation observed in the ion trap is a surface-induced dissociation process, this has yet to be proven.

Sensitivity and detection limits achieved with electrospray ionization combined with ion trap mass spectrometry have been demonstrated to be equivalent to, or better than, those obtained with beam instruments.

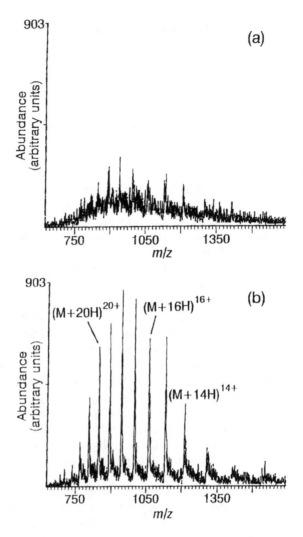

FIGURE 3.3

(a) The electrospray mass spectrum of myoglobin without using resonant excitation to de-solvate the analyte ions, and (b) the electrospray mass spectrum of myoglobin using resonant excitation to de-solvate the analyte ions in the ion trap. (Reprinted from Ref. 38 with permission.)

In the initial work, low femtomole amounts of sample for peptides and proteins ranging in mass from 555 Da (leucine enkephalin) to 66,000 Da (bovine albumin) were detected.[37] These levels cannot be considered detection limits because the experiments were performed by continuous infusion of sample. They do, however, demonstrate the potential of the technique, provided that level of sample can be introduced into the ion

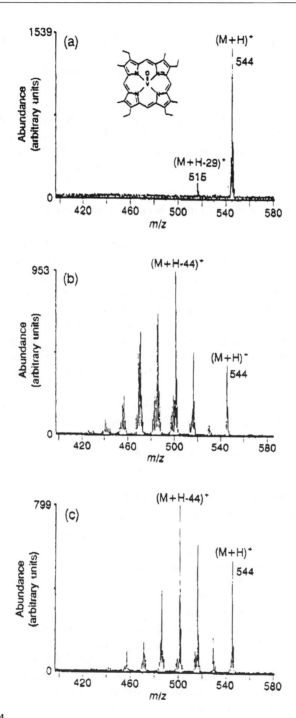

FIGURE 3.4
The electrospray mass spectra of vanadyl etioporphyrin III obtained by: (a) low voltages in the electrospray interface region and an ion trap low-mass cut-off of 50 Da, (b) using higher voltages in the electrospray interface region to effect interface CID, and (c) with low voltages in the interface region but an ion trap low-mass cut-off of 150 Da to effect injection-induced dissociation. (Reprinted from Ref. 40 with permission.)

trap over an appropriate period of time. One method to introduce the sample into the ion trap over a short period of time is flow injection. A study using flow injection demonstrated detection of low femtomole amounts of porphyrins.[40]

High-performance liquid chromatography has been combined with ion spray/ion trap analysis. In this study, sub-picomole amounts of sample injected onto the column were detected.[41] The combination of liquid chromatography/electrospray/ion trap will likely be one of the most important applications of the ion trap in the future.

5. Fast Atom Borbardment/Liquid Secondary Ionization Mass Spectrometry

For the last decade, fast atom bombardment and liquid secondary ionization mass spectrometry (LSIMS) have been the mainstay ionization technique for the analysis of involatile and thermally labile polar molecules. LSIMS has been combined with an ion trap and the results reported thus far are encouraging.[42,43] By using an ion beam as the bombarding species rather than a neutral atom beam, the bombarding beam can be deflected readily from the sample. Thus, when using an ion trap, sample can be conserved by arresting bombardment of the sample during stages of analysis when ions are not being stored in the ion trap.

While only a little work has been done combining FAB/LSIMS with the ion trap, that which has been performed has been impressive. Mass spectra of sub-femtomole and MS/MS spectra of low femtomole amounts of analyte loaded onto the sample probe have been reported.[43]

6. RF-Glow Discharge for Elemental Analysis

The ion trap has also been interfaced with an RF-glow discharge source for elemental analysis.[44] A unique feature of this combination is that matrix ions (for example, ArH^+, Ar^+, Ar_2^+) either are dissociated upon injection into the ion trap or react with neutral compounds present in the ion trap (for example, residual water vapor). These reaction products, (for example, m/z 18 or 19) have a mass/charge ratio below the low-mass cut-off. These ions are ejected, therefore, and thus do not contribute to space charge in the ion trap. Water was found also to react with oxides, forming hydroxides. The hydroxides are easier to dissociate than oxides and MS/MS efficiencies of close to 100% were observed, demonstrating the possibility of transforming or removing interfering polyatomic ions. By using mass-selective ion injection techniques, detection limits in the low parts per million were achieved.

III. OPERATION OF THE ION TRAP

As previously mentioned, the development of the current generation of quadrupole ion traps began with the invention of the mass-selective instability mode of operation by Stafford et al.[1] In this mode of operation, all ions with a mass/charge greater than a certain value (often referred to as the low-mass cut-off) are contained in the ion trap by applying an RF voltage only (no DC voltage is applied) to the ring electrode. In this mode of operation, the low-mass cut-off is determined by the axial (z) boundary of the stability region as defined by the Mathieu equations. With no DC applied, the pertinent equation is:

$$q_z = \frac{8eV}{m(r_0^2 + 2z_0^2)\Omega^2}$$

(3.1)

in which, q_z is the Mathieu parameter, V is the amplitude of the RF, Ω is the radial frequency of the RF, r_0 the radius of the ion trap, e the charge of the ion, and m the mass of the ion.

The q_z-axis of the stability region intersects the z boundary ($\beta_z = 1$ boundary of Fig. 2.1 (Chapter 2)) of the stability region at $q_z = 0.908$ and thus, nominally, any ions with q_z values less than 0.908 have stable trajectories within the ion trap volume and, thus, are trapped. From Eq. (3.1), the q_z value of an ion is inversely proportional to its mass and proportional to the amplitude of the applied RF voltage. By increasing the RF amplitude, ions of higher mass/charge ratios are destabilized sequentially in the z-direction and ejected from the ion trap. By placing a detector such that these ejected ions can be detected, a mass spectrum can be obtained by simply ramping the fundamental RF amplitude.

A. Resonant Ejection

Resonant ejection, or axial modulation, in the standard ion trap scan, is replacing the use of mass-selective instability for obtaining mass spectra because it enhances sensitivity and mass resolution.[45] In resonant ejection, a supplemental RF is applied across the end-cap electrodes of the ion trap with a typical amplitude of 5 to 20 $V_{(p-p)}$. When an ion has a secular frequency of motion (vide infra) that corresponds to that of the supplemental RF potential, the ion will absorb power. This power absorption causes the ion to gain kinetic energy which can lead, in the limit, to the ion being ejected from the ion trap. The secular frequency, ω_z, of an ion is determined by the following equation:

$$\omega_z = \frac{\beta_z \Omega}{2} \tag{3.2}$$

β_z is a function of the Mathieu parameters a_z and q_z and can range from 0 to 1. Thus, ions can have axial secular frequencies up to half that of the RF drive frequency. In the mode of resonant ejection known as axial modulation, β_z is close to one; thus, the supplemental RF frequency at which axial modulation normally occurs is slightly less than half of the RF drive frequency.

B. Mass Range Extension

The mass range of the ion trap can be extended by performing resonant ejection at β_z values much lower than one, altering r or Ω, or scanning the DC, or the DC and RF voltages.[46-48]

The factor by which the mass range is increased by resonant ejection at lower β_z values depends on the resonant ejection frequency.[42] This mass range extension factor can be calculated by the following equation:

$$F_{mr} = \frac{0.908}{q_{z,re}} \tag{3.3}$$

in which F_{mr} is the mass range extension factor and $q_{z,re}$ is the Mathieu q_z value at which resonant ejection is effected. The 0.908 comes from the $q_{z,max}$ associated with the mass-selective instability mode of operation. Thus, the normal mass range of 650 Da per charge for the Finnigan ion trap, operated in the mass-selective instability mode is increased to 1300 Da per charge if resonant ejection is effected at a $q_{z,re}$ of 0.454, and is increased to 10,0000 Da per charge at a $q_{z,re}$ of 0.059.

From Eq. (3.1), the mass/charge ratio with a q_z value of 0.908 can be increased by decreasing r or Ω, and thus the mass range can be extended when operating in the mass-selective instability mode or using resonant ejection.[42,46] Reducing Ω is preferred because making smaller electrodes to reduce the radius amplifies field imperfections, which can cause deterioration of performance. Good performance can be obtained by reducing the frequency, within a limited range.[47]

As an alternative to resonant ejection or decreasing r or Ω and rather than using the crossing of the $\beta_z = 1$ boundary as point of instability, instability at the $\beta_z = 0$ boundary has been investigated as a means to extend the mass range.[48] Four different methods have been explored.

These methods are known as *capscan, downscan, reverse,* and *anglescan*. The *capscan* method is the simplest and involves scanning a DC voltage that is applied to one of the end-cap electrodes of the ion trap assembly; however, this mode of operation gave mass spectra of the poorest quality. In the *downscan* method, a DC voltage that is applied to the ring electrode is ramped downwards. This scan mode gives better spectra than the *capscan* but suffers from ejection of ions due to nonlinear resonances. These same nonlinear resonances are observed in the *reverse* scan in which the RF and DC potentials applied to the ring electrode are scanned such that the ratio of a_z/q_z is constant. (This mode is similar to the operation of a linear quadrupole except that, in the *reverse* scan, the scan line intersects a much wider portion of the stability region and the RF and DC voltages are decreased during the scan.) The most promising of the scan modes using instability at the $\beta_z = 0$ boundary is the *angle scan*. Here the RF and DC are both scanned, but in opposite directions. By appropriate choice of RF and DC values, the scan lines intersect at close to right angles with the $\beta_z = 0$ boundary, thereby increasing the resolution.

From the various studies of the methods to extend the mass range of the ion trap, it appears that resonant ejection is the preferred method; it is the easiest to implement, it gives the greatest mass range, and provides the best sensitivity. If the fundamental frequency, Ω, can be varied easily, this method may be used also, and in combination with resonant ejection.

C. High Mass Resolution

The quadrupole ion trap is often considered from the same point of view as the quadrupole mass filter, and so it may be assumed that, like the quadrupole mass filter, only modest mass resolution can be obtained. However, from considerations of the relative pathlengths of ions in the two instruments, it is evident that much higher mass resolution should be obtainable with the ion trap and, indeed, much higher mass resolutions with the ion trap have been demonstrated recently.

Preliminary work demonstrated that varying the scan speed led to variations in resolution.[47] Subsequently, it was shown that by slowing down the scan speed substantially in the resonant ejection mode of operation (by decreasing the rate at which the RF drive potential amplitude is ramped), high mass resolution could be obtained with a quadrupole ion trap.[49,50] Alternatively, the amplitude of the RF drive potential can be held constant while the resonant ejection *frequency* is scanned slowly.[51] The ability to perform high mass-resolution experiments is a significant advancement in ion trap technology. Now, experiments can be performed on the ion trap that previously required the resolution of a sector or Fourier transform ion cyclotron resonance (FT-ICR) instrument, for example, the

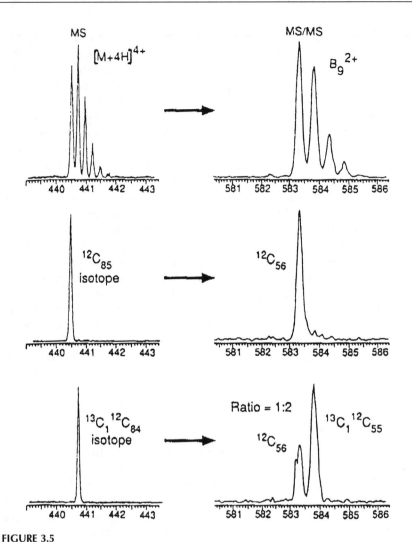

FIGURE 3.5
Demonstration of the high resolution capabilities of the ion trap. The upper spectra show the isotope-resolved cluster (left) of the quadruply-protonated renin substrate (MW 1757.9) and a region of the MS/MS product ions spectrum (right) from these ions. The middle spectra show the isolated ^{12}C peak from the isotope cluster (left) and the resulting B_9^{2+} product ion. The bottom spectra are the same as the middle except that a parent ion containing one ^{13}C is isolated for MS/MS. (Reprinted from Ref. 54 with permission.)

separation of carbon isotopes in electrospray mass and MS/MS spectra, as shown in Fig. 3.5.

According to a theory developed to explain the effects observed when the scan rate is varied,[52] a slight shift in peak position should be (and is) observed as the scan rate is decreased.[51] The validity of the theory has been supported by experiment.[53] This experimental work demonstrated that the shape of the curve of resolution *versus* scan rate matches that predicted by the theory, although the absolute value of the experimental resolution is higher than that predicted by the theory. This difference in absolute value of the resolution is believed to arise from the fact that the theory is based on lineshapes of ion motion *within* the ion trap, whereas the observed resolution depends upon ions which are *ejected* and impinge on the detector; that is, it depends on the apex of the lineshape.

It is noteworthy that, since the final analytical scan during which ions are ejected and detected is independent of any prior manipulations of the ions, high mass resolution can be obtained for MS or MS/MS spectra.[49] In addition, parent ion selection can be performed at high mass resolution for MS/MS experiments.[54]

The best mass resolution, defined as the ratio of the mass to the full width at half height, demonstrated to date with an ion trap is over twelve million at m/z 614.[53] A significant difference exists between high mass resolution on an ion trap and a FT-ICR even though the resonant absorption process leading to high mass resolution is somewhat analogous to FT-ICR. When the experimental parameters are held constant in FT-ICR, the resolution decreases as the mass increases, while resolution in an ion trap increases with mass.

D.　Mass Accuracy

Coupling ionization techniques such as electrospray ionization and laser desorption to the ion trap have created the need to perform accurate mass measurements on high-molecular weight molecules such as proteins and peptides. This need has become a current challenge in the development of ion trap technology. Recently, it was revealed that the ion trap manufactured by Finnigan is not a pure quadrupolar geometry.[55,56] The geometry of the ion trap was stretched to alleviate mass accuracy problems observed with a pure quadrupolar geometry,[55] although such problems are still sometimes observed.[57] The background story to the "stretching" of the ion trap is recounted in Chapter 4 of Volume 1 of this series.[58]

Recently, a peak-matching procedure[59] was developed, using CsF cluster ions as internal calibrants. This peak-matching procedure provides mass accuracies of better than 0.007% (Table 3.2). The procedure is, however, awkward to implement on a routine basis. Thus, other ways to improve mass accuracy will undoubtly be developed.

E.　Higher-Order Multipole Fields

The vast majority of work that has been done since the development of mass-selective instability has used Finnigan ion traps. Due to the stretched

TABLE 3.2

Accurate mass-measurements of protonated Gramicidin S molecules*

Calibrated m/z	Theoretical m/z	Mass Accuracy
1141.79	1141.72	0.006%
1142.79	1142.72	0.006%
1143.80	1143.72	0.007%
1144.70	1144.73	0.0009%
average mass accuracy		0.005%

* Table derived from Ref. 59.

geometry of the Finnigan ion trap, the trapping field is not a pure quadrupole field. The actual field can be described as a quadrupole field with superimposed higher-order multipole fields.[60] Whereas the quadrupole field provides a linear restoring force as a function of the distance from the center of the field, these higher-order fields are non-linear and recent results indicate that they can affect the performance of the ion trap. Nonlinear fields have been discussed at length in Chapter 3 of Ref. 58.

It has been observed that MS/MS product ions with a q_z value of 0.78 can be lost from the ion trap prior to mass analysis.[61] This behavior has been attributed to an overtone of the fundamental RF frequency.[62] An expanded investigation of this observation showed that the $a_z = 0$, $q_z = 0.78$ point is just one point along a loci of points in the Mathieu a_z, q_z space along which ion loss is observed.[63] Recent work has provided a theoretical basis for the experimental observations.[64] This work determined that hexapole fields are present due to the holes in the end-cap electrodes and octapole fields which occur as a result of the stretched geometry.

F. Ion Temperature

While a thermal equilibrium does not exist in the ion trap, the concept of the effective ion "temperature," based on kinetic and internal energies, has been of interest. In an attempt to measure effective ion temperatures, ion/molecule reactions with known dependencies on kinetic energy have been used as probes of the ion kinetic temperature. The initial study used the reaction of O_2^+ with CH_4 and estimated a temperature of 600 to 700 K.[65] A subsequent re-examination of this reaction and the reaction of Ar^+ with N_2 concluded that the ion kinetic temperatures were in the range of 1700 to 3300 K.[66] There are a number of assumptions made in these temperature determinations and ion temperature will no doubt continue to be an area open to debate.

It is generally accepted that the ions in an ion trap are cooled kinetically by collisions with the buffer gas. One possible channel for kinetic cooling is conversion into internal energy, which would mean that the ion internal temperature is different from that of the bath gas temperature. Measurements of the dissociation rates of protonated water and methanol clusters have been made to investigate internal temperatures.[67] It was found that the internal temperature exceeds that of the bath gas and is, typically, in the range of 350 to 700 K. It was observed also that the temperature is related to bath gas pressure, decreasing with increasing bath gas pressure.

G.   Trajectory Simulations

To increase the understanding of experimental results related to the operation of the ion trap, a number of groups are performing ion trajectory calculations. One group of programs has evolved to allow increasingly complicated simulations and these have been applied to problems such as resonant excitation.[68,69] Simpler simulations have also been performed to investigate the resonant excitation process.[3] Other computer programs have been developed which allow simulations of complex operating modes of the ion trap and include effects such as collisions with the He buffer gas and space charge.[70] These latter programs have provided theoretical support to experimental observations such as the kinetic energy of ions ejected from the ion trap.[71]

A potential caveat to the simulations done to date is that most simulations have used pure quadrupole fields and thus potential effects of the higher-order nonlinear fields are neglected. One exception to this is a study investigating the effect of intentionally-added higher-order fields.[60]

IV.   TANDEM MASS SPECTROMETRY

A.   Fundamentals of Tandem Mass Spectrometry, MS/MS

Perhaps the most impressive attribute of the quadrupole ion trap is its ability to effect MS/MS experiments. There are at least four features of the MS/MS experiment in an ion trap that standout compared to what can be accomplished with beam instrument mass spectrometers. These features are the collision-induced dissociation (CID) MS/MS efficiency, the ability to perform MS^n experiments, the high efficiency of ion/molecule reactions, and the upper mass/charge limit for CID.

The MS/MS efficiency is probably the best known feature of the ion trap. The most commonly assumed reason for this high efficiency is that ions do not have to be transported from one analyzer to a collision region to a second analyzer as in beam instruments. While this "transmission" advantage of the ion trap undoubtedly contributes to its MS/MS efficiency, a less well recognized, but perhaps more important, factor is the time scale of the experiment. Ions typically have at least two to three orders of magnitude longer time to dissociate in the ion trap than in beam instruments. This long time scale for dissociation means that the kinetic shift is reduced and thus less internal energy is required to induce dissociation. An alternative perspective is that for a given internal energy a far greater fraction of ions will dissociate on the time scale of the ion trap MS/MS experiment than for a beam instrument MS/MS experiment.

Most of the CID experiments performed in the ion trap have been effected by resonant excitation. Resonant excitation is the same as resonant ejection except that a resonant frequency of much lower amplitude is used. The lower amplitude causes the rate of power absorption to be slower. Thus, an ion can undergo many collisions with the helium buffer gas, leading to dissociation, before it acquires sufficient kinetic energy to be ejected from the ion trap.

A subtle but potentially important difference exists between collisional activation by resonant excitation in an ion trap and collisional activation in a beam instrument. In a beam instrument, an ion undergoes a collision and converts some of its kinetic energy to internal energy. Thus, there is a finite limit in the number of inelastic collisions an ion can undergo before it loses all of its kinetic energy so that it is no longer able to pass into the next analyzer. In the ion trap, however, the ion is re-accelerated between collisions by the resonant excitation and hence a limit does not exist on the number of inelastic collisions an ion can undergo. Dissociation is, however, a competitive process with resonant ejection (in addition to other forms of internal energy dissipation) and, under certain conditions ejection can be the dominant process even when low resonant frequency amplitudes are used.[72]

Parameters important in the resonant excitation CID experiment in the ion trap include the q_z value of the ion being excited resonantly, the duration and amplitude of the resonant excitation, and the pressure and composition of the bath/target gas with which the ions collide. Initial studies of the influence of some of these parameters on MS/MS in the ion trap[72-74] demonstrated that the q_z value of the ion being excited is important because it determines the trapping well-depth. At low q_z values, the ion is more easily ejected than at higher q_z values. Thus, ejection can occur before the ion undergoes sufficient collisions to activate it to a level at which it may dissociate. Reduced activation levels can lead to reduced MS/MS efficiency and discrimination against competitive dissociation pathways with higher critical energies.

The duration of the resonant excitation affects the amount of internal energy deposited into the parent ion as does the amplitude of the resonant excitation.[73] These parameters interact with the q_z value and bath/target gas pressure in determining the relative rates of activation/dissociation *versus* ejection.[72] Raising the amplitude of the resonant excitation voltage increases the rate of gain of kinetic energy of the parent ion, which can result in fewer collisions occurring before the ion has sufficient kinetic energy to be ejected from the ion trap. Ion ejection thus occurs to a greater relative extent than ion dissociation when large resonant excitation amplitudes are used.

Increasing the bath gas pressure increases the number of collisions, thereby slowing the gain in kinetic energy of the parent ion; in turn, dissociation is favored over ejection, although other factors associated with the bath gas pressure, such as reduced sensitivity due to scattering losses, must be considered also.

The maximum amount of kinetic energy that can be converted to internal energy increases with target gas mass. Thus, target gases heavier than helium, such as argon or xenon, are used typically in low energy CID experiments on beam instruments. However, because helium is used as the bath gas with the ion trap, it is also generally the target gas used in CID experiments performed on the ion trap. Use of heavier gases as the bath gas can lead to complete loss of signal,[27] although this is not necessarily the case for all geometries of ion traps.[75] A mixture of up to a few percent of a heavier target gas with helium can increase the amount of internal energy deposited into a parent ion during resonant excitation.[76,77] However, since the ion undergoes hundreds of collisions, most of these collisions still occur with helium. Thus, while the addition of a heavier target gas increases the rate of dissociation relative to ejection, higher internal energy states are accessed only if the parent ion undergoes a collision with the heavier target gas after it has been excited to near its dissociation threshold.[76] The addition of even a few percent of a heavier target gas can, however, affect sensitivity by increasing scattering losses just as increasing the bath gas pressure does.

One of the limitations of resonant excitation is that the resonant frequency is dependent upon the number of ions in the ion trap. Thus, typically, some operator tuning is necessary to optimize the resonant frequency. A number of different approaches to address this problem have been demonstrated. One approach is to use quick forward and reverse resonant excitation frequency scans so as to determine the optimum resonant frequency.[78] An alternative approach employs a 1.7 kHz bandwidth resonant excitation frequency.[78] Both of these approaches require *a priori* knowledge of the parent ion mass/charge ratio; the former approach also requires that the analyte concentration remains fairly constant, although only over a short period of time. A third approach, which does not require any *a priori* knowledge of the number or nature of the ions in the ion trap, is to use noise to effect broadband resonant excitation.[79] As a knowledge of the species in the ion trap is not required, broadband resonant excitation using noise is a potentially very useful method with ionization techniques that produce only pseudo-molecular ions.[80] A consequence of using a broadband signal is that product ions also are excited resonantly when they are formed. These product ions can dissociate subsequently, giving an MS/MS spectrum that contains a greater variety of product ions. This additional information can be an asset when structural information is desired to identify compounds.

A useful MS/MS experiment in quadrupole-based systems is energy-resolved mass spectrometry (ERMS). In this experiment, the relative intensity of product ions is measured as a function of collision energy. ERMS can be accomplished in an ion trap by varying either the amplitude of the resonant excitation or the duration of irradiation.[81] This study using ERMS concluded that lower ion internal energies were accessed in the ion trap compared to a triple quadrupole instrument. Data exists, however, which indicates that both these instruments can provide comparable information needed to differentiate isomers.[82]

B.   Nonresonant Excitation for Collision-Induced Dissociation

Another means to effect CID, other than resonant excitation, is to apply a suitable DC voltage along with the RF drive potential to move the parent ion to a working point near a boundary of the stability region. When an ion is at a working point near the boundaries of the stability region, its kinetic energy will increase which can lead to CID.[83,84] This recent observation needs further work in order to determine the potential utility of this method. It should, however, be less sensitive to the number of ions present than resonant excitation, while still being mass dependent, like the 1.7 kHz bandwidth resonant excitation discussed above.[78]

The fact that ions can be collisionally activated when they are at a working point close to the boundary of the stability region can have a negative consequence. Most ion isolation procedures in the ion trap move the parent ion close to a stability region boundary during the ejection of lower or higher mass ions. If the ion being isolated has a barrier to isomerization lower than any dissociation pathways, isomerization can occur during ion isolation as a result of collisional activation due to the increased kinetic energy of the ion.[85] Therefore, the ion being probed in an MS/MS experiment may have a different structure than the original ion that was generated for the neutral compound.

C.   Alternative Activation/Dissociation Methods

Two alternative dissociation methods to CID that have been demonstrated in the ion trap are photodissociation and surface-induced dissociation. Early work on photodissociation used the ion trap as an ion storage device, and mass-analyzed the product ions by ejection into another mass analyzer. Details of these types of experiments can be found in a recent monograph.[25] Given this past history, it is somewhat surprising that little photodissociation has been reported in the current generation of ion traps in which the whole experiment, including mass analysis, is performed in the ion trap. A fiber optic has been used to introduce laser light into an ion trap to perform photodissociation, which was compared to CID.[86] Dissociation efficiencies up to 95% were obtained and greater amounts of internal energy were deposited into the parent ion in the photodissociation experiment. A fiber optic has also been used to introduce a laser beam into the ion trap to perform a gas chromatography/tandem mass spectrometry (GC/MS/MS) experiment (see also Chapters 2 and 7).[87]

Recently, a method was developed to implement surface-induced dissociation in an ion trap.[88,89] The SID experiment is performed by applying a fast DC pulse to the end-caps of the ion trap. This causes the ions to become unstable in the radial direction and subsequently collide with the ring electrode. Relatively large amounts of internal energy can be deposited into the parent ion in this experiment and the energy deposition is related to the amplitude of the DC pulse. However, the MS/MS efficiency is poorer than for CID, especially at high energy deposition. It was also observed that the application of the DC pulse has a phase dependence with the fundamental RF for best efficiency.

D. Ion/Molecule Reactions in Tandem Mass Spectrometry

The ion trap also affords the capability to incorporate ion/molecule reactions into the MS/MS experiment. This can be as simple as the proton or electron transfer reactions in selected reagent ion chemical ionization[14] or more complex reactions that can be used to probe ion structure. The advantage of using ion/molecule reactions lies in the fact that additional energy is not deposited into the system and thus isomerization will not be induced. Employing an ion/molecule reaction was the method used to demonstrated that the ion isolation process can cause isomerization.[85] In that example, 100% of the $C_7H_7^+$ ions (m/z 91) generated by methane CI of 3-fluorotoluene reacted with dimethylether when m/z 91 was not isolated, indicating that these ions all had the tolyl structure.[90] When these same ions were isolated before undergoing the ion/molecule reaction, less than 100% of them reacted (the amount that reacted depended upon the isolation procedure). The unreactive ions had isomerized to either the benzyl or tropylium structure. These two structures can also be differentiated by an ion/molecule reaction.[91] However, CID of the three different structures in the ion trap gives the identical spectrum and thus is not useful for differentiating these ion structures. It is expected that ion/molecule reactions for structural analysis will be an important area of future research with the ion trap.

E. Multiple Stages of Tandem Mass Spectrometry (MS^n)

One last aspect of MS/MS in the ion trap that is particularly noteworthy is the ability to do multiple stages of MS/MS (MS^n). While MS^n ion trap experiments have become fairly routine in some laboratories, two studies devoted to this topic provided the framework for MS^n experiments. The first of these studies demonstrated the capability of doing MS^n experiments.[92] This report illustrated the ability to perform six consecutive stages of CID to generate an MS^7 spectrum and the combination of CID with ion/molecule reactions to effect an MS^{12} experiment. The other report focusing on the implementation of MS^n experiments dealt with the chemical and instrumental conditions necessary for such experiments.[93] Two factors lead to the ability of the ion trap to perform routinely MS^n experiments. The first of these is the instrumental efficiency of the ion trap. The second is that the CID method tends to access only one or a couple of dissociation channels. Thus, a large percentage of the initial parent ion population is often contained in a single product ion. This combination of efficiency and limited dissociation pathways provides the needed sensitivity to perform the MS^n experiment. However, it also necessitates such experiments to obtain sufficient information for structural determination. Also noted in the latter MS^n study is the role that ion/molecule reactions are likely to play in MS^n experiments. Unique reactant ions can be generated by dissociation MS/MS techniques, ion structure can be probed by ion/molecule reactions, and dissociation MS/MS techniques can be used to help elucidate structures of ion/molecule reaction products.

V. CONCLUSION

The quadrupole ion trap has undergone tremendous development as a mass spectrometer since it was introduced as a GC detector. Two of the important areas of development have been (1) the coupling of external ionization techniques, and (2) the tandem mass spectrometry capabilities. While there is obviously more development yet to occur in these two areas, a great growth in the application of these techniques to solve fundamental and applied research problems lies on the horizon.

While the application of the ion trap is growing rapidly, the understanding of the operation of the ion trap is still in its infancy. Effects of higher-order fields, space-charge, bath gas, scan speed, and many other parameters associated with the ion trap operation are yet to be understood, both empirically and theoretically, in their entirety. It may well be that the current geometry and modes of operation of the ion trap will become obsolete in the future as the influences of the various ion trap parameters become better understood.

Overall, the future of the ion trap looks very promising and exciting. The ion trap is already capable of performing a greater variety of experiments than any other mass spectrometer. It will become almost certainly a major mass spectrometry tool for fundamental and applied research as systems with the demonstrated capabilities become available commercially.

REFERENCES

1. Stafford, G. C., Jr.; Kelley, P. E.; Syka, J. E. P.; Reynolds, W. E.; Todd, J. F. J. *Int. J. Mass Spectrom. Ion Processes.* 1984, *60*, 85.
2. Paul, W.; Steinwedel, H. U. S. Patent No. 2,939,952 (1960).
3. Pedder, R. E.; Yost, R. A. *Proc. 36th ASMS Conf. Mass Spectrometry and Allied Topics.* San Francisco, 1988, p. 632.
4. Hart, K. J.; McLuckey, S. A.; Glish, G. L. *Proc. 39th ASMS Conf. Mass Spectrometry and Allied Topics.* Nashville, 1991 p. 1495.
5. Eckenrode, B. A.; McLuckey, S. A.; Glish, G. L. *Int. J. Mass Spectrom. Ion Processes.* 1991, *106*, 137.
6. Fulford, J. E.; March, R. E. *Int. J. Mass Spectrom. Ion Processes.* 1978, *26*, 155.
7. Creaser, C. S.; Mitchell, D. S.; O'Neill, K. E. *Int. J. Mass Spectrom. Ion Processes.* 1991, *106*, 21.
8. Buttrill, Jr., S. E.; Shaffer, B.; Karnicky, J.; Arnold, J. T. *Proc. 40th ASMS Conf. Mass Spectrometry and Allied Topics.* Washington, DC, 1992, p. 1015.
9. Hanson, C. D.; Castro, M. E.; Kerley, E. L.; Russell, D. H. *Anal. Chem.* 1990, *62*, 1352.
10. Marshall, A. G.; Wang, T.-C. L.; Ricca, R. L. *J. Am. Chem. Soc.* 1985, *107*, 7893.
11. McLuckey, S. A.; Glish, G. L.; Asano, K. G.; *Anal. Chem.* 1988, *60*, 2312.
12. Bonner, R. F.; Lawson, G.; Todd, J. F. J. *J. Chem. Soc. Chem. Commun.* 1972, 1179.
13. Brodbelt, J. S.; Louris, J. N.; Cooks, R. G. *Anal. Chem.* 1987, *59*, 1278.
14. Glish, G. L.; Van Berkel, G. J.; McLuckey, S. A. *Advances in Mass Spectrometry.* Vol. 11A, P. Longevialle (ed.), Heyden and Sons, 1989, p. 596.
15. Berberich, D. W.; Hail, M. E.; Johnson, J. V.; Yost, R. A. *Int. J. Mass Spectrom. Ion Processes.* 1989, *94*, 115.
16. Glish, G. L.; Van Berkel, G. J.; Asano, K. G.; McLuckey, S. A. *Proc. 36th ASMS Conf. Mass Spectrometry and Allied Topics.* San Francisco, 1988, p. 1112.
17. Strife, R. J.; Keller, P. R. *Org. Mass Spectrom.* 1989, *24*, 201.
18. Van Berkel, G. J.; Glish, G. L.; McLuckey, S. A.; Tuinman, A. A. *J. Am. Chem. Soc.* 1989, *111*, 6027.
19. Brodbelt, J.; Liou, J.; Donovan, T. *Anal. Chem.* 1991, *63*, 1205.
20. McLuckey, S. A.; Glish, G. L.; Kelley, P. E. *Anal. Chem.* 1987, *59*, 1670.
21. Eckenrode, B. A.; Glish, G. L.; McLuckey, S. A. *Int. J. Mass Spectrom. Ion Processes.* 1990, *99*, 151.

22. Lifshitz, C.; Goldenberg, M.; Malinovich, Y.; Peres, M. *Org. Mass Spectrom.* 1982, *17*, 453.

23. Goeringer, D. E.; Whitten, W. B.; Ramsey, J. M. *Int. J. Mass Spectrom. Ion Processes.* 1991, *106*, 175.

24. Goeringer, D. E.; Glish, G. L.; McLuckey, S. A. *Anal. Chem.* 1991, *63*, 1186.

25. March, R. E.; Hughes, R. J. *Quadrupole Storage Mass Spectrometry.* Chemical Analysis Series, Vol. 102, John Wiley & Sons, New York, 1989.

26. Louris, J. N.; Amy, J. W.; Ridley, T. Y.; Cooks, R. G. *Int. J. Mass Spectrom. Ion Processes.* 1989, *88*, 97.

27. McLuckey, S. A.; Glish, G. L.; Asano, K. G. *Anal. Chim. Acta.* 1989, *225*, 25.

28. Hart, K. J.; McLuckey, S. A.; Glish, G. L. *J. Amer. Soc. Mass Spectrom.* 1992, 3 549.

29. Pedder, R. E.; Yost, R. A. *Proc. 37th ASMS Conf. Mass Spectrometry and Allied Topics.* Miami Beach, 1989, p. 468.

30. McLuckey, S. A.; Glish, G. L.; Asano, K. G.; Grant, B. C. *Anal. Chem.* 1988, *60*, 2220.

31. Williams, J. D.; Reiser, H.-P.; Kaiser, Jr., R. E.; Cooks, R. G. *Int. J. Mass Spectrom. Ion Processes.* 1991, *108*, 199.

32. Glish, G. L.; Goeringer, D. E.; Asano, K. G.; McLuckey, S. A. *Int. J. Mass Spectrom. Ion Processes.* 1989, *94*, 15.

33. Heller, D. N.; Lys, I.; Cotter, R. J.; Uy, O. M. *Anal. Chem.* 1989, *61*, 1083.

34. Chambers, D. M.; Goeringer D. E.; McLuckey, S. A.; Glish, G. L. *Anal. Chem.* 1993, *65*, 14.

35. Schwartz, J. C.; Bier, M. E. *Rapid Commun. Mass Spectrom.* 1993, *7*, 27.

36. Jonscher, K.; Currie, G.; McCormack, A. L.; Yates, III, J. R. *Rapid Commun. Mass Spectrom.* 1993, *7*, 20.

37. Van Berkel, G. J.; Glish, G. L.; McLuckey, S. A. *Anal. Chem.* 1990, *62*, 1284.

38. McLuckey, S. A.; Glish, G. L.; Van Berkel, G. J. *Anal. Chem.* 1991, *63*, 1971.

39. Loo, J. A.; Udseth, H. R.; Smith, R. D. *Rapid Commun. Mass Spectrom.* 1988, *2*, 207.

40. Van Berkel, G. J.; McLuckey, S. A.; Glish, G. L. *Anal. Chem.* 1991, *63*, 1098.

41. McLuckey, S. A.; Van Berkel, G. J.; Glish, G. L.; Wang, E. C.; Henion, J. D. *Anal. Chem.* 1991, *63*, 375.

42. Kaiser, Jr., R. E.; Louris, J. N.; Amy, J. W.; Cooks, R. G. *Rapid Commun. Mass Spectrom.* 1989, *3*, 225.

43. Kaiser, Jr., R. E.; Cooks, R. G.; Syka, J. E. P.; Stafford, Jr., G. C. *Rapid Commun. Mass Spectrom.* 1990, *4*, 30.

44. McLuckey, S. A.; Glish, G. L.; Duckworth, D. C.; Marcus, R. K. *Anal. Chem.* 1992, *64*, 1606.

45. Weber-Grabau, M.; Kelley, P. E.; Bradhaw, S. C.; Hoekman, D. J. *Proc. 36th ASMS Conf. Mass Spectrometry and Allied Topics*, San Francisco, 1988, p. 1106.

46. Kaiser, Jr., R. E.; Cooks, R. G.; Moss, J.; Hemberger, P. H. *Rapid Commun. Mass Spectrom.* 1989, *3*, 50.

47. Kaiser, Jr., R. E.; Cooks, R. G.; Stafford, Jr. G. C.; Syka, J. E. P.; Hemberger, P. H. *Int. J. Mass Spectrom. Ion Processes.* 1991, *106*, 79.

48. Todd, J. F. J.; Penman, A. D.; Smith R. D. *Int. J. Mass Spectrom. Ion Processes.* 1991, *106*, 117.

49. Schwartz, J. C.; Syka, J. E. P.; Jardine, I. *J. Amer. Soc. Mass Spectrom.* 1991, *2*, 198.

50. Williams, J. D.; Cox, K. A.; Cooks, R. G.; Kaiser, Jr., R. E.; Schwartz, J. C. *Rapid Commun. Mass Spectrom.* 1991, *5*, 327.

51. Goeringer, D. E.; McLuckey, S. A.; Glish, G. L. *Proc. 39th ASMS Conf. Mass Spectrometry and Allied Topics.* Nashville, 1991, p. 532.

52. Goeringer, D. E.; Whitten, W. B.; Ramsey, J. M.; McLuckey, S. A.; Glish, G. L. *Anal. Chem.* 1992, *64*, 1434.

53. Londry, F. A.; Wells, G. J.; March, R. E. *Rapid Commun. Mass Spectrom.* 1993, *7*, 43.

54. Schwartz, J. C.; Jardine, I. *Rapid Commun. Mass Spectrom.* 1992, *6*, 313.

55. Louris, J.; Schwartz, J.; Stafford, G.; Syka, J.; Taylor, D. *Proc. 40th ASMS Conf. Mass Spectrometry and Allied Topics.* Washington, DC, 1992, p. 1003.

56. Julian, Jr., R. K.; Reiser, H.-P.; Cooks, R. G. *Proc. 40th ASMS Conf. Mass Spectrometry and Allied Topics.* Washington, DC, 1992, p. 1777.

57. Traldi, P.; Catinella, S.; Bortolini, O. *Org. Mass Spectrom.* 1992, *27*, 927.

58. *Practical Aspects of Ion Trap Mass Spectrometry,* R. E. March and J. F. J. Todd (eds.), Vol. 1, "Fundamentals and Instrumentation," Modern Mass Spectrometry series, CRC Press, Roca Baton, FL, 1995.

59. Williams, J. D.; Cooks, R. G. *Rapid Commun. Mass Spectrom.* 1992, *6*, 524.

60. Franzen, J. *Int. J. Mass Spectrom. Ion Processes.* 1991, *106*, 63.

61. Guidugli, G.; Traldi, P. *Rapid Commun. Mass Spectrom.* 1991, *5*, 343.

62. Morand, K. L.; Lammert, S. A.; Cooks, R. G. *Rapid Commun. Mass Spectrom.* 1991, *5*, 491.

63. Eades, D. M.; Yost, R. A. *Rapid Commun. Mass Spectrom.* 1992, *6*, 573.

64. Wang, Y.; Franzen, J.; Wanczek, K. P. *Int. J. Mass Spectrom. Ion Processes.* 1993, *124*, 125.

65. Nourse, B. D.; Kenttamaa, H. I. *J. Phys. Chem.* 1990, *94*, 5809.

66. Basic, C.; Eyler, J. R.; Yost, R. A. *J. Amer. Soc. Mass Spectrom.* 1992, *3*, 716.

67. McLuckey, S. A.; Glish, G. L.; Asano, K. G.; Bartmess, J. E. *Int. J. Mass Spectrom. Ion Processes.* 1991, *109*, 171.

68. March, R. E.; McMahon, A. W.; Londry, F. A.; Alfred, R. L.; Todd, J. F. J.; Vedel, F. *Int. J. Mass Spectrom. Ion Processes.* 1989, *95*, 119.

69. March, R. E.; McMahon, A. W.; Allinson, T. E.; Londry, F. A.; Alfred, R. L.; Todd, J. F. J.; Vedel, F. *Int. J. Mass Spectrom. Ion Processes.* 1990, *99*, 109.

70. Reiser, H.-P.; Julian, Jr., R. K.; Cooks, R. G. *Int. J. Mass Spectrom. Ion Processes.* 1992, *121*, 49.

71. Reiser, H.-P.; Kaiser, R. E.; Savickas, P. J.; Cooks, R. G. *Int. J. Mass Spectrom. Ion Processes.* 1991, *106*, 237.

72. Charles, M. J.; McLuckey, S. A.; Glish, G. L. *Proc. 40th ASMS Conf. Mass Spectrometry and Allied Topics.* Washington, DC, 1992, p. 57.

73. Louris, J. N.; Cooks, R. G.; Syka, J. E. P.; Kelley, P. E.; Stafford, Jr., G. C.; Todd, J. F. J. *Anal. Chem.* 1987, *59*, 1677.

74. Gronowska, J.; Paradisi, C.; Traldi, P.; Vettori, U. *Rapid Commun. Mass Spectrom.* 1990, *4*, 306.

75. Franzen, J.; Gabling, R. H. *Proc. 40th ASMS Conf. Mass Spectrometry and Allied Topics.* Washington, DC, 1992, p. 1009.

76. Glish, G. L.; McLuckey, S. A.; Goeringer, D. E.; Van Berkel, G. J.; Hart, K. J. *Proc. 39th ASMS Conf. Mass Spectrometry and Allied Topics.* Nashville, 1991, p. 536.

77. Morand, K. L.; Cox, K. A.; Cooks, R. G. *Rapid Commun. Mass Spectrom.* 1992, *6*, 520.

78. Yates, N. A.; Yost, R. A. *Proc. 39th ASMS Conf. Mass Spectrometry and Allied Topics.* Nashville, 1991, p. 132.

79. McLuckey, S. A.; Goeringer, D. E.; Glish, G. L. *Anal. Chem.* 1992, *64*, 1455.

80. Goeringer, D. E.; Chambers, D. M.; McLuckey, S. A.; Glish, G. L. *Proc. 40th ASMS Conf. Mass Spectrometry and Allied Topics.* Washington, DC, 1992, p. 1749.

81. Brodbelt, J. S.; Kenttamaa, H. I.; Cooks, R. G. *Org. Mass Spectrom.* 1988, *23*, 6.

82. Evans, C.; Traldi, P.; Mele, A.; Sottani, C. *Org. Mass Spectrom.* 1991, *26*, 347.

83. Paradisi, C.; Todd, J. F. J.; Traldi, P.; Vettori, U. *Org. Mass Spectrom.* 1992, *27*, 251.

84. Curcuruto, O.; Fontana, S.; Traldi, P. *Rapid Commun. Mass Spectrom.* 1992, *6*, 322.

85. Hart, K. J.; McLuckey, S. A.; Glish, G. L. *J. Amer. Soc. Mass Spectrom.* 1992, *3*, 680.

86. Louris, J. N.; Brodbelt, J. S.; Cooks, R. G. *Int. J. Mass Spectrom. Ion Processes.* 1987, *75*, 345.

87. Creaser, C. S.; McCoustra, M. R. S.; O'Neill, K. E. *Org. Mass Spectrom.* 1991, *26*, 335.

88. Lammert, S. A.; Cooks, R. G. *J. Amer. Soc. Mass Spectrom.* 1991, *2*, 487.

89. Lammert, S. A.; Cooks, R. G. *Rapid Commun. Mass Spectrom.* 1992, *6*, 528.

90. Heath, T. G.; Allison, J.; Watson, J. T. *J. Amer. Soc. Mass Spectrom.* 1991, *2*, 270.

91. Ausloos, P.; Jackson, J.-A. A.; Lias, S. G. *Int. J. Mass Spectrom. Ion Phys.* 1980, *33*, 269.
92. Louris, J. N.; Brodbelt-Lustig, J. S.; Cooks, R. G.; Glish, G. L.; Van Berkel, G. J.; McLuckey, S. A. *Int. J. Mass Spectrom. Ion Processes.* 1990, *96*, 117.
93. McLuckey, S. A.; Glish, G. L.; Van Berkel, G. J. *Int. J. Mass Spectrom. Ion Processes.* 1991, *106*, 213.

Part 2

PRACTICAL ION TRAP TECHNOLOGY

Chapter 4

PRACTICAL ION TRAP TECHNOLOGY: GC/MS AND GC/MS/MS

*Nathan A. Yates, Matthew M. Booth, James L. Stephenson Jr.,
and Richard A. Yost*

CONTENTS

0-8493-8251-3/95/$0.00+$.50

121

I. INTRODUCTION

In this chapter, the current capabilities of the ion trap for gas chromatography/mass spectrometry (GC/MS) are presented and illustrated, and new advances which may be incorporated into the next generation of commercial instruments are highlighted. Although many aspects of ion trap performance enhancement have been reported in the literature, such as parent and neutral loss scans,[1,2] high resolution,[3] and accurate mass measurement,[4,5] few of these methods have been reduced to practice; these features are not covered here. Rather, this chapter emphasizes the role of the ion trap in GC/MS in which it excels, and examines both the present limitations of the ion trap and possible future capabilities.

In recent years, the quadrupole ion trap has grown from a novel ion storage device into a conventional mass analyzer used in GC/MS. The development of the mass-selective instability scan in 1983 by Stafford *et al.* revolutionized the use of the ion trap as a mass analyzer.[6] Commercial instruments that are based on this convenient method of mass analysis exhibit unit resolution, a mass range of 650 u, and high sensitivity.[7,8] Several ion trap applications include the monitoring of priority pollutants in environmental chemistry,[9] the identification of drugs and their metabolites from biological matrices in clinical chemistry,[10] and the analysis of rock extracts in geochemistry.[11] With some 3500 instruments sold worldwide, the ion trap has become an important mass analyzer for GC/MS.

Although GC/MS has been the driving force behind the commercial development of the quadrupole ion trap, many expect that the use of the ion trap for gas chromatography/tandem mass spectrometry (GC/MS/MS) will become important as well. The ion trap's tandem mass spectrometric (MS/MS) capabilities were demonstrated first by Louris *et al.*[12] Many research groups have further developed this technique in the last several years.[13–21] Efficient fragmentation and detection are primary advantages of ion trap MS/MS. Full-scan product ion spectra have been obtained for 1000 times less sample than could be observed on a triple quadruple tandem mass spectrometer.[22] These techniques, which are powerful characterization and identification tools for the chemist, are not in general use for gas chromatographic applications.[23] However, the development of new instrumental methods that make GC/MS/MS practical is ongoing. With the commercialization of new ion isolation and excitation techniques, the next generation of GC/ion trap instruments is expected to incorporate powerful MS/MS capabilities into a benchtop gas chromatographic detector.

This chapter is structured in the following way: a short discussion of the commercial ion trap instruments is presented to provide a basis for those who may be unfamiliar with each instrument's features and methods of operation. Next, several GC/MS and GC/MS/MS application examples are summarized. The operational principles and characteristics of GC/MS ion traps are described, with attention given to the underlying differences between the ion trap and the quadrupole mass filter. Following this, new methods are presented that reduce space charge effects, decrease the extent of ion/molecule reactions, and further the applicability of GC/MS and GC/MS/MS. Finally, GC/MS/MS results from two unique ion traps, are presented. The aim of this discussion is to convey the key issues underlying GC/MS and GC/MS/MS on the ion trap, and to provide an appreciation of the applicability of both techniques.

II. INSTRUMENTATION AND METHODS OF OPERATION

A. The Ion Trap Detector, ITD700®

The ITD700® (Finnigan) was the first ion trap mass spectrometer available commercially.[7] Capable of acquiring full-scan and selected-ion electron ionization (EI) mass spectra over a mass range of some 650 u, the ITD700 shared similar features with other low cost GC/MS detectors, such as its small size and simple software-controlled operation.[7] A heated transfer line coupled the GC to the ion trap *via* an open-split interface. Sample effluent passed through this interface and into the internal volume of the ion trap. The scan sequence was made up of three steps that were performed in rapid succession: (1) ionization and storage of all sample ions above a particular mass/charge ratio, (2) ejection of background ions below a low-mass cut-off, and (3) the ejection and detection of ions in order of increasing mass/charge ratio to produce a mass spectrum. A diagram of this scan program sequence, also referred to as a *scan function,* is shown in Fig. 4.1. This scan program produced mass spectra for ions between m/z 60 and m/z 100. Full-scan mass spectra covering m/z 60 to m/z 650 were acquired using a segmented EI scan program which is discussed later in this section. The execution of a single scan program is referred to as a *microscan.* During a GC/MS acquisition, the data for each microscan are summed, and an average spectrum is returned to the data system and recorded as a single *analytical scan.*[6]

The ion trap is operated with 1 mTorr of helium *buffer gas* that serves to dampen the motion of trapped ions as well as to enhance the mass resolution and sensitivity.[8] The buffer gas is introduced through the GC column and made up by an open-split interface. The number of ions that are created and trapped for a given sample pressure is controlled by the ionization time. In Fig. 4.2 are compared four profile mass spectra that were acquired with (a) 50 µs, (b) 350 µs, (c) 1 ms, and (d) 5 ms ionization times. For ionization times of 350 µs, 1 ms, and 5 ms, the loss of resolution observed for low mass/charge ions is due to the presence of too many ions in the ion trap. Resolution is improved for high mass/charge ions because those of low mass/charge are scanned out of the trap prior to detection of the high mass/charge ions. These *space-charge effects* which impair the performance of the ion trap are discussed below. Calibration curves generated with the ITD700 typically exhibited a linear dynamic range of two to three orders of magnitude. Although the introduction of the ITD700 represented a breakthrough in ion trap technology, the manual adjustment of the ionization time and the narrow linear dynamic range were clearly drawbacks for routine GC/MS.

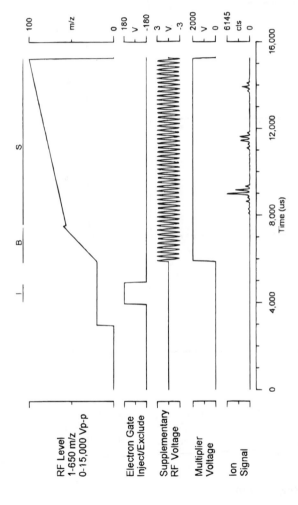

FIGURE 4.1

Ion trap scan program used to acquire EI mass spectra for m/z 60 to m/z 100. Ions are generated during period I, background ions are ejected at point B, and the mass analysis scan is performed during stage S. The RF level determines the ion of lowest mass/charge ratio stored in the trap. Ions detected during stage B are not acquired by the data acquisition software. A supplementary RF voltage frequency, denoted as supplementary AC voltage in the diagram, of 530 kHz on the ITMS and 485 kHz on the ITS40 is applied during period S to effect axial modulation (frequency not drawn to scale). Multiplier settings of 0 and 2000 V represent ON and OFF values, and do not correspond to the true analog voltages applied. The ion signal shown corresponds to the actual ion signal observed for 1-octene; the peak widths and data acquisition rate are drawn to scale.

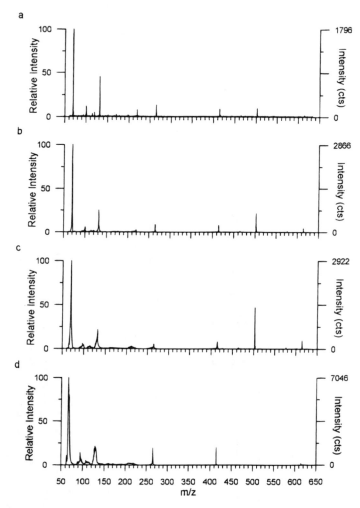

FIGURE 4.2
EI/MS profile spectra of ionized perfluorotributylamine $(C_4F_9)_3N$ (MW 671) acquired with ionization times of (a) 50 μs, (b) 350 μs, (c) 1 ms, and (d) 5 ms. Intensity (cts) correspond directly to the ion current measured by the electrometer circuit without the application of a scaling factor for ionization time.

B.   The Ion Trap Detector, ITD800®

The ITD800® (Finnigan) is an upgraded version of the ITD700 that includes chemical ionization (CI) capabilities and improved on several of the ITD700's most apparent limitations. ITD800s also were sold with the ATD50/8000 nameplate by Perkin Elmer, and upgrade kits were available for ITD700 users. Positive CI is carried out by forming an excess of CI reagent ions that can react further to produce protonated sample ions.[24,25]

Fig. 4.3 compares the background-subtracted full-scan spectra for 10 ng of amobarbital (MW = 226) using (a) EI and (b) low pressure methane CI. The CI scan program is shown in Fig. 4.4. Note that the CI scan begins with an EI ionization event that produces the reagent ions which react further to form $[M + H]^+$ ions. The number of $[M + H]^+$ ions produced in a CI experiment is controlled by the duration of the CI reaction period. Importantly, the EI spectrum shows a much larger $[M + H]^+$ peak, m/z 227, than an M^+ peak due to self-CI.

In addition to the ITD800's CI capabilities, automatic ionization control routines were introduced that eliminated the manual adjustment of the ionization time. Automatic gain control (AGC) and automatic reaction control (ARC) set the ionization and reaction times for EI and CI analyses, respectively.[26,27] The AGC and ARC pre-scans, shown in Fig. 4.5(a) and Fig. 4.5(b), determine the total ion signal for a fixed ionization period. Using this measurement, a microprocessor calculates an appropriate ionization time for the next microscan. To correct for the effect of variable ionization times, all ion intensity data are normalized to an ionization time of 25 ms. AGC and ARC allow sample amounts ranging from the low picogram to the high nanogram levels to be injected without the requirement for manual adjustment of the ionization time. Fig. 4.6 shows a calibration curve for hexachlorobenzene that was acquired with AGC. Note that the linear dynamic range extends over six orders of magnitude. With the addition of CI and the automatic ionization control techniques, ITD800 users found that the sensitivity, spectral quality, dynamic range, and ease of operation were improved significantly compared to the ITD700.[28]

C. The Ion Trap Mass Spectrometer, ITMS®

The ITMS® (Finnigan) is a multi-purpose research instrument; approximately 50 ITMS® instruments have been produced. The main use of the ITMS has been in the development and characterization of new ion trap capabilities, although some GC/ITMS applications have been demonstrated. MS/MS,[15,17,18,29–31] ion injection,[32–37] CI,[24,25,38,39] mass range extension,[40,41] and high resolution[3,42–44] are several examples of the "high performance" features that have been studied using the ITMS. The ion trap analyzer, electronics, and GC/MS transfer-line interface are similar to those of the ITD700 and ITD800. Hardware differences include a multiport UHV vacuum system, a probe lock, a heated solids probe, and a programmable DC supply and frequency synthesizer.

1. Ion Trap Tandem Mass Spectrometry

In addition to the standard EI and CI scan programs shown in Figs. 4.1 and 4.4, the ITMS provides versatile software for creating custom scan

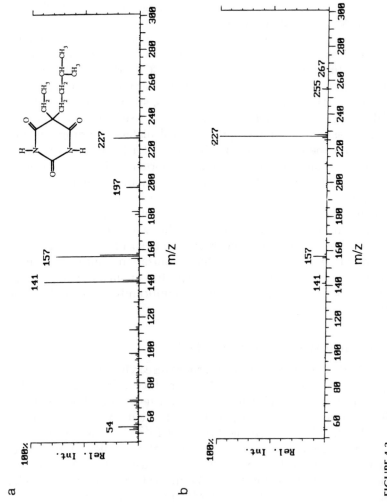

FIGURE 4.3

Background-subtracted full-scan spectra for the injection of 10 ng amobarbital (MW 226) with (a) EI, and (b) methane CI.

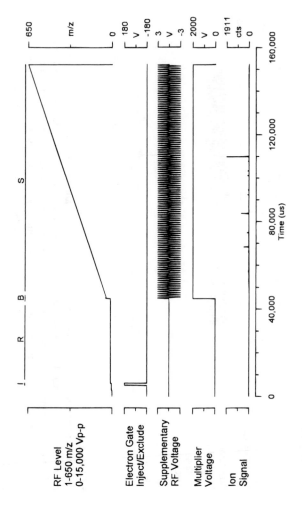

FIGURE 4.4

Ion trap scan program used to acquire CI mass spectra for m/z 50 to m/z 650. CI reagent ions are generated during period I, [M + H]+ ions form during interval R, background ions are ejected at point B, and the mass-analysis scan is performed during stage S.

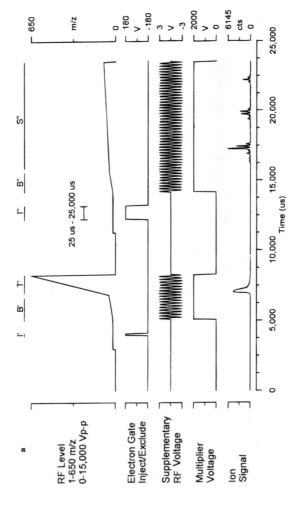

FIGURE 4.5

Automatic ionization control scan-program for (a) EI (AGC) and (b) CI (ARC). Pre-scan ionization is performed during stage I′, background ions are ejected at point B′, and the total ion current is collected during interval T′. The ion signal obtained during T′ is used to set the ionization time (I″) for the next microscan. B″ and S″ denote the background ejection and mass-analysis stage for the microscan. R′ and R″ indicate the CI reaction periods for the ARC pre-scan and microscan, respectively.

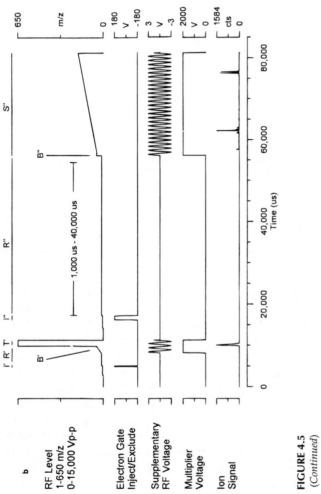

FIGURE 4.5 (*Continued*)

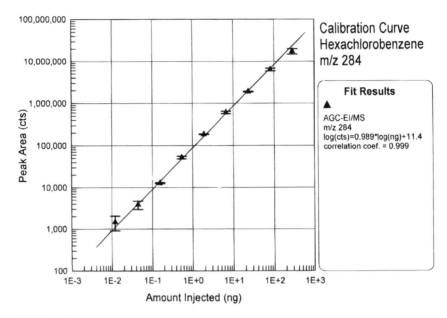

FIGURE 4.6

EI/MS calibration curve for m/z 284 of hexachlorobenzene (MW 284). Each datum point
(▲) represents three replicate injections. Error bars correspond to ±2 standard deviations.
The best fit line was determined by linear regression using the data points between 0.02
ng and 100 ng.

programs. Fig. 4.7 shows a custom MS/MS scan program that was created
using the ITMS *scan editor* software. The ion trap MS/MS experiment has
been described in detail previously.[13,14,29] Briefly, MS/MS is performed in
four steps; (a) ionization of the sample molecules, (b) isolation of the
parent ion, (c) activation of the parent ion to promote collision-induced
dissociation (CID), and (d) collection of the resulting product ion spectrum. The isolation step (b) is accomplished by the application of a combination of RF and DC voltages to the ring electrode of the ion trap.[14,45–47]
The CID step (c) is achieved by applying a supplementary RF voltage
across the end-cap electrodes of the ion trap.[13,48] The frequency of the
supplementary RF voltage is adjusted to match the fundamental axial
secular frequency of ion motion. With this method of *resonant excitation*
ions are accelerated axially. CID occurs when the collisions between the
accelerated ions and the helium buffer gas impart sufficient internal
energy to the bonds within individual ions. Fig. 4.8 shows the MS/MS
product ion spectrum for the molecular ion m/z 134, of *n*-butylbenzene.

2. Axial Modulation

The supplementary RF voltage is applied also during the mass-selective instability scan. This method of *axial modulation* improves the resolu-

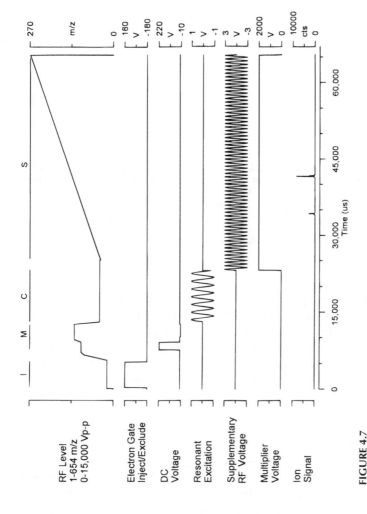

FIGURE 4.7

Ion trap scan program used to acquire EI/MS/MS product ion spectra for m/z 134. Periods I, M, C, and S correspond to ionization, mass selection, CID, and mass analysis, respectively. Positive and negative DC voltages are applied to the ring electrode to isolate the ion of interest using the $\beta_z = 0$ and $\beta_z = 1$ stability boundaries, respectively. Typical values for the resonant excitation signal are 100 to 200 kHz and 0.5 to 2.0 $V_{(p-p)}$ (frequency not drawn to scale).

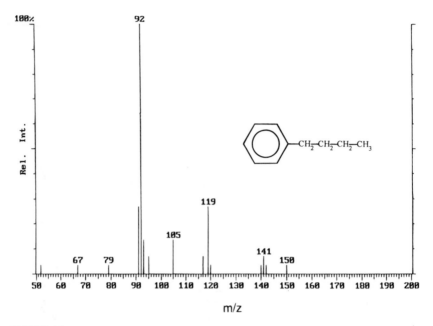

FIGURE 4.8
Background subtracted EI/MS/MS product ion spectrum for $M^{+\cdot}$ ion (m/z 134) of n-butylbenzene (MW 134); 1 pg injected.

tion and dynamic range by increasing the efficiency with which ions are ejected from the ion trap.[47] In Fig. 4.9 are compared two spectra that were acquired with and without the use of axial modulation. Note that peak widths of unit resolution are observed with axial modulation, and that the peak height observed for m/z 69 is five times greater than the peak height recorded without the use of axial modulation.

3. Problem Areas

Although the ITMS is a versatile instrument for tandem mass spectrometry, several problems limit the use of MS/MS with gas chromatographic sample introduction. One problem is that the AGC and ARC routines, which improve GC/MS performance, have not been integrated into the scan editor software that is used to create MS/MS scan programs. Tuning MS/MS scan programs for GC analyses is very time-consuming since multiple sample injections are required. Although some GC/MS/MS applications have appeared in the literature,[15,22,30,49,50] the ITMS is best suited for fundamental studies where sample pressures can be controlled and maintained constant.

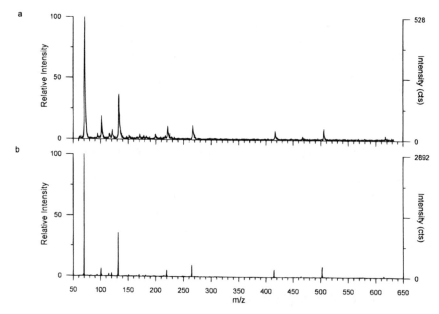

FIGURE 4.9

EI/MS profile spectra of ionized perfluorotributylamine $(C_4F_9)_3N$ (MW 671) (a) without axial modulation and (b) with axial modulation of 6.0 $V_{(p - p)}$ at 530 kHz. A fixed ionization time of 150 μs was used to acquire both spectra. Intensity (cts) correspond directly to the ion current measured by the electrometer circuit without the application of a scaling factor for ionization time.

D. The Ion Trap System, ITS40®, MAGNUM®3 and SATURN®

The ITS40® ion trap (Finnigan) is a benchtop GC/MS system designed for routine analyses. The hardware for the MAGNUM® (Finnigan) and SATURN® (Varian) instruments is functionally identical to that of the ITS40. The open-slit interface used on the ITD and ITMS instruments is replaced by a heated transfer line that directs all of the column effluent into the ion trap analyzer. Additionally, both manufacturers use an open-split interface for air monitoring work, and a jet separator for volatiles. A fixed-frequency crystal oscillator and balun transformer supply the axial modulation signal that improves ion detection, reproducibility, and ion statistics. The ITS40's scan acquisition processor (SAP) board centroids the profile spectra before transferring them to the data system. This procedure reduces the time that the instrument spends communicating with the data system, resulting in a higher scan repetition rate. Both systems use a new data system which integrates the control of an optional autosampler, the GC, and the mass spectrometer into an easy-to-use package. The standard EI and CI scan programs are supported, but the ability to create custom scan programs is not available. Both AGC and ARC are included.

In addition, the MAGNUM and SATURN III instruments use new techniques that remove EI-produced ions having mass/charge ratios greater than that of the CI reagent ions. In summary, the ITS40-based instruments are refined GC/MS ion traps that are capable of generating full-scan EI and CI spectra over a wide dynamic range with very high sensitivity.

III. GC/MS AND GC/MS/MS APPLICATIONS

Of the commercial ion traps sold, greater than 97% are dedicated GC/MS instruments. It is not possible to present here a complete summary of the many applications in which GC/MS ion traps have been used. This problem, therefore, has been mitigated by selecting examples that demonstrate the use of the ion trap in several areas of chemistry.

A.   GC/MS Ion Trap Applications

Although ion trap MS is a relatively new technique compared to the more conventional MS methods, the use of the ion trap for GC/MS has been widespread. Firstly, the driving force behind this growth has been the ability to obtain full-scan spectra with high sensitivity. Full-scan spectra can improve the certainty of an identification relative to single ion monitoring (SIM) methods. Secondly, switching from EI and CI is accomplished through software, and does not require hardware changes such as swapping ion volumes. The alternating EI and CI scan capability provides a powerful method for obtaining molecular weight and structural information in a single experiment.[51] A current limitation that has been observed in environmental and clinical GC/MS applications is the loss in sensitivity that occurs when background chemical interferences are high. Additionally, negative ion CI cannot be performed with the standard configuration of the ion trap.

Several reports have noted differences between ion trap and quadrupole spectra exist, these variations are analogous to the differences that are observed commonly for spectra obtained on different types of instruments.

1.   Priority Pollutants

The identification of priority pollutants in complex matrices has been one of the primary applications of GC/MS. Ion trap GC/MS is employed for many environmental applications,[9,55-59] including the U.S. Environmental Protection Agency (U.S. EPA) groundwater monitoring methods (624, 625) and drinking water methods (524.2, 525). The determination of trace levels of herbicides and their degradation products in surface and groundwaters was demonstrated by Pereira *et al.*[60] Calibration curves were linear from 0.06 to 6 μg/mL for the analytes of interest. The lower level of detection of the analytes was 60 pg (with a signal/noise ratio, S/N, greater than 10:1). Full-scan mass spectra were obtained routinely for 1 ng of analyte; the author noted that the ion trap is rapid, specific, sensitive, inexpensive, and could be automated easily as a screening method. Groundwater samples were observed to be less difficult to analyze than surface water samples because of reduced matrix interference problems. Fig. 4.10 shows the ion trap mass spectra obtained for approximately 1 ng of the compounds selected for this study. The spectra demonstrate the ability of this technique to provide absolute and unambiguous identification of environmental organic contaminants. Canela and Muehleisen applied headspace GC/MS to the measurement of industrial air emissions.[57] Detection with an ITD improved the identification and determination of priority pollutants with LODs of 0.50 and 0.12 mg/m³ for passive and active adsorption, respectively.

2. Clinical and Pharmaceutical Studies

GC/MS ion traps have been applied also to an assortment of clinical and pharmaceutical studies.[10,15,28,39,61–66] De Boer compared the performance of an ITD to a Hewlett-Packard Mass Selective Detector (MSD) for the analysis of TMS-enol, TMS-ether, methoxime TMS, and pentafluoropropionic ester derivatives of the main urinary metabolite of a nandrolone, norandrostone.[67] Fig. 4.11 shows the ITD chromatographic data and full-scan mass spectrum for derivatized norandrosterone. The full-scan sensitivity and selective CI capabilities of the ITD are observed to be its two main advantages in this application. However, the Hewlett-Packard MSD gave better sensitivity, in the SIM mode, for norandrosterone extracted from urine because of the limited ability of the ITD to handle background interferences. Genin and Lehrer used the ITD for antidoping studies to determine low levels of drugs and their metabolites in biological matrices.[68] The ITD allowed complete CI mass spectra to be obtained from less sample than that required by other MS methods. Full-scan analyses of seven drugs resulted in improved sensitivity and specificity compared to multiple ion detection methods. Horman *et al.* have demonstrated routine GC/MS analysis of fatty acid methyl esters using the ITD.[54] They obtained full-scan EI mass spectra that gave good correlation with literature reference spectra for sample amounts from 2 pg to 225 ng. Enhanced [M + 1]⁺ ions are formed at levels above 50 ng due to self-CI, but this did not prevent identification.

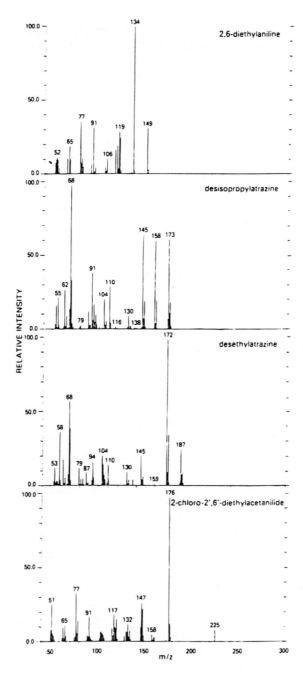

FIGURE 4.10

Ion trap mass spectra of selected compounds taken from the determination of herbicides in surface and groundwaters.[56]

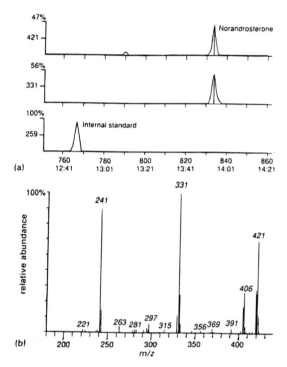

FIGURE 4.11

GC/MS data of the trimethylsilyl-enol trimethylsilyl derivative of norandrosterone isolated from a urine sample and analyzed by the ion trap detector in the positive-ion CI mode: (a) mass chromatograms at m/z 421 and m/z 331, and (b) full-scan mass spectrum.[57]

B. GC/MS/MS Ion Trap Applications

Compared to GC/MS, relatively few GC/MS/MS ion trap applications appear in the literature.[15,22,23,30,49,50,69–71] One reason for this discrepancy is that less than 2% of all ion traps sold before 1994 have MS/MS capabilities. Of these instruments, few have been used for GC/MS/MS applications. The second reason for such a small number of GC/MS/MS applications is that several problems currently limit the usefulness of GC/MS/MS on the ion trap. Johnson compared GC/MS/MS carried out with the ITMS with that carried out with a triple quadrupole mass spectrometer (TQMS).[22] Full-scan product ion spectra were obtained for the injection of 15 pg and 1.5 ng of diisopropyl methylphosphonate (DIMP) standards on the ITMS and TQMS, respectively. The maximum efficiency of CID was observed to be 90% on the ITMS (compared to 56% on the TQMS), but the overall MS/MS efficiency, including the efficiency of isolation of the parent ion, was not determined. MS/MS log-log calibration curves obtained for DIMP on the ITMS show a slope of 0.53 (compared to the ideal of 1.0) demonstrating the first limitation of GC/MS/MS on the ITMS. Strife and Simms evaluated the

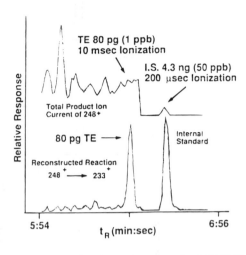

FIGURE 4.12
GC full-scan MS/MS analysis of an extract of plasma initially spiked at 1 ng/g. The ionization time for tebufelone is 10 ms. The upper trace is a sum of all MS/MS product ion current as a function of time. The lower trace is reconstructed from product ions of m/z 233 only.[50]

quantitative analysis of tebufelon (TE) in crude plasma extracts by MS/MS with an ITMS in an effort to determine the analytical utility of the technique.[50] Quantitation, using an internal standard method previously described by Strife and Simms, resulted in a correlation coefficient (r^2) of 1.000 for the TE calibration curve. The authors noted that the only practical choice for GC/MS/MS quantitation on the ITMS is the use of internal standards, due to the ITMS' nonlinear external calibration curves. RF/DC isolation was observed to be important for the analysis of real samples because it reduces space charge caused by the presence of the matrix ions. Fig. 4.12 shows the analysis of a plasma extract for a sample spiked at 1 ppb of TE. Note that the upper trace is the sum of the product ion current, which would be equivalent to the ion signal for a selected ion using SIM. The reconstructed ion current for the reaction of m/z 248$^+$ → m/z 233$^+$ demonstrates the enhancement in S/N that can be obtained with MS/MS relative to SIM. A limitation for practical GC/MS/MS, as noted by Strife and Simms[50] and by Johnson et al.,[22] is that the instrument software does not support the automatic ionization control techniques that improved significantly the usefulness of ion trap GC/MS when the ITMS is operated in the MS/MS mode. Several examples of GC/MS/MS using automatic ionization control have been demonstrated on modified ITS40 instruments.[69,72,73] Practical methods for performing GC/MS/MS on the ion trap are under investigation at the present time. Clearly, the development of an affordable and easy-to-use benchtop GC/MS/MS ion trap is expected to expand use of the ion trap to a new class of research applications.

IV. OPERATIONAL PRINCIPLES FOR ION TRAP GC/MS

A. Space-Charge Effects

Space charge, defined as the electrostatic forces that exist between ions that are held in close proximity to each other, can degrade severely the performance of the ion trap as a mass spectrometer.[47,74-77] The effects of space charge are observed commonly as losses in mass resolution and/ or mass shifts.[78,79] Although space charge can degrade the performance of most types of mass spectrometers, space-charge effects are apparent at lower sample pressures with the ion trap.

Some simple models may help to explain space-charge effects in the ion trap. Confinement of the ions within the ion trap increases the effect of space charge. In many ways, this situation is analogous to tossing a handful of marbles into a mixing bowl, where they will be trapped. Compare this situation with that of a linear quadrupole, which can be likened to inserting marbles into one end of a length of tubing; the marbles will roll through the tube and be gone. The marbles (ions) that pass through the tube (quad) interact with the other marbles that are immediately in front of and behind them in line. In the bowl (ion trap), all of the marbles (ions) can interact with one another for an extended period of time. Clearly, the storage of ions exacerbates the susceptibility of the ion trap to space-charge effects. In addition, friction will dampen the marbles' trajectories toward the bottom of the bowl. Similarly, collisions with buffer gas (He) will "frictionally" cool the ions so that they migrate towards the center of the trap, bringing the ions together and increasing space charge.

Furthermore, space-charge effects are more noticeable with the ion trap because, typically, all of the ions above a particular mass/charge ratio are stored at the same time. To illustrate how this situation contributes to space charge, consider a crowded elevator that is carrying people from the basement of a building up to their respective floors. Initially, people getting off at the lower levels must push past the others to get out of the elevator. As the car travels upward the people continue to get off, however, it becomes easier to get out of the elevator. The crowding in this example is analogous to the space charge that exists when the ion trap is operated in the mass-selective instability mode. Initially, all of the ions that are stored and available for detection are formed during the ionization event, and space charge is high. Ions of low mass/charge are scanned out of the ion trap under conditions of high space charge due to the presence of ions of higher mass/charge so their detection may be perturbed. By the time ions of high mass/charge are themselves scanned out of the ion trap, the ion trap has been emptied partially and space charge has diminished. Fig. 4.2 shows clearly how resolution improves with increasing mass/ charge as space charge is reduced during an analytical scan.

Although space charge is associated normally with losses in resolution, it can affect the ion trap operation in a variety of ways. For example, low levels of space charge can bring about a slight shift in an ion's secular frequency.[22,48,70,80,81] March et al. used such frequency shifts as a sensitive method of measuring the number of ions in the ion trap.[48] At moderate space-charge levels, peak broadening (2 to 3 u) is observed, particularly

a

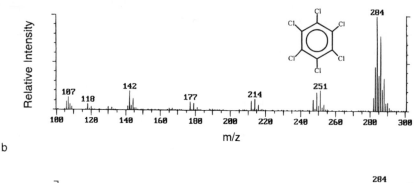

b

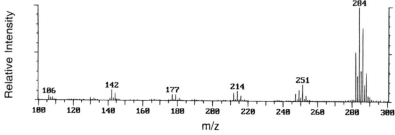

FIGURE 4.13

Background-subtracted centroid mass spectra for (a) 2.0 and (b) 81 ng of hexachlorobenzene. Note the differences in the relative intensity of m/z 282 and m/z 283 for the two spectra. Ionization times for the displayed spectra were set by AGC.

for ions of low mass/charge. Good GC/MS performance is maintained under moderate space-charge conditions when axial modulation is used. Gross space-charge effects may be observed as very broad peaks, with widths ranging from 5 to 25 u. Since space charge limits the ability of the ion trap to produce identifiable mass spectra, it is difficult to determine how gross space charge affects other ion trap operations. Space charge decreases the efficiency with which additional ions can be stored in the ion trap, and leads ultimately to a maximum ion density within the device. For example, Eades *et al.* demonstrated recently that space charge increases the magnitude of "black holes" that augment ion losses due to nonlinear resonances.[82]

The ability to recognize space-charge effects in GC/MS data is complicated because the current commercial instruments are limited to the acquisition of centroided mass spectra. Profile spectra can be viewed when the instrument is tuned, but they cannot be collected to a data file. Once profile data have been converted to centroid data, it is difficult to characterize changes in resolution, peak shape, and mass assignment that may be caused by space charge. Fig. 4.13 shows EI mass spectra for the injection of 2.0 and 81 ng of hexachlorobenzene. For both amounts, the hexachlorobenzene spectra are easily identified. However, it is difficult to differentiate between unit resolved peaks and peak broadening/shifts that are caused by space charge (compare the relative intensity of m/z 283 and m/z 284 in both spectra). For GC/MS/MS experiments, space charge and the lack of profile acquisition make it difficult to assess how well the parent ion has been mass-isolated prior to CID. Calibration curves are often the best indication of space charge for GC/MS experiments. Nonlinearities at high concentrations show when the efficiency of ion detection begins to drop off and provide the only method of estimating the mass resolution and peak shape for samples of lower concentration.

B. Space Charge and Automatic Ionization Control

Automatic ionization control, which we refer to here as AGC,[26,27] is a method of reducing space charge by limiting the number of ions that are created at different points along a chromatographic run. When the AGC pre-scan returns a large total ion current (TIC) measurement, the ionization time for the subsequent microscan is reduced. Conversely, when the AGC pre-scan TIC measurement is low, the ionization time for the next microscan is increased. Thus, AGC improves the linear dynamic range of the ion trap in two ways. First, it minimizes space charge when sample pressures are high, and, second, it enhances sensitivity when sample pressures are low.

Although the development of AGC significantly improved the utility of the ion trap for GC/MS, background interferences can cause AGC to reduce the ionization time which, in turn, can diminish sensitivity for the analyte. When the ratio of background molecules to analyte molecules in the ion trap is high, the majority of the ions formed during the AGC pre-scan are background ions. When the background ions are present in large amounts, the AGC pre-scan shortens the ionization time to prevent space charge caused by the background ions. This reduction in ionization time also decreases the number of analyte ions that are formed and detected. Similar problems can occur when co-eluting internal standards are used for quantitation. Large discrepancies in the quantities of deuterated internal standard and analyte amounts can adversely affect detection of the compound present in lesser amounts. Thus, a high background-to-analyte ion ratio can reduce the sensitivity of GC/MS on the ion trap. It is important to note that the use of internal standards that are structural analogs helps to eliminate these problems because co-elution is less common. The use of sequential RF/DC isolation steps has been demonstrated as a suitable method for obtaining a linear calibration curve for an analyte and its isotopically-labeled internal standard that co-elute with vastly different partial pressures.[83]

C. EI Scan Segments

Full-scan EI spectra covering a wide mass range normally are acquired with a segmented EI scan program. Fig. 4.14 shows the scan program diagram for a segmented EI scan over a scan range of m/z 60 to m/z 650. This scan program produces a complete mass spectrum by combining the spectra obtained from each of the four scan segments. Each scan segment resembles a separate scan program, including the ionization, low mass ion ejection, and data acquisition steps. Although a single scan segment could be used to obtain full scan EI data, the segmented scan program helps to reduce self-CI ions that can be observed in EI spectra due to ion/ molecule reactions. Breaking long scan programs into four shorter scan segments reduces the residence time of the ion in the ion trap, that is, between ion formation and ion ejection to the detector. By minimizing the time available for ion/molecule reactions, the intensity of ion/molecule reaction products in EI spectra may be reduced.

In addition, the segmented scan program allows different ionization settings to be used for different sections of the mass spectrum. To alter the intensity of ions in a particular scan segment, the scaling factor and/or RF ionization level can be adjusted. The scaling factors allow shorter ionization levels to be used during the low mass scan segments and longer ionization times to be used for the high mass scan segments. This procedure reduces the number of ions that are in the ion trap when low mass ions are scanned out, yet increases the number of high mass ions that are detected. The adjustment of the RF ionization level corrects for the interdependence between the efficiency of ionization and the magnitude of mass/charge.[84] Together, the proper adjustment of the scaling factor and the RF ionization level for each segment of the EI scan program can improve the sensitivity of the ion trap for GC/MS.

D. Scan Repetition Rate

Here we define the scan repetition rate (scans/s) as the number of scans that can be stored to disk in one second. Our justification for the use of repetition rate instead of scan rate is that scan rate usually describes the speed of the acquisition step for a scan program (u/s). In the ion trap software, the term *scan rate* is used incorrectly to refer to the time required for a scan to be performed (s/scan), which is not a rate but a period time. To avoid further confusion, we will use the term *repetition rate* to describe the number of scans that are acquired each second (for example, 1.3 scans/s).

For GC/MS, the repetition rate is important because it defines the number of scans that can be acquired over a GC peak. A minimum repetition rate of 1 scan/s is required for most GC/MS applications.

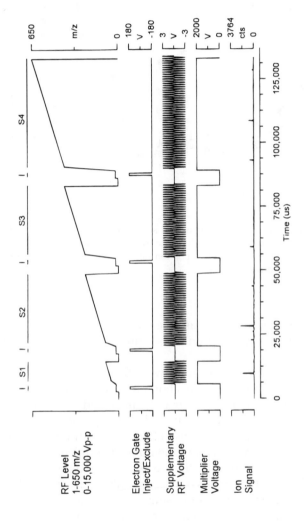

FIGURE 4.14

Ion trap scan program used to acquire EI mass spectra from m/z 60 to m/z 650 using four scan segments. Periods labeled I correspond to the ionization step for each of the four scan segments S1 to S4. The mass/charge values for the segment breaks can be set manually; the default values are shown. The supplementary RF voltage is available on ITMS and ITS40 based ion traps only (frequency not drawn to scale). The ion signal shown corresponds to the actual ion signal observed for perfluorotributylamine and the peak widths and data acquisition rate are drawn to scale.

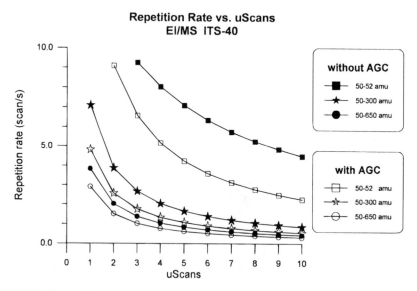

FIGURE 4.15
Scan repetition rate (scan/s) for an EI/MS segmented scan program as a function of the number of microscans. Unfilled data points include the extra scan time for the AGC pre-scan, and solid data points correspond to the repetition rate without AGC. All scan rates were calculated by the ITS40 software.

However, higher repetition rates increase the number of data points that define the shape of the GC peak profile and can improve quantitation precision.

In Fig. 4.15 is shown the relationship between the scan repetition rate, the number of microscans, and the acquisition range for an ITS40 EI scan program. In general, the repetition rate may be increased in one of two ways: reducing the number of microscans or decreasing the duration of the microscan. Communication delays also affect the repetition rate, but they cannot be reduced easily. Reducing the number of microscans that are averaged to produce a spectrum will increase the scan repetition rate. However, a loss in S/N and mass spectral reproducibility is observed. In practice, the signal-averaging benefits of using three or more microscans outweigh the gains that are made in repetition rate by using one microscan. The scan repetition rate can be increased also by reducing the duration of the scan program, which can be accomplished by eliminating the AGC pre-scan or reducing the acquisition range. By judicious choice of an acquisition range that minimizes the number of scan segments, large gains in repetition rate can be affected. For example, the repetition rate increases dramatically when the end-mass value is changed from m/z 200 to m/z 199 because the number of scan segments is reduced from three to two.

The largest gains in scan repetition rate are obtained when a single scan segment is used. Fig. 4.15 shows that repetition rates approach 10 scans/s when a $2u$ wide acquisition range is used. Also, scan repetition rates are usually higher for CI and MS/MS scan programs because these scan modes use a single scan segment.

E. Single Ion Monitoring *versus* Full Scan

The trade-offs between full-scan and single ion monitoring (SIM) modes for the ion trap are different from those for the quadrupole mass filter instruments. SIM improves sensitivity for beam-type mass spectrometers by collecting current continuously for a single mass, thus enhancing the S/N. For example, when a quadrupole mass filter is scanned over a 200 u mass range every second, 0.005 s is spent collecting the ion current for each mass. By changing to SIM and monitoring the ion current for a single mass continuously, the observed S/N would be enhanced by a factor of $\sqrt{200} = 14.4$. For the ion trap, reducing the acquisition range to cover a single mass only shortens one part of the scan program; thus, the reduction in scan time is small. Since SIM on the ion trap does not monitor a single mass continuously, only modest gains in S/N are observed.

In practice, switching from full-scan to SIM on the ion trap eliminates the information available in full-scan spectra for only a slight gain in sensitivity. Fig. 4.16 shows how the number of microscans that can be acquired changes with acquisition range for an AGC-EI/MS scan program of 1 s. When full-scan data are acquired over a m/z range of 650, 3 microscans can be acquired in 1 s. For an acquisition m/z range of 1, 23 microscans can be acquired. By switching from full-scan to SIM, the ion current observed for SIM would be expected to be 8 times greater, which results in a S/N enhancement of 2.7. While this modest gain in S/N may be beneficial in some cases, the additional information that would have been provided by a full-scan spectrum is, typically, more significant. Although the limited sensitivity enhancement afforded by SIM on the ion trap appears to be a significant drawback for GC/MS, such is not the case. The inherent high sensitivity of the ion trap allows full-scan spectra to be obtained at lower levels than can be detected with SIM on conventional instruments.

Ion trap operation in the SIM mode benefits in several other ways. When the acquisition range is reduced, SIM increases the scan repetition rate which, as we have seen in the previous section, improves quantitation by reproducing more accurately the GC peak profile. In addition, several SIM methods in which background interference ions are ejected from the ion trap prior to mass analysis have been discussed in the literature. Such a procedure serves to enhance sensitivity, as longer ionization times may

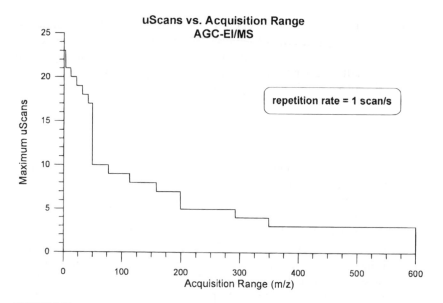

FIGURE 4.16

Relationship between the acquisition range and the maximum number of microscans that can be averaged for an AGC-EI/MS scan program with a scan repetition rate of 1 scan/s. A fixed start mass (m/z 50) was used for each acquisition range and the large decreases in the number of microscans at acquisition ranges of 50, 200, and 350 correspond to the segment breaks at m/z 100, m/z 250, and m/z 400, respectively.

be employed without space-charge effects. One group observed 1000 times more signal using this approach.[85] The use of SIM to eliminate background interferences is covered later in this chapter.

F. Ion/Molecule Reactions

The problem of the occurrence of ion/molecule reactions during GC/MS analysis with an ion trap has been addressed by several researchers.[78,79,86,87] The occurrence of such reactions, which results in excessive $[M + 1]^+/M^+$ ratios and adduct peaks for many classes of compounds that are analyzed typically by GC/MS (see Fig. 4.17), is commonly termed *self-induced chemical ionization* or *self-CI*. The presence of these spectral artifacts can complicate the unambiguous identification of a compound. Thus, it is important to recognize the presence of such artifacts and to understand their origin.

At relatively low sample pressures, self-CI effects are observed for those compounds which can either donate or accept a proton, such as aliphatic molecules containing oxygen or nitrogen, that is, aliphatic alcohols, amines, ketones, and esters; the alkyl fragment ions for these types

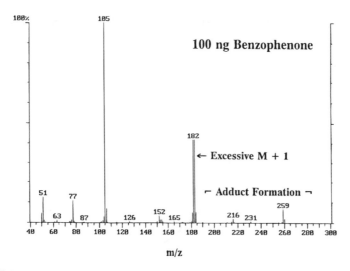

FIGURE 4.17
Mass spectrum for 100 ng of benzophenone analyzed using standard GC/ion trap MS configuration, showing excessive [M + 1]⁺ abundance and adduct formation due to ion/molecule reactions.

of compounds have relatively high gas-phase acidities. For example, Ratnayake et al.[87] observed increased [M + 1]⁺/[M]⁺ ratios and other differences in the spectra of certain lipids including fatty acid esters, when performing ITD analyses.

As discussed by Pannell et al.,[78] proton transfer reactions typically are observed for compounds which form alkyl fragments having conjugate bases of low proton affinity. Even-electron fragment ions react with neutral compounds to form even-electron products through proton and alkyl group transfer (leading to the presence of self-CI product ions), while odd-electron fragment ions react with neutral compounds through charge exchange (leading to M⁺ ions). In the case of methyl decanoate, Pannell et al. found that the tendency for even-electron fragment ions to donate a proton to the neutral molecule was greatest for the lower mass fragment ions (Fig. 4.18).[78] In the case of the m/z 41 fragment ion, they also observed subsequent fragmentation of the [M + H]⁺ ion formed from the proton transfer reaction due to the very low proton affinity of the conjugate base for the m/z 41 ion. Losses of H_2 and methanol were also observed for this reaction.

McLuckey et al. have also described the conditions for which self-CI is observed in the ion trap.[79] Although the partial pressure of neutral molecules [M] varies with the GC peak profile under GC/MS conditions, it does so relatively slowly during each individual scan; a standard rate equation was used to approximate the extent of self-CI occurring during a typical GC peak width of the order of 1 to 5 s. They concluded that a

Methyl Decanoate 50ng Injected on DB-1

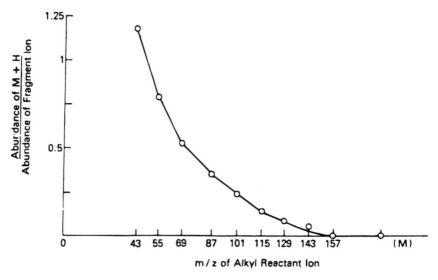

FIGURE 4.18
Tendency of fragment ions from methyl deconate to donate a proton to the neutral ester. A 50 ng sample of methyl deconate was injected onto a DB-1 column.

50 ng sample of a substance capable of reacting with itself or its fragment ions with a rate constant of *ca.* 3×10^{-9} cm^3 per (molecule s) and with a reaction time of *ca.* 50 ms in the ion trap would show about 13% excess $[M + 1]^+/M^+$ as it elutes from a typical GC column.

The authors referred to immediately above made the important point that AGC has no effect in reducing the potential for self-CI. The relative degree of self-CI is governed by the partial pressure of neutrals and the reaction time for a fixed number of ions. The use of AGC only helps to reduce the degree of self-CI because of the use of segmented scan programs which, in turn, leads to a reduction in reaction time.

G. Mass Spectral Skewing

The ion trap does not exhibit the mass spectral distortion (skewing) that results from changes in the partial pressure of the sample in the ion source. Mass spectral skewing is observed when the ions are formed continuously in the ion source and the time required to obtain a mass spectrum is comparable to the GC peak width.[88] Since the ion trap concurrently creates all of the ions that are available for detection, the short (1 to 25 ms) ionization period produces an accurate snapshot of the sample molecules that are present in the ion trap at the beginning of the scan. Thus, changes in the partial pressure of the sample that occur during mass analysis do not cause mass spectra skewing because no ions are being formed during this period.

V. ADVANCES IN ION TRAP GC/MS AND GC/MS/MS

Although the ion trap is considered by many to be the most sensitive mass spectrometer for EI and positive ion CI GC/MS, the development of new techniques that enhance further instrument performance are being pursued actively. Of the many new developments in ion trap mass spectrometry, techniques that reduce background interferences, minimize ion/ molecule reactions, and improve on-line MS/MS, appear to hold the greatest potential for GC/MS applications. This section summarizes some of the recent advancements that have been made in these areas.

A. Methods for Reducing Background Interferences

The problem of background ions that can reduce the sensitivity and dynamic range of ion trap GC/MS experiments has been addressed by several researchers.[23,47,89–92] The presence of high levels of background ions normally requires that shorter ionization times be used so as to avoid space charging. However, shortening the ionization time also reduces the number of analyte ions that are available for detection, resulting in a loss of sensitivity and dynamic range. Space-charge effects caused by background interference ions can be reduced by ejecting these ions from the ion trap prior to a mass selective instability scan. Generally, it is preferable to remove these ions during ionization, or shortly thereafter, since gross space charge may affect also the storage efficiency of the ion trap. Although many of these methods have been developed using a variety of inlet systems and ionization techniques, they are equally useful for GC/MS. These methods reduce the effects of background interferences and, in certain cases, can boost sensitivity by as much as three orders of magnitude.[85]

1. Single-Frequency Resonance Ejection

Single-frequency resonance ejection removes ions of a particular mass/ charge ratio by applying a sine wave to the endcap electrodes at the fundamental axial secular frequency of ion motion. Such ions absorb power from the applied field and, at the limit, are ejected axially from the ion trap. Although single-frequency resonant ejection is mass selective, ions over a range of mass/charge can be ejected by varying either the excitation frequency or the trapping conditions so as to bring other ion species into resonance.

McLuckey *et al.* have described several single-frequency resonance ejection techniques that use a combination of RF voltages and swept-frequency resonant ejection to effect selective ion isolation/ejection over a broad mass range (see Chapter 2).[89] For example, matrix ions were removed from a spectrum of 2,4,6-trinitrotoluene (TNT) using swept frequencies to eject ions over the range m/z 46 to m/z 120 during ion injection from an air sampling source. The ejection amplitude in these studies was 30 $V_{(p-p)}$. The authors also observed that the use of swept-frequency resonance ejection during ion accumulation does not completely remove all of the matrix ions, since ions can accumulate once the frequency sweep has passed beyond resonance.

To effect simultaneous ejection over a wide mass/charge range during ionization, Eades *et al.* demonstrated a broadband excitation technique that switched rapidly between discrete frequencies within a range of 300 kHz.[90] The programmable frequency synthesizer and special instrument control software of the ITMS were used to generate 300 kHz-wide broadband excitation signals.[93] Compared to swept-frequency resonant ejection, broadband ejection over a 300 u mass range (375 to 75 kHz at 6 $V_{(p-p)}$) during ionization does not allow ions to reform because the entire frequency range is excited in rapid succession.

These studies demonstrate that single-frequency resonant ejection can reduce space-charge effects by ejecting background interference ions both during and after ionization. However, only modest gains in sensitivity are observed when single-frequency ejection methods are used in conjunction with long ionization times. McLuckey *et al.* noted that the analyte signal for TNT could be increased by a factor of 8 only with a 150-fold increase in ion accumulation time, even though frequency sweeps were used to prevent gross space-charge affects.[89] Eades *et al.* observed also a nonlinear relationship between ionization time and detected ion signal using the broadband ejection methods.[90] These results suggest that the efficiency with which single-frequency methods can be used to remove ions over a wide mass/charge range may not be sufficient to produce large gains in sensitivity for a single ion. This result is logical since the frequency sweep and broadband excitation methods excite ion species in succession; thus, each ion species is subjected to the resonant excitation for only a fraction of the irradiation period. Single-frequency ejection methods are better suited to the elimination of a single ion species.

2. Multiple-Frequency Resonance Ejection

Multiple-frequency resonance ejection methods use nonsinusoidal waveforms containing a spectrum of frequencies. These methods eject one or more ion species simultaneously. Excitation waveforms are created by numerical methods that provide greater control over the excitation process. These digital waveforms are then converted into analog signals by means of a high speed digital-to-analog converter (DAC).

a.   Arbitrary Waveform Generator Commercial arbitrary (ARB) waveform generators offer a convenient method for producing these types of analog signals. The main advantage of multiple-frequency resonance ejection over single-frequency techniques is that they excite the selected ions continually, and prevent all of the interfering species from accumulating to any appreciable extent.

b.   Stored-Waveform Inverse Fourier Transform (SWIFT) Stored-waveform inverse Fourier transform (SWIFT) broadband excitation, as developed by Marshall and co-workers, is a powerful tool for the manipulation of ions that are held in ion cyclotron resonance (ICR) or quadrupole ion traps.[94] SWIFT on the ICR has been used to isolate/eject ions with high resolution and efficiency. Fig. 4.19 demonstrates the use of SWIFT on the quadrupole ion trap for the elimination of matrix ions that prevented the detection of 37 pmol of substance P using desorption ionization.[91] A limitation of the SWIFT approach is that it is computationally intensive. Thus, the time required to calculate and download a single SWIFT waveform restricts the use of multiple waveforms during a GC analyses. Also, because SWIFT pulses have a finite duration, the length of excitation is limited. As an example, an 8192-point SWIFT pulse that is output at a rate of 2 μs per point lasts approximately 16 ms. During the 16 ms pulse, ions are excited for *ca.* 12 ms due to the reduced amplitude during the leading and tailing edges of the waveform. Multiple SWIFT excitation pulses can be linked together to form longer ejection periods, but it is important to note that the excitation is not continuous. Thus, ions formed between SWIFT pulses are not ejected immediately from the ion trap. The time-varying nature of SWIFT waveforms may also affect the resolution with which an ion can be isolated since the resolution of other resonance ejection techniques is observed to improve as the length of the excitation pulse is increased.[95]

c.   Filtered Noise Field (FNF) A second type of multiple-frequency resonant ejection applies a filtered noise field (FNF) to the ion trap to isolate/eject ions of specific *m/z*.[85] The FNF method uses a single excitation waveform containing all frequencies to excite all of the ions in the ion trap. Then, before this waveform is converted to an analog signal, rapid digital filtering methods are applied to remove selected frequencies from the noise spectrum. By selecting filters that mask the mass/charge ratio of interest, FNF removes all of the other ions from the ion trap. FNF has

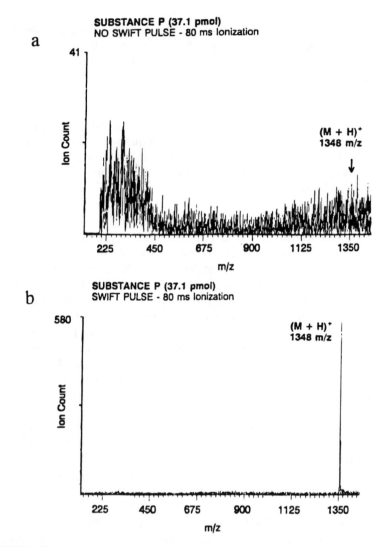

FIGURE 4.19
Resonant ejection mass-selective instability scan of a 37.1 pmol sample of substance P. Ions are created *via* desorption ionization from glycerol–thioglycerol matrix. The sample was bombarded by Cs$^+$ beam of 7 keV for 80 ms during which time ions were injected into the trap: (a) no SWIFT pulse applied; this spectrum shows the severe effect of space charge, with low parent ion intensity; (b) SWIFT pulse applied during ion injection.[91]

been shown to enhance GC/MS sensitivity by as much as 1000 times over GC/MS alone.[85] Fig. 4.20 shows two sets of ion chromatograms for the molecular and fragment ions of cocaine present in a mixture of polycyclic aromatic hydrocarbons. An advantage of the FNF approach is that the digital filters can be exchanged rapidly, allowing different ions to be

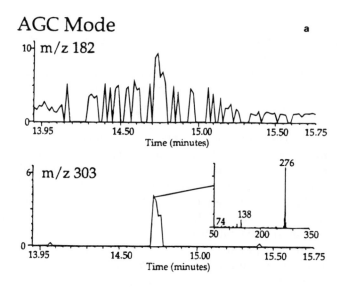

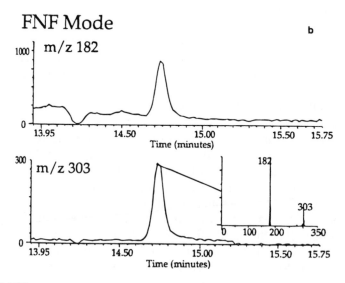

FIGURE 4.20

Comparison of GC/MS data for 200 pg of cocaine (MW 303) in the presence of a 100 ng of PAH background, using (a) AGC and (b) FNF selected ion monitoring modes.

isolated/ejected at different points during the chromatogram. Fig. 4.20 also demonstrates that the FNF technique can be used to isolate multiple mass/charge windows, allowing more than a single ion to be monitored at high sensitivity. The FNF electronics are available commercially and can be retro-fitted to existing ion trap instruments. (See also Chapter 2, Section IV.B.6.b.)

d. Waveform Generators The commercial introduction of ion isolation/ ejection methods that reduce background interferences will improve further ion trap performance for GC/MS and GC/MS/MS. A similar method which employs a waveform generator also uses complex waveforms; this method has been demonstrated by Varian Associates and is available for the SATURN instruments.[92] Alternatively, the availability of inexpensive arbitrary waveform generators, digital function generators, and personal computers offers a practical means for implementing the SWIFT and single-frequency resonant ejection methods on existing ion trap hardware. While these multiple-frequency techniques appear to improve the capabilities of the ion trap, detailed studies that characterize the isolation efficiency, ejection efficiency, simplicity, sensitivity, and linear dynamic range for GC/MS have not been carried out as yet.

3. RF/DC Methods

RF/DC methods can be used to remove rapidly interfering ions from the ion trap by operating at RF and DC voltages that induce instability into the trajectories of such ion species. The rapid nature of RF/DC ejection would appear to make it a more efficient method for eliminating matrix ions. However, the large RF and DC voltages that are associated commonly RF/DC methods can reduce the efficiency with which analyte ions are created and trapped. Nevertheless, RF/DC methods have been shown to be useful for reducing certain types of background interferences.

The application of a small positive DC voltage (5 to 20 V) during electron ionization can be used to remove efficiently high mass/charge ions from the ion trap. For CI experiments, the ejection of high mass EI ions that are formed at the same time that the CI reagent ions are formed can improve CI sensitivity and eliminate the EI ions from CI spectra. Eades showed that the application of a small positive DC voltage could be used to eject all ions above m/z 50.[90] This procedure allowed the ionization time to be increased from 1 to 50 ms and resulted in a 20-fold increase in ion signal for m/z 41.

Several RF/DC methods have been demonstrated that reduce background interferences by isolating the ion of interest after the ionization event. Strife showed that the sequential ionization and isolation of an analyte and its internal standard improved dramatically the ability of the ion trap to detect compounds that co-elute with vastly different concentrations.[23] In subsequent work, sequential RF/DC isolation steps were used to obtain a calibration curve for an analyte and its isotopically-labeled internal standard for sample amounts that spanned six orders of magnitude.[83]

B.   External Ionization GC/MS

In the standard ion trap configuration (see Fig. 4.21a), GC effluent is introduced into the body of the ion trap; electrons emitted from a filament located externally to the ion trap are gated into the trap. Ions formed by electron ionization are stored and mass analyzed subsequently by ramping the RF voltage applied to the ring electrode. During this procedure, neutral molecules are present in the same region as the ions that are being stored and analyzed. As reaction times of *ca.* 50 ms are encountered typically, ion/molecule reactions can occur, leading to self-CI, as previously discussed. Both proton transfer (creating excessive $[M + 1]^+$ abundances) and alkyl group transfer (creating adduct ion peaks) can be seen clearly in the ion trap mass spectrum of benzophenone (see Fig. 4.17).

A solution to the problem of self-CI during GC/ITMS analyses is to reduce the number density of neutral molecules (either analyte, solvent, or contaminants) in the ion trap during storage and scanning by creating the ions externally, in a conventional ion source, and injecting them into the ion trap. The injection of externally-generated ions into an ion trap has been used by several researchers since the original work of Louris *et al.*[32] Some of this work includes atmospheric sampling glow discharge ionization,[96] particle beam liquid chromatography,[97] supercritical fluid chromatography,[98] and electrospray ionization.[99]

Fig. 4.21(b) shows an external ion-source/ion-trap GC/MS interface that was described recently.[100] For this work, a Finnigan 4500 ion source was coupled directly to a Finnigan ITMS-style ion trap. Injection of ions was accomplished by grounding the ion source and floating the end-cap and ring electrodes of the ion trap by a variable potential. A standard ITS40 electron gate was used as the ion gate for this system. The electron gate TTL signal produced by the ion trap electronics was inverted and amplified using a custom circuit and a variable power supply to produce the ion gate voltage. For this work, a standard EI AGC scan program was used.

Since benzophenone forms both the protonated molecule (m/z 183) and adduct ions (m/z 216 and 259) under EI conditions, it was used to evaluate the GC/MS performance of this system. Fig. 4.22 shows the total ion chromatogram and mass spectrum of 1 ng benzophenone analyzed on the GC/ion injection ion trap system. Fig. 4.23 compares the

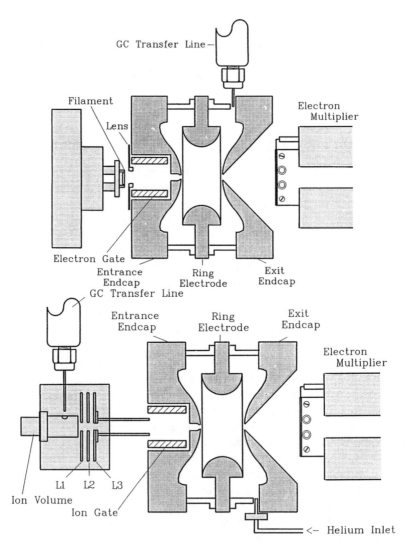

FIGURE 4.21

Schematic of (a) standard GC/ion trap MS configuration using internal ionization and (b) external ion source used for GC/ion trap MS. Helium is introduced directly into the trap (a) *via* the GC and (b) by an additional gas line. The external ion source was constructed from a 4000 series ion source. Lens 3 (L3) has a tube lens that helps pass ions from the ion source to the entrance end-cap. The pressure in the ion trap was maintained at a pressure ten times greater than the surrounding vacuum system *via* the conductance spacers that are located between the ring and end-cap electrodes.

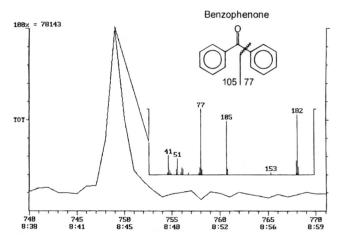

FIGURE 4.22

Total ion chromatogram and mass spectrum for 1 ng of benzophenone analyzed on the GC/ion trap MS system using external ionization. Ion injection times were controlled by AGC.

(M + 1)/(M) Ratio vs. Amount Injected

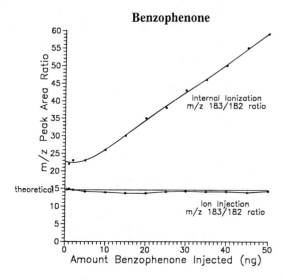

FIGURE 4.23

Comparison of the [M + 1]+/[M+] ratio for the analysis of benzophenone using both internal and external ionization systems. The theoretical peak area ratio of 14.6% assumes that only ^{13}C contributes to [M + 1]+.

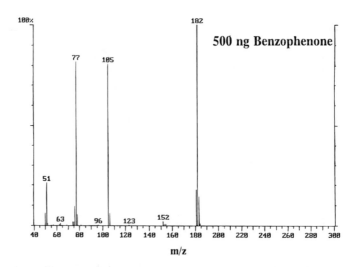

FIGURE 4.24

Mass spectrum for 500 ng of benzophenone analyzed using GC/ion trap MS with external ionization, showing correct [M + 1]$^+$ abundance and no adduct formation.

[M + 1]$^+$/[M]$^+$ ratio for the analysis of benzophenone using both internal and external ionization systems. As can be seen from this figure, the [M+1]$^+$/[M]$^+$ ratio remains constant and very close to the theoretical value for the GC/ion injection ion trap system. The data for internal ionization, in contrast, show that even at 0.5 ng benzophenone, the proper [M+1]$^+$/[M]$^+$ ratio is never reached. At low concentrations, this may be due to lower mass ions from the column bleed that were present during the analysis. These ions also have relatively high gas-phase acidities, and can donate protons to neutral benzophenone. At the higher concentrations of benzophenone, self-protonation occurs, leading to a linear relationship between the [M+1]$^+$/[M]$^+$ ratio and the amount of benzophenone injected.

Fig. 4.17 shows the mass spectrum for 100 ng of benzophenone analyzed on the GC/internal ionization ion trap system. At this concentration, both protonated benzophenone molecule (*m/z* 183) and adduct ions (*m/z* 216 and 259) can be seen. Fig. 4.24 shows the mass spectrum for 500 ng of benzophenone analyzed on the GC/ion injection ion trap system. Even at this high concentration, the correct [M + 1]$^+$/[M]$^+$ ratio is observed, and no adduct ion species are observed in this spectrum.

It is important to note that only minor modifications to a standard ITS40 GC/MS ion trap system were needed in order to use an external ion source. For this system, differential pumping was not required, a tube lens was added to the third lens of the ion source, and helium was added directly into the ion trap.

Although this system does show that it is capable of delaying the onset of ion/molecule reactions, it is important to realize that all mass spectrometers are susceptible to ion/molecule reactions. Besides reducing ion/molecule reactions, another potential benefit in using an external ion source is that negative ion chemical ionization (NICI) may be performed. NICI cannot be performed when using internal ionization because of the lack of thermal electrons due to the presence of the ion trap's high voltage RF storage field. This mode of operation is particularly useful when analyzing compounds with very electronegative functional groups, such as halogenated pesticides and PCBs.

The major drawback to using an external ion source is the added complexity of operation. The external ion source requires both a vacuum system somewhat larger than the standard configuration and the necessary power supply to drive the lenses. Additionally, changing between EI and CI ionization modes involves hardware adjustments with an external ionization source, in contrast to the minor software modifications that are required for internal ionization. Even with these limitations, the development of commercial external ionization source GC/ion trap systems will expand the use of ion traps for applications that require additional ionization modes and for which ion/molecule reactions must be avoided.

C. Tuning Methods for GC/MS/MS

Creating and tuning a MS/MS scan program is currently a time-consuming process that can restrict the usefulness of GC/MS/MS on the ion trap. Several parameters may be adjusted, such as ionization time, RF level for ionization, RF and DC voltages for isolation, isolation times, RF level for CID, excitation frequency, excitation time, excitation voltage, etc. For GC sample introduction, tuning time is limited to the width of the chromatographic peak, typically 1 to 5 s over the course of the GC run. Thus, if 50 manual adjustments are required, 50 repetitive sample injections would also need to be performed. Repetitive injections require a lot of time, and can be impractical even if only a few injections are needed. Additionally, either the sample must be of sufficient concentration to allow tuning to be performed or standards must be available. Alternatively, samples of sufficient volatility can be introduced *via* a variable leak valve to allow for off-line tuning of the scan program. Clearly, the amount of tuning that is required currently to analyze a single compound, present at a known concentration in a clean sample matrix, makes general use of ion trap GC/MS/MS impractical.

Todd *et al.* have demonstrated a sophisticated set of programs with which MS/MS scan program parameters are adjusted automatically.[101] With the use of the dynamically-programmed scan method, complete

MS/MS fragmentation data maps were constructed with a single multi-scan sequence. However, the rate at which these types of experiments were performed would limit their applicability on a GC time scale.

New software developed for the ITMS simplifies the process of setting up MS/MS scan programs by scaling automatically a template of an MS/MS scan program for a parent ion of interest.[70,93] This software adjusts the RF voltages, segment durations, and scan program structure without tuning. Tuning the resonant excitation frequency is obviated by the use of broadband excitation (discussed in the following section). Additionally, the DC voltage used for isolation is calculated empirically. After proper calibration of the DC voltage, these methods can isolate automatically a window some 2 to 3 u wide in the range m/z 70 to m/z 300. Fine tuning of the isolation voltages to isolate a single ion species still requires manual adjustment and remains the biggest limitation for tuning MS/MS scan programs for GC analyses. The use of a scan program template also restricts the versatility of the software to particular types of MS/MS scan programs. Regardless of these limitations, this software simplifies many of the tuning requirements for MS/MS scan programs and can save a considerable amount of time in comparison with that required to create new scan programs.

Griffin and co-workers have developed several sophisticated programs that tune MS/MS scan programs automatically for an ITS40 modified for GC/MS/MS experiments.[72,91] This software uses the Mathieu parameters a_z and q_z (see Chapter 1) and the modified field equations for the stretched geometry of the ion trap to calculate the RF and DC voltage for each stage of the scan.[102] This procedure allows scan tables to be scaled for a particular ion species of interest. Additionally, versatile automatic ionization control techniques for EI and CI are included that eliminate adjustment of the ionization time. Creation of an MS/MS scan program for m/z 303 begins with the creation of a MS/MS scan program for m/z 264 of perfluorotributylamine (PFTBA). PFTBA is employed as a calibration compound that is introduced readily into the ion trap *via* a leak valve. Once the a_z, q_z, and table time parameters have been adjusted manually for PFTBA, the scan program can be scaled to the ion species of interest with a software toggle. Although the isolation of a single ion species can be erratic, windows of mass/charge ranges approximately 2 to 3 u wide can be isolated with good reproducibility. In most cases, these programs allow MS/MS scan programs to be created without the need for repetitive injections.

The process of performing GC/MS/MS on the ion trap is limited, in part, by the current complexity of tuning MS/MS experiments. However, the development of more accurate electronics and sophisticated tuning methods should greatly simplify the conversion from MS to MS/MS.

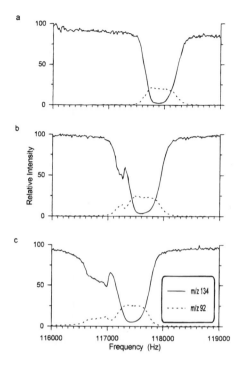

FIGURE 4.25
Comparison of three frequency breakdown curves for m/z 134 of n-butylbenzene at ionization times of (a) 5 ms, (b) 40 ms, and (c) 85 ms. Resonant excitation parameters include amplitude of 400 mV$_{(0-p)}$, duration of 10 ms, and a frequency range of 116,000 to 119,000 Hz at 23 Hz increments.

D. Frequency Shifts, Resonant Excitation Methods, and CID

Shifts in the secular frequency of ion motion that occur when the number of ions in the ion trap changes have been investigated by several groups.[22,48,70,80,81] Fig. 4.25 contains three frequency breakdown curves for the molecular ion of n-butylbenzene, m/z 134. The minimum of each curve corresponds to the most efficient resonant excitation frequency, and may be interpreted as the secular frequency of ion motion in the axial direction. Interestingly, the secular frequency shifts to lower values as the number of ions is increased by extending the ionization time from 5 to 85 ms. These frequency shifts complicate single-frequency resonant excitation experiments, since the excitation frequency must match the secular frequency of ion motion. To illustrate the effects of irradiation with an excitation frequency that does not match the secular frequency of ion motion, Fig. 4.26 shows three product ion spectra for m/z 134 of n-butylbenzene. These spectra were acquired using ionization periods of (a) 5 ms, (b) 40 ms, and (c) 80 ms, and an excitation frequency of 117.900 kHz. Note that, as the ionization (and thus the number of ions in the ion trap) increases, the efficiency of CID for m/z 134 decreases due to shifts in the secular frequencies of ion motion.

For GC/MS/MS, the connection between CID efficiency and the number of ions in the ion trap can be particularly bothersome since the number

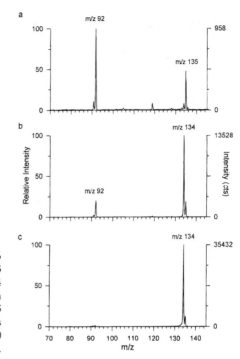

FIGURE 4.26
Comparison of three MS/MS product ion spectra for m/z 134 of n-butylbenzene at ionization times of (a) 5, (b) 40, and (c) 85 ms. Resonant excitation parameters of 400 $mV_{(0-p)}$, 117.9 kHz, and 10 ms were used.

of ions changes continually during a GC run. Tuning single-frequency resonant excitation scan programs for GC can also be complicated because it is difficult to determine from a CID spectrum whether the applied excitation frequency was too high or low. To simplify CID on the GC time scale, several methods have been developed that compensate for the effects of frequency shifts.

1. Frequency-Assignment Pre-Scan

The frequency-assignment pre-scan corrects for frequency shifts by performing two rapid pre-scans prior to each analytical scan that determine the secular frequency of parent ion motion.[70] Each pre-scan employs a frequency sweep of 10 ms duration that determines the secular frequency of parent ion motion by ejecting the parent ion to the detector. By correlating the observed ion signal with the frequency sweep rate, the secular frequency of motion may be determined.

Fig. 4.27 shows the frequency breakdown curve and frequency sweep data for m/z 134 of n-butylbenzene. Note that peaks observed for the positive and negative slope frequency sweeps are equivalent to the secular frequency of the ion as determined by the minima of the frequency breakdown curve. The 50 ms required for the two pre-scans is 10,000 times

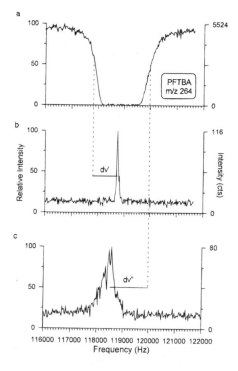

FIGURE 4.27

Comparison of (a) the frequency breakdown curve and (b,c) swept-frequency ejection curves for m/z 264 of perfluorotributylamine. A frequency sweep rate of ± 840 kHz/s and amplitude of 6.0 $V_{(p-p)}$ were used to eject the mass-isolated m/z 134 ions stored at $q_z = 0.3$. Labels dv' and dv'' designate the delay associated with energy absorption and ion ejection for the two frequency sweeps.

shorter than the several minutes that are needed to collect the data for a frequency breakdown curve.

Fig. 4.28 shows an MS/MS scan program that incorporates the frequency-assignment pre-scans. When executed, the pre-scans ionize, isolate, and measure the secular frequency of parent ion motion. Next, the instrument calculates the average for the two frequency measurements and uses this value as the resonant excitation frequency during the MS/MS scan.

Fig. 4.29 shows three product ion spectra that were acquired using the frequency-assignment pre-scan. Note that the ionization times used were (a) 5, (b) 80, and (c) 200 ms for the top, middle, and bottom mass spectra, respectively. For each ionization time, the frequency-assignment pre-scan determined an appropriate excitation frequency that was used to produce a reliable product ion spectrum.

The frequency-assignment pre-scan is the only method of determining the proper resonant excitation frequency on the millisecond time scale. A limitation of this method is that the isolated parent ion must be of sufficient intensity to allow detection by the pre-scan. Also, the extra 50 to 70 ms required for the pre-scan to be performed may be a limitation for some applications. Although the scan program used by this method is complex, the technique has been very reliable in determining the appro-

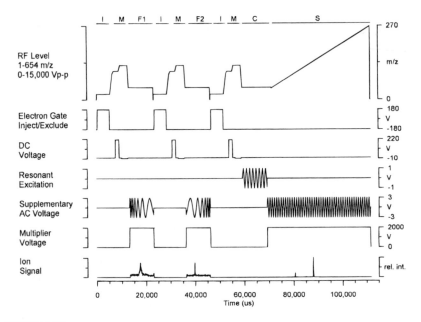

FIGURE 4.28

A frequency-assignment pre-scan, steps I to F2, is performed prior to the MS/MS product ion scan to determine automatically an appropriate resonant excitation frequency to be used in step C. Ions are generated during the periods labeled I, parent ions are mass-selected using two-step isolation during the stages labeled M, the secular frequency of parent ion motion is determined with two frequency sweeps periods F1 and F2, the parent ions are dissociated during stage C, and mass analysis of the product ions is performed during period S. Increasing and decreasing frequency sweeps F1 and F2 (not drawn to scale) cover a 10 kHz range centered at 118 kHz. A supplementary RF voltage frequency of 530 kHz is applied during period S to affect axial modulation. The ion signals observed during the pre-scan and MS/MS scan are not shown on the same relative intensity scale.

priate resonant excitation frequencies for ions in the range m/z 50 to m/z 650.

2. Broadband Excitation

Broadband excitation compensates for frequency shifts by exciting a range of frequencies that cover the expected magnitude of the frequency shift.[70] On the ITMS, broadband excitation may be implemented by programming the frequency synthesizer to switch between selected frequencies in a phase-continuous manner. A typical broadband excitation pulse switches between 1400 discrete frequencies covering a 10 kHz-wide range. Centering this broadband excitation signal at 118 kHz excites a 1.2 u wide mass/charge range at $q_z = 0.3$ on the ITMS.

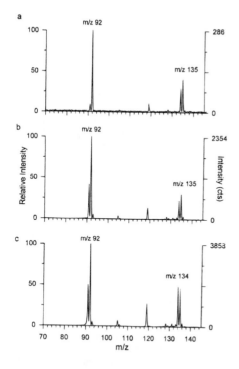

FIGURE 4.29

Comparison of three MS/MS product ion spectra for m/z 134 of n-butylbenzene at ionization times of (a) 5, (b) 85, and (c) 200 ms. Resonant excitation amplitude and durations were 400 $mV_{(0-p)}$ and 10 ms, respectively. The excitation frequency used for resonant excitation was determined automatically *via* the frequency-assignment pre-scan.

Fig. 4.30 shows a chromatogram for 800 pg of n-butylbenzene that was acquired with an MS/MS scan program and broadband excitation. The product ion spectra observed at the top and near the base of the peak correlate well with each other and to reference product ion spectra for n-butylbenzene. No adjustment of the excitation frequency was required. However, when the experiment was repeated with single-frequency resonant excitation, no product ions were observed.

Qualitatively, the CID efficiencies observed for broadband excitation are 10 to 40% less than are observed with optimized single-frequency resonant excitation. Although the frequency pattern that is used to create the broadband signal is designed to distribute the excitation power evenly over the 10 kHz wide excitation envelope, the true power spectrum may contain irregular power distribution that can lead to ion losses and reduced CID efficiencies. Broadband excitation also requires a 40 ms excitation time and 1 to 2 $V_{(0-p)}$ amplitudes in order to obtain spectra similar to those observed with a 10 ms excitation time and 300 to 700 $mV_{(0-p)}$ amplitude using single-frequency excitation. The need for longer excitation times and higher voltages is a result of the excitation power being spread over a wider frequency range; thus, less power is put into each frequency. Regardless of the reduced CID efficiencies and the need for increased excitation times and amplitudes, broadband excitation may

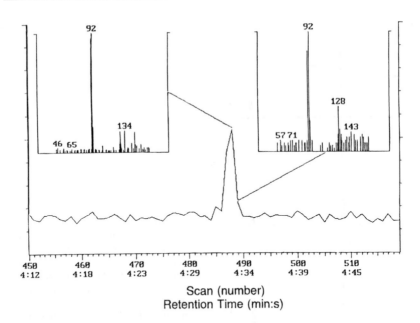

FIGURE 4.30

GC/MS/MS product ion data for m/z 134 of n-butylbenzene, obtained using broadband excitation over a 10 kHz frequency range centered at 118 kHz. An excitation time and amplitude of 40 ms and 1.2 $V_{(p-p)}$, respectively, were used. The insets show the MS/MS product ion spectra observed at the top and bottom of the chromatographic peak.

result in higher CID efficiencies for GC/MS/MS because the overall CID efficiency remains constant as the secular frequency of ion motion shifts.

Broadband excitation over a 10 kHz frequency range can be achieved also with the use of SWIFT waveforms. While SWIFT methods have been demonstrated for the CID, isolation, and ejection of ions on the ion trap,[91] SWIFT broadband excitation is particularly useful for GC/MS/MS experiments because it is unaffected by shifts in the secular frequency of ion motion.[72] Unlike frequency modulated methods, the power spectrum for SWIFT broadband excitation exhibits flat power response over the excited frequency range. SWIFT effects continuous excitation, meaning that ion motion is excited in a phase-continuous manner for the duration of the SWIFT pulse. As a result, the CID efficiency for SWIFT broadband excitation is comparable to single-frequency resonant excitation.[72] The overall MS/MS efficiency including the SWIFT broadband excitation step is characterized in Section VI.

Qualitatively, SWIFT broadband excitation is simple to adjust and produces reliable CID spectra for GC/MS/MS experiments. The adjustment of the SWIFT excitation amplitude gives the same control over excitation energy that is observed for single-frequency CID. Although the duration of the SWIFT excitation waveform is determined by the

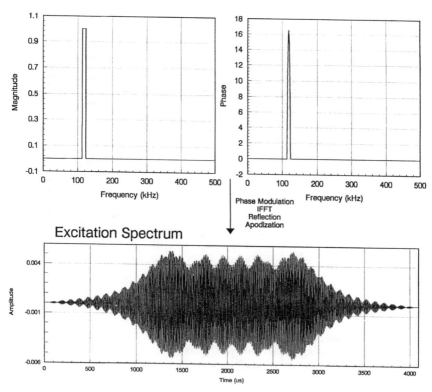

Excitation Spectrum

FIGURE 4.31

Diagram of the computation procedure for stored waveform inverse Fourier transform (SWIFT) broadband excitation. The broadband frequency domain spectrum is subject to quadratic phase modulation prior to being converted to a time domain signal excitation signal. The inverse fast Fourier transform (IFFT) is used to convert between the frequency and time domain. The resulting time domain data are reflected and apodized as discussed by Marshall prior to being downloaded to the arbitrary waveform generator.[94]

digitization rate and the number of data points in the SWIFT waveform, these values can be altered to increase or decrease the excitation time. Fig. 4.31 outlines the calculation of the SWIFT waveform for broadband excitation. Typical values for a 10 kHz wide SWIFT broadband excitation period are: digitization rate = 500 kHz, data points = 8192, waveform bursts = 3, excitation amplitude 200 to 1000 mV$_{(0-p)}$. Interestingly, changing to a digitization rate of 1 MHz, 1024 data points, and a burst count of 39 appears to have little effect on the CID spectra that are observed. The manner in which ion activation in the ion trap is affected by these parameters has yet to be characterized. As a general rule, SWIFT settings that excite an ion for 10 to 50 ms are observed to produce the highest CID efficiencies for GC/MS/MS.[72]

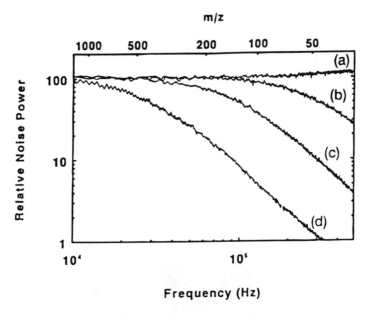

FIGURE 4.32

Relative noise power *versus* frequency as a function of –3 dB roll-off frequency for several low pass filters: (a) no filter, (b) $f_c = 300$, (c) $f_c = 100$, (d) $f_c = 30$ kHz. Mass/charge values are indicated also for a low-mass cut-off corresponding to m/z 35.[103]

3. *Random Noise Excitation*

Random noise excitation is a simple method for exciting ions resonantly over a wide mass/charge range that is also unaffected by frequency
shifts. The technique can be implemented with a commercial noise generator that produces a random noise signal with a flat power spectrum from
10 Hz to 500 kHz.[103] Although low-pass filters can be used to alter this
power spectrum, the technique is not mass selective. Fig. 4.32 shows the
power spectrum of a random noise source that has been attenuated by
several low-pass filters. Note that, for a low-mass cut-off of 35, a frequency
range from 10 to 530 kHz corresponds to a mass range of m/z 35 to
m/z 1000. Thus, random noise excitation results in the simultaneous activation of all of the ions in the ion trap.

Since random noise excitation is not mass selective, product ions
formed by CID of parent ions are excited also and, in turn, fragment to
form other product ions, etc. Thus, random noise may produce more
fragment ions and more structural information in a single experiment
than is observed for mass-selective CID methods. However, additional
fragmentation does not always provide more structural information
because formation of low mass/charge fragment ions is often less useful
for structure elucidation studies. Additional fragmentation may also lead

to the production of ions below the low-mass cut-off of the ion trap. In this case, both the CID efficiency and sensitivity are reduced due to ejection of such ions from the ion trap.

Increased ion losses due to resonant ejection may also be observed when random noise is used for CID. Although the average power spectrum of a random noise source exhibits a flat response, discrete segments of the random noise signal may exhibit irregularities in time domain. Individual ions may be unaffected, excited, or ejected, depending on the instantaneous phase and frequency of the noise source and the velocity, position, and mass/charge ratio of the ion. Thus, random noise excitation may be more erratic than single-frequency methods which exhibit a smooth activation process. In practice, ion losses are 40 to 60% greater for random noise relative to single-frequency resonant excitation.[103]

4. Filtered Noise Field

The filtered noise field (FNF) method that is used for the removal of background matrix ions can be also applied as an excitation method for GC/MS/MS. FNF broadband excitation can be implemented by applying a digital filter to the noise field that masks all of the frequencies except for a 10 kHz-wide window centered at the frequency of motion for the parent ion of interest. Digital methods also allow the phase relationship of the unfiltered noise field to be controlled so that ion activation is a smooth process rather than an erratic one. Although many of the details of the FNF method have yet to be described, the ion isolation and broadband excitation capabilities of FNF appear to make it a powerful technique for GC/MS/MS.

5. DC Pulse Axial Activation

DC pulse axial activation can be used to affect ion activation and dissociation in the quadrupole ion trap.[104] The method employs short (2 μs) DC pulses that are applied to the end-cap electrodes of the analyzer to accelerate the ions in the axial direction. Excitation is not mass selective and must be combined with isolation to affect MS/MS experiments. Interestingly, the maximum internal energy that can be deposited has been shown to be considerably higher than is possible with resonant excitation methods. Dissociation efficiencies are less than those observed typically for resonant excitation, ranging from 50 to 5% as the amplitude of excitation is increased.

Wang et al. demonstrated a similar nonresonant ion activation method that employs low frequency (100 to 500 Hz) excitation signals.[105] Ions were excited by applying, in dipolar mode, a potential oscillating at 200 Hz across the end-cap electrodes for a period of 10 ms; the amplitude of the potential was 56 $V_{(0-p)}$. This signal corresponds to 2 DC pulses of alternating polarity and each of 5 ms duration. High CID efficiencies and evidence for considerable internal energy deposition was obtained. The detection of 100 pg of *n*-butylbenzene was demonstrated using nonresonant excitation GC/MS/MS. Interestingly, the internal energy deposition was noticeably less for the GC/MS/MS results, as indicated by the intensity ratio for the product ions m/z 91 and m/z 92. Although these DC pulse axial activation methods are relatively new tools for affecting CID, their simplicity, energy deposition, and lack of dependence on the secular frequency of parent ion motion augur well for their future application in GC/MS/MS experiments.

VI. BENCHTOP GC/MS/MS INSTRUMENTS

Based on the present understanding of the GC/MS and MS/MS capabilities of the ion trap, several criteria can be identified which must be fulfilled for practical GC/MS/MS analysis.

The first criterion is the ability to analyze sample amounts that extend over a wide dynamic range without manually adjusting the ionization time. Automatic ionization control methods have been effective for GC/MS and would appear to be a reasonable approach for limiting space charge, maximizing parent ion intensity, and enhancing dynamic range for GC/MS/MS experiments. External ionization would help to reduce ion/molecule reactions and allow selective ionization modes like electron capture negative CI to be used.

The second criterion is the ability to analyze samples regardless of high levels of chemical interferences. The incorporation of isolation methods that reduce background interferences by ejecting matrix ions from the ion trap should reduce space-charge effects from complex samples. Consequently, isolation of an ion species of interest is a necessary first step for any MS/MS experiment. Thus, MS/MS analyses should be less affected by space charge because of the reduced number of ions that are analyzed compared to MS analyses.

The third criterion is that the method of ion activation yields high CID efficiencies regardless of the number of ions in the ion trap. Many of the broadband excitation methods that have been developed appear to be viable methods for affecting CID over a GC peak.

Perhaps the most important criterion is simplicity of operation. By using improved methods of operation, the need for repetitive sample injections can be reduced. The development of software that automatically handles the creation and tuning of scan programs would reduce the level of sophistication that is required currently to take advantage of the MS/MS capabilities of the ion trap.

Although none of the ion traps incorporate currently all of these features, several groups have strived to improve the usefulness of GC/MS/MS on the ion trap in order to demonstrate the analytical utility of the technique. Van Berkel *et al.* recently demonstrated the use of random noise excitation on a modified ITS40.[69] Product ion spectra were obtained by triggering the random noise signal during the CI reaction period of a ARC-CI/MS scan program. CID spectra for GC peaks that correspond to amphetamine and methamphetamine were obtained. The authors pointed out that while the random noise technique is simple to implement, the ITS40 software was intended for GC/MS experiments and can be restrictive for GC/MS/MS analyses. For instance, the ITS40 EI segmented scan program does not hold the ions at a fixed RF level for an extended period of time; thus, random noise CID experiments cannot be performed using EI.

An important advantage of random noise CID is observed when complete chromatographic separation is achieved. That is, the same scan program can be used to produce product ion spectra for each compound. Additionally, toggling the random noise on or off with each scan would allow molecular weight and structural information to be obtained in a single experiment.

A disadvantage of random noise CID on the ITS40 is that the parent ions are not mass selected prior to CID; such a procedure is not a true MS/MS experiment. This type of CID/MS experiment can perform only "orphan scans", because there is no way at this time of determining the mass/charge of the parent ion from the product ion spectra. Most importantly, the analytical utilities of CID/MS and MS/MS are not the same. In contrast to MS/MS, CID/MS produces product ions from all of the ions in the ion trap, including matrix ions. Thus, chemical noise is not reduced and no gains in selectivity or signal/noise ratio are observed.

GC/MS/MS was demonstrated recently on a modified ITS40 that employs automatic ionization control (AGC and ARC), RF/DC two-step isolation, SWIFT broadband excitation, and new instrument control software that scales the MS/MS scan programs automatically for the parent ion of interest. Fig. 4.33 shows a diagram of the major components that make up the GC/MS/MS ion trap system. The main hardware modifications include an arbitrary waveform generator (ARB) that is used to produce the SWIFT waveforms, and an active dipolar-excitation mixing circuit that was developed to connect the SWIFT and axial modulation signals to the instrument. The MS/MS software is integrated into the standard ITS40 data system so that control of the sample injection, chromatographic analysis, and mass analysis could be automated completely.

Fig. 4.34 shows an example of an AGC-EI/MS/MS scan program that can be executed on the modified ITS40. ARC-CI/MS/MS scan programs (not shown) can be implemented also. The scan programs shown here

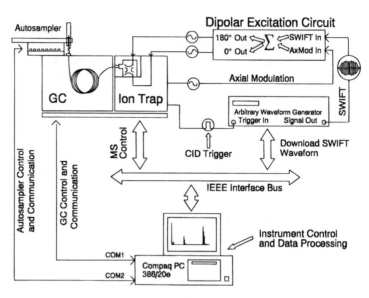

FIGURE 4.33

Block diagram of a modified ITS40 GC/MS/MS ion trap. Samples are injected into the gas chromatograph by an A200S autosampler. Following chromatographic separation, samples elute into the ion trap where they are mass-analyzed. MS/MS product ion scans are performed by the ion trap's embedded microprocessor and MS/MS software. The arbitrary waveform generator is used to produce the SWIFT broadband excitation signals that effect CID in the ion trap. The dipolar excitation circuit sums the SWIFT and axial modulation signals and splits the result into two signals that have a phase separation of 180°. These signals are connected to the ion trap end-cap electrodes. A personal computer provides a user interface to control the instrument and to perform data processing.

use the same type of AGC and ARC pre-scans that are used for GC/MS on the ITS40. However, modified AGC pre-scans have been developed that adjust the ionization time as the intensity of a selected parent ion changes.[72] The parent ion is selected with a two-step RF/DC isolation technique that removes ions of mass/charge both greater and less than the parent ion by moving to the working points $q_z = 0.70$, $a_z = -0.23$ and $q_z = 0.85$, $a_z = 0.06$ in consecutive steps.[46,106] Although a single ion species can be isolated with good efficiency using two-step isolation, greater than 90% of the ion of interest can be lost due to improper adjustment of the isolation voltages. In addition, tuning is complicated because it is difficult to determine just how well a single ion species is isolated using the centroid spectra that are stored by the data system. To reduce ion losses and lengthy tuning procedures, an isolation window of 2 to 3 u is normally used. In most cases, the software can calculate proper voltages to isolate a ion of interest in the range m/z 50 to m/z 300. SWIFT broadband excitation was implemented using a 8192-point waveform that was digi-

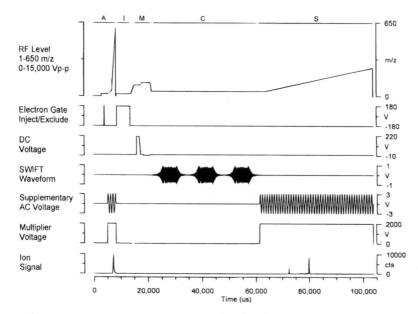

FIGURE 4.34

Ion trap scan program used to acquire GC/MS/MS EI mass spectra on a modified ITS40. The AGC pre-scan (A) sets the ionization time used during period I, the parent ion is mass-selected using two-step RF/DC isolation during period M, CID is accomplished during period C with three SWIFT broadband excitation bursts that excite a 10 kHz range of frequencies from 102 to 112 kHz, and the resulting product ions are mass-analyzed during period S. A supplementary RF voltage frequency of 485 kHz is applied during period S to effect axial modulation.

tized at a rate of 500 kHz. This waveform excites a 10 kHz-wide frequency range that is centered on 112 kHz, which corresponds to a $q_z = 0.3$ for the ITS40. Three SWIFT bursts are emitted by the ARB, corresponding to an excitation time of approximately 40 ms. Excitation amplitudes of *ca* 300 m$V_{(0-p)}$ and 1500 m$V_{(0-p)}$ are sufficient usually to promote CID and ion ejection, respectively; however, for CID, the most appropriate setting depends on the stability of the ion structure that is to be dissociated. The SWIFT signal passes through a dipolar-excitation mixing circuit where it is inverted to produce two waveforms that are 180° out-of-phase; these waveforms are applied to the end-cap electrodes of the ion trap. Detection of the product ion spectrum is performed with a mass-selective instability scan that uses axial modulation at 485 kHz for ion ejection.

The instrumental performance of the modified ITS40 was characterized using benzophenone standards. Fig. 4.35 shows the background-subtracted product spectrum of the M$^{+\bullet}$ ion (m/z 182) that was observed for 3.56 ng of benzophenone. The main product ions result from the

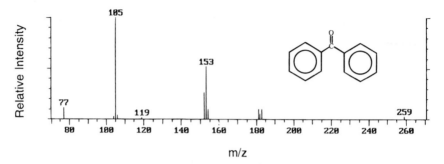

FIGURE 4.35
Background-subtracted GC/MS/MS EI m/z 182 product ion spectrum of 3.56 ng of benzo-phenone (MW 182).

cleavage of the phenyl-keto bond. The m/z 153 ion has been identified as the biphenylene ion that is formed by the loss of CO and rearrangement of the phenyl rings.[107] The parent ion was selected with an isolation window approximately 2 to 3 u wide that allowed some m/z 181 and m/z 183 to remain in the ion trap. Additional mass selection is afforded using broadband excitation, and preferential excitation of m/z 182 over m/z 181 and m/z 183 is observed. However, it is difficult to differentiate products of m/z 182 and m/z 183 because they both fragment to form m/z 105.

Qualitatively, the product ion spectra obtained over a wide range of sample amounts show good correlation to one another. However, near the limit of detection, product ions of low abundance are often absent from the product spectra. For larger sample amounts, product ions may be observed that result from ion/molecule reactions. The product ion m/z 77 of the $M^{+\bullet}$ of benzophenone can react with neutral benzophenone molecules to form m/z 259, as seen in Fig. 4.35.

Fig. 4.36 shows three calibration curves for the analysis of benzophe-none using EI/MS, EI/ISO/MS, and EI/MS/MS scan modes, where ISO refers to the isolation of the parent ion. EI/MS data were collected using the standard ITS40 segmented EI scan program. EI/ISO/MS and EI/MS/MS data were collected using the EI/MS/MS scan program shown in Fig. 4.34 both with and without the application of the SWIFT broadband excitation signal. Five replicate injections were performed for each sample amount for each scan mode, for a total of 180 sample injections. The experiment required 36 h to complete during which no adjustment of the scan program or instrument settings was made.

The slopes of the EI/MS, EI/ISO/MS, and EI/MS/MS calibration curves were determined by linear regression to be 0.96, 0.97, and 1.05, respectively; the linearity of these log-log calibration curves are reflected in the small differences from an ideal slope of 1. Interestingly, the slope for the EI/MS/MS data shows a positive curvature while the EI/MS

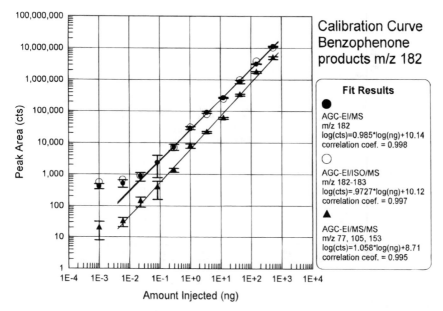

FIGURE 4.36
Calibration curve for benzophenone (MW 182). Each datum point represents five replicate injections. Error bars correspond to ±2 standard deviations. AGC-EI/MS data (●) correspond to the peak area observed for m/z 182 without isolation or CID. AGC-EI/ISO/MS data (○) correspond to the isolated peak for m/z 182 without CID. AGC-EI/MS/MS data (▲) correspond to total peak area for the product ions m/z 77, m/z 105, and m/z 153 of m/z 182.

and EI/ISO/MS show a negative trend. The change in slope is being investigated currently; however, several explanations can be offered here. First, the overall CID efficiency may decrease with decreasing amounts of sample. Alternatively, detection efficiency for EI/MS/MS may be higher than for EI/MS at higher sample concentrations since fewer ions are in the ion trap. Differences in quantitation methods have also been observed to produce significant changes in calibration slope. Differences in the $[M]^+/[M + H]^+$ ratio observed at various sample amounts suggests that self-CI is affecting the MS/MS calibration curve. Perhaps a combination of these effects is occurring. Nevertheless, slopes of 0.96, 0.97, and 1.05 show reasonably good linearity for MS, ISO/MS, and MS/MS.

The efficiency of isolation and CID can be determined from Fig. 4.36 by comparing the difference in peak height that separates the three curves. Since the EI/MS and EI/ISO/MS curves closely overlap one another, the isolation efficiency for m/z 182 is nearly 100%. The high efficiency is due certainly, in part, to the isolation of a 2 to 3 u window. The overall CID efficiency, as estimated by the separation between the EI/ISO/MS and EI/MS/MS curves, varies between 15% and 60% depending on the amount of

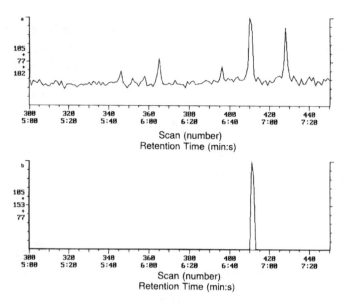

FIGURE 4.37
Comparison of (a) GC/MS and (b) GC/MS/MS chromatogram for 23 pg and benzophenone.
Selected ion chromatogram for the three predominant EI and product ions are plotted for
AGC-EI/MS and AGC-EI/MS/MS scan modes.

sample. Although CID efficiencies approaching 100% have been observed
previously with the ion trap, CID efficiencies are compound dependent
since they depend on the structure of the ion that is to be dissociated.

A linear dynamic range (LDR) from 0.022 to 85 ng was observed for
EI/MS, EI/ISO/MS, and EI/MS/MS scan modes as shown in Fig. 4.36.
For EI/MS/MS, the LDR extends down to 0.008 ng of benzophenone
injected. However, the selected-ion chromatogram and background-
subracted spectra observed at these levels show only the main product
ion m/z 105. In general, good agreement in the LDR has been observed
between the GC/MS and GC/MS/MS calibration curves for both EI and
CI modes.

Comparable precision is observed for GC/MS and GC/MS/MS peak
areas. The error bars for each data set in Fig. 4.36 represent ±2 standard
deviations for five replicate injections. Although the heights of the error
bars increase as the peak area decreases, the relative error at a constant
amount injected is the same for MS, ISO/MS, and MS/MS. This precision
demonstrates that GC/MS/MS can produce reliable results when the
instrument is operating properly.

Fig. 4.37 compares the S/N ratio of the GC/MS and GC/MS/MS
chromatograms for 0.022 pg of benzophenone. Note the S/N enhancement
that is afforded by monitoring the selective MS/MS fragmentation of m/z

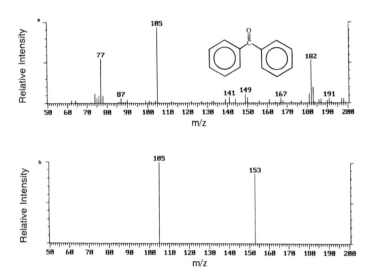

FIGURE 4.38
Comparison of the background-subtracted GC/MS full-scan and GC/MS/MS product ion spectra for 23 pg of benzophenone.

182 relative to the nonmass-selective EI/MS fragmentation. The background-subtracted mass spectra for the MS and MS/MS chromatographic peaks are shown in Fig. 4.38. Although the chemical noise that is present in the standards is low, MS/MS improves dramatically the selectivity of detection over MS alone.

The comparisons of GC/MS and GC/MS/MS that have been made so far have relied on AGC to set the ionization time and have used, therefore, the same ionization times at each concentration. Fig. 4.39 compares the S/N ratio of the GC/MS and GC/MS/MS chromatogram for 1 pg of benzophenone using 5 ms and 150 ms fixed ionization times, respectively. Longer ionization times can be used without causing gross space-charge effects because the isolation step removes most of the ions from the ion trap prior to CID and mass analysis. Note that the added sensitivity gained by the 150 ms ionization time significantly increases the peak area and S/N ratio observed for the products of m/z 182. Although long ionization times can be used to enhance the detection of analytes that are present in very small quantities, direct correlation of the peak area with sample amounts is complicated by gross nonlinearities in the calibration curves. The LDR for these analyses extended for 0.1 to 1.2 ng of benzophenone injected and is a clear limitation of long ionization times.

Although the incorporation of AGC, RF/DC isolation, SWIFT broadband excitation, and automated software helps to facilitate GC/MS/MS on the ion trap, further advancements are needed before these methods will be suitable for routine analyses. In particular, the automated isolation

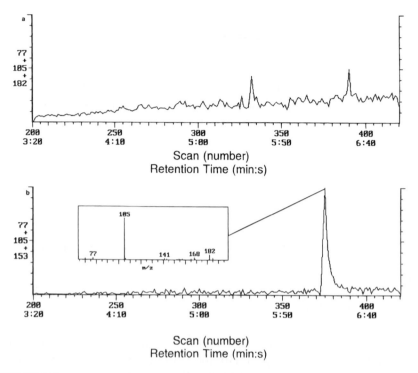

FIGURE 4.39
Comparison of fixed ionization time (a) GC/MS and (b) GC/MS/MS chromatogram for 8 pg of benzophenone. Selected ion chromatogram for the three predominant EI and product ions are plotted for EI/MS and EI/MS/MS scan modes using ionization times of 5 ms and 150 ms, respectively. Inset shows the MS/MS product ion spectra observed for m/z 182 and m/z 183 of benzophenone.

of a single ion species presents the biggest limitation of the current approach. Secondly, ion/molecule reactions are a source of chemical noise for MS/MS experiments. As an example, Fig. 4.40 shows the GC/MS/MS data for the analysis of n-butylbenzene in unleaded gasoline. Unexpectedly, the product spectrum for m/z 134 for n-butylbenzene was not observed. Instead, the products of ion/molecule reactions dominated the spectrum. The main improvement needed for SWIFT is the ability to rapidly switch waveforms and amplitudes during a GC analysis. This limitation is due largely to the commercial hardware that was used for these studies and not the SWIFT approach. SWIFT broadband excitation is an extremely reliable method for affecting CID over a GC peak. Excitation of a 10 kHz range eliminates the adjustment of the excitation frequency so it need be set only once. During the course of these experiments, the same frequency settings were used without adjustment of the SWIFT

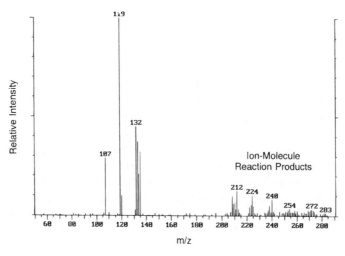

FIGURE 4.40
Background-subtracted GC/MS/MS product ion spectrum of m/z 134 for gas oil sample. The ions observed above m/z 200 are the result of ion/molecule reactions that occur following the isolation of m/z 132 to m/z 135.

waveform for the analysis of approximately 500 samples over a period of three months.

VII. SUMMARY AND CONCLUSION

The quadrupole ion trap has undergone a renaissance from a novel ion storage device to a conventional mass analyzer used for GC/MS. Although GC/MS has been the driving force behind the commercial development of the quadrupole ion trap, many expect that the use of the ion trap for GC/MS/MS will become important also. In this chapter, the features and methods of operation of commercial ion trap instruments have been reviewed, and examples of GC/MS and GC/MS/MS applications have been summarized.

Several new methods of operation of the ion trap are being developed currently with the aim of improving its GC/MS capabilities. The use of single-frequency and multi-frequency excitation to isolate selectively an ion species of interest while ejecting interference ions shows a strong potential for improving the ability of the ion trap to analyze trace analytes that co-elute from complex mixtures. The development of ion activation methods such as SWIFT and FNF broadband excitation that are capable of dissociating ions without a loss in efficiency or the need for extensive manual tuning will also improve the capabilities of the ion trap for GC/MS/MS. DC pulse axial excitation appears to be a promising method for accessing higher fragmentation energies. The benefits of external ionization and ion ejection may prove also to be very useful for complex GC analyses.

Finally, the development of sophisticated ion trap systems that are not only powerful mass spectrometers, but also simple to operate, is now well under way with the result that the ion trap's high performance capabilities are becoming accessible to a wide range of users. Current commercial instruments, which have been targeted at the single quadrupole (GC/MS) market, are now being superceded by ion trap instruments which offer full-scan sensitivity and tandem mass spectrometry (GC/MS/MS) operation. Perhaps the greatest gains in ion trap mass spectrometry can be made by taking advantage of the ion trap's unique abilities for ion storage and consecutive stages of mass selectivity while appreciating the underlying limitations of this marvelous ion storage device.

REFERENCES

1. Johnson, J. V.; Pedder, R. E.; Yost, R. A. *Int. J. Mass Spectrom. Ion Processes.* 1991, *106*, 197–212.
2. Johnson, J. V.; Pedder, R. E.; Yost, R. A.; Story, M. S. *U.S. Patent No. 5,075,547*, 1991, 1–8.
3. Schwartz, J. C.; Syka, J. E. P.; Jardine, I. *J. Am. Soc. Mass. Spectrom.* 1991, *2*, 198–204.
4. Williams, J. D.; Cooks, R. G. *Rapid Commun. Mass Spectrom.* 1992, *6*, 524–527.
5. Williams, J. D.; Cox, K. A.; Cleven, C. D.; Cooks, R. G. *Proc. 40th ASMS Conf. Mass Spectrometry and Allied Topics.* San Francisco, 1993, p. 697a.
6. Stafford, G. C.; Kelley, P. E.; Stephens, D. R. *U.S. Patent No. 4,540,884*, 1985, 1–14.
7. Stafford, G. C.; Kelley, P. E.; Bradford, D. C. *Int. Lab.* 1983, *13*, 84–92.
8. Stafford, G. C.; Kelley, P. E.; Syka, J. E. P.; Reynolds, W. E.; Todd, J. F. J. *Int. J. Mass Spectrom. Ion Processes.* 1984, *60*, 85–98.
9. Eichelberger, J. W.; Bellar, T. A.; Donnelly, J. P.; Budde, W. L. *J. Chromatogr. Sci.* 1990, *28*, 460–467.
10. Lim, H. K.; Sakashita, C. O.; Foltz, R. L. *Rapid Commun. Mass Spectrom.* 1988, *2*, 129–131.
11. Philp, R. P.; Lewis, C. A.; Campbell, C.; Johnson, E. *Org. Geochem.* 1989, *14*, 183–187.
12. Louris, J. N.; Amy, J.; Ridley, T.; Kascheres, C.; Cooks, R. G. *Proc. 40th ASMS Conf. Mass Spectrometry and Allied Topics.* Denver, 1987, p. 769–770.
13. Syka, J. E. P.; Louris, J. N.; Kelley, P. E.; Stafford, G. C.; Reynolds, W. E. *U.S. Patent.* 1988, 4,736,101.
14. Louris, J. N.; Brodbelt, J. S.; Cooks, R. G.; Glish, G. L.; Van Berkel, G. J.; McLuckey, S. A. *Int. J. Mass Spectrom. Ion Processes.* 1990, *96*, 117–137.
15. Strife, R. J.; Simms, J. R. *Anal. Chem.* 1989, *61*, 2316–2319.
16. Todd, J. F. J.; Penman, A. D.; Thorner, D. A.; Smith, R. D. *Rapid Commun. Mass Spectrom.* 1990, *4*, 108–113.
17. Morand, K. L.; Cox, K. A.; Cooks, R. G. *Rapid Commun. Mass Spectrom.* 1992, *6*, 520–523.
18. Brodbelt, J. S.; Liou, C. C., Donovan, T. *Anal. Chem.* 1991, *63*, 1205–1209.

19. McLuckey, S. A.; Van Berkel, G. J.; Glish, G. L.; Huang, E. C.; Henion, J. D. *Anal. Chem.* 1991, *63*, 375–383.
20. Glish, G. L.; Goeringer, D. E.; Asano, K. G.; McLuckey, S. A. *Int. J. Mass Spectrom. Ion Processes.* 1989, *94*, 15–24.
21. McLuckey, S. A.; Glish, G. L.; Van Berkel, G. J. *Int. J. Mass Spectrom. Ion Processes.* 1991, *106*, 213–235.
22. Johnson, J. V.; Yost, R. A.; Kelley, P. E.; Bradford, D. C. *Anal. Chem.* 1990, *62*, 2162–2172.
23. Strife, R. J.; Simms, J. R.; Lacey, M. P. *J. Am. Soc. Mass Spectrom.* 1990, *1*, 265–272.
24. Brodbelt, J. S.; Louris, J. N.; Cooks, R. G. *Anal. Chem.* 1987, *59*, 1278–1285.
25. Louris, J. N.; Syka, J. E. P.; Kelley, P. E. *U.S. Patent No. 4,686,367,* 1987, 1–14.
26. Weber-Grabau, M.; Bradshaw, S. C.; Syka, J. E. P. *U.S. Patent No. 4,771,172,* 1988, 1–7.
27. Shetty, H. U.; Daly, E. M.; Greig, N. H.; Rapoport, S. I.; Soncrant, T. T. *J. Am. Soc. Mass Spectrom.* 1991, *2*, 168–173.
28. De Jong, E. G.; Maes, R. A. A.; Van Rossum, J. M. *J. Pharm. Biomed. Anal.* 1988, *6*, 987–993.
29. Louris, J. N.; Cooks, R. G.; Syka, J. E. P.; Kelley, P. E.; Stafford, G. C.; Todd, J. F. J. *Anal. Chem.* 1987, *59*, 1677–1685.
30. Strife, R. J.; Kelley, P. E.; Weber-Grabau, M. *Rapid Commun. Mass Spectrom.* 1988, *2*, 105–109.
31. Kaiser, R. E.; Cooks, R. G.; Syka, J. E. P.; Stafford, G. C. *Rapid Commun. Mass Spectrom.* 1990, *4*, 30–33.
32. Louris, J. N.; Amy, J. W.; Ridley, T. Y.; Cooks, R. G. *Int. J. Mass Spectrom. Ion Processes.* 1989, *88*, 97–111.
33. Williams, J. D.; Reiser, H.-P.; Kaiser, R. E.; Cooks, R. G. *Int. J. Mass Spectrom. Ion Processes.* 1991, *108*, 199–219.
34. Louris, J. N.; Brodbelt-Lustig, J. S.; Kaiser, R. E.; Cooks, R. G. *Proc. 36th ASMS Conf. Mass Spectrometry and Allied Topics.* San Francisco, 1988, p. 968.
35. Schwartz, J. C.; Cooks, R. G. *Proc. 36th ASMS Conf. Mass Spectrometry and Allied Topics.* San Francisco, 1988, p. 634.
36. McLuckey, S. A.; Glish, G. L.; Van Berkel, G. J. *Proc. 38th ASMS Conf. Mass Spectrometry and Allied Topics.* Tucson, 1990, p. 512.
37. Pedder, R. E.; Yost, R. A.; Weber-Grabau, M. *Proc. 37th ASMS Conf. Mass Spectrometry and Allied Topics.* Miami Beach, 1989, p. 468.
38. Eckenrode, B. A.; Glish, G. L.; McLuckey, S. A. *Int. J. Mass Spectrom. Ion Processes.* 1990, *99*, 151–167.
39. Fales, H. M.; Sokoloski, E. A.; Pannell, L. K.; Quan Long, P.; Klayman, D. L.; Lin, A. J.; Brossi, A.; Kelley, J.A. *Anal. Chem.* 1990, *62*, 2494–2501.
40. Kaiser, R. E.; Cooks, R. G.; Moss, J.; Hemberger, P. H. *Rapid Commun. Mass Spectrom.* 1989, *3*, 50–53.
41. Kaiser, R. E.; Louris, J. N.; Amy, J. W.; Cooks, R. G. *Rapid Commun. Mass Spectrom.* 1989, *3*, 225–229.
42. Williams, J. D.; Cox, K.; Morand, K. L.; Cooks, R. G.; Julian, R. K., Jr. *Proc. 39th ASMS Conf. Mass Spectrometry and Allied Topics.* Nashville, 1991, p. 1481.
43. Cooks, R. G.; Cox, K.; Williams, J. D.; Morand, K. L.; Julian, R. K., Jr. *Proc. 39th ASMS Conf. Mass Spectrometry and Allied Topics.* Nashville, 1991, p. 469.
44. Goeringer, D. E.; Whitten, W. B.; Ramsey, J. M.; McLuckey, S. A.; Glish, G. L. *Anal. Chem.* 1992, *64*, 1434–1439.
45. Weber-Grabau, M. *U.S. Patent No. 4,818,869,* 1989, 1–8.
46. Yates, N. A.; Yost, R. A.; Bradshaw, S. C.; Tucker, D. B. *Proc. 39th ASMS Conf. Mass Spectrometry and Allied Topics.* Nashville, 1991, p. 1489.
47. Tucker, D. B.; Hameister, C. H.; Bradshaw, S. C.; Hoekman, D. J.; Weber-Grabau, M. *Proc. 36th ASMS Conf. Mass Spectrometry and Allied Topics.* San Francisco, 1988, p. 628.

48. Fulford, J. E.; Hoa, D. N.; Hughes, R. J.; March, R. E.; Bonner, R. F.; Wong, G. J. *J. Vac. Sci. Technol.* 1980, *17*, 829–835.
49. Creaser, C. S.; McCoustra, M. R. S.; O'Neil, K. E. *Org. Mass Spectrom.* 1991, *26*, 335–338.
50. Strife, R. J.; Simms, J. R. *J. Am. Soc. Mass Spectrom.* 1992, *3*, 372–377.
51. Creaser, C. S.; Mitchell, D. S.; O'Neill, K. E.; Trier, K. J. *Rapid Commun. Mass Spectrom.* 1990, *4*, 217–221.
52. Galletti, G. C.; Traldi, P.; Bonaga, G. *Rapid Commun. Mass Spectrom.* 1989, *3*, 241–243.
53. Galletti, G. C.; Marotti, M.; Piccaglia, R.; Pelli, B.; Traldi, P. *Flavour Fragrance J.* 1989, *4*, 155–159.
54. Horman, I.; Traitler, H. *Biomed. Environ. Mass Spectrom.* 1989, *18*, 1016–1022.
55. Arnold, N. S.; Kalousek, P.; McClennen, W. H.; Gibbons, J. R.; Maswadeh, W.; Meuzelaar, H. L. C. *Proc. 38th ASMS Conf. Mass Spectrometry and Allied Topics.* Tucson, 1990, p. 613.
56. Bruns Weller, E.; Tillmanns, U. *GIT Fachz. Lab.* 1989, *33*, 1159–1162.
57. Canela, A. M.; Muehleisen, H. *J. Chromatogr.* 1988, *456*, 241–249.
58. Fuerst, P.; Krueger, C.; Meemken, H. A.; Groebel, W. *J. Chromatogr.* 1987, *405*, 311–317.
59. Evans, S.; Smith, R. D.; Wellby, J. K. *Int. J. Mass Spectrom. Ion Processes.* 1984, *60*, 239–249.
60. Pereira, W. E.; Rostad, C. E.; Leiker, T. J. *Anal. Chim. Acta.* 1990, *228*, 69–75.
61. De Boer, D.; de Jong, E. G.; Maes, R. A. A. *Rapid Commun. Mass Spectrom.* 1990, *4*, 181–185.
62. Horman, I.; Traitler, H.; Aeschlimann, J. J. *High Resolut. Chromatogr.* 1989, *12*, 308–315.
63. Aggazzotti, G.; Predieri, G.; Fantuzzi, G.; Benedetti, A. *J. Chromatogr.* 1987, *24 Apr 1987*, *60*, 125–160, 130.
64. Anon., *Anal. Rep.* 1987, 16–18.
65. Leloux, M. S.; Maes, R. A. A. *Biol. Mass Spectrom.* 1991, *20*, 382–388.
66. Leloux, M. S.; De Jong, E. G.; Maes, R. A. A. *J. Chromatogr.* 1989, *24 Mar 1989*, *80*, 357–380, 367.
67. De Boer, D.; De Jong, E. G.; Maes, R. A. A. *Rapid Commun. Mass Spectrom.* 1990, *4*, 181–185.
68. Genin, E.; Lehrer, M. *Analysis.* 1988, *16*, LIV–LVII.
69. Van Berkel, G. J.; Goeringer, D. E. *Anal. Chim. Acta.* 1993, *277*, 41–54.
70. Yates, N. A.; Yost, R. A.; Bradshaw, S. C.; Tucker, D. B. *Proc. 39th ASMS Conf. Mass Spectrometry and Allied Topics.* Nashville, 1991, p. 132.
71. Chappel, C. G.; Creaser, C. S.; Stygall, J. W.; Shepherd, M. J. *Biol. Mass Spectrom.* 1992, *21*, 688–692.
72. Yates, N. A.; Griffin, T. P.; Yost, R. A. *Proc. 41st ASMS Conf. Mass Spectrometry and Allied Topics.* San Francisco, 1993, p. 444a.
73. Schachterle, S.; Brittain, R. D. *Proc. 41st ASMS Conf. Mass Spectrometry and Allied Topics.* San Francisco, 1993, p. 638a.
74. Stafford, G. C.; Taylor, D. T.; Bradshaw, S. C.; Syka, J. E. P.; Uhrich, M. *Proc. 35th ASMS Conf. Mass Spectrometry and Allied Topics.* Denver, 1987, p. 775.
75. Todd, J. F. J.; Waldren, R. M.; Mather, R. E. *Int. J. Mass Spectrom. Ion Phys.* 1980, *34*, 325–349.
76. Vedel, F.; Andre, J. *Int. J. Mass Spectrom. Ion Processes.* 1985, *65*, 1–22.
77. Chattopadhyay, A. P.; Ghosh, P. K. *Int. J. Mass Spectrom. Ion Phys.* 1983, *49*, 253–263.
78. Pannell, L. K.; Pu, Q. L.; Fales, H. M.; Mason, R. T.; Stephenson, J. L. Jr. *Anal. Chem.* 1989, *61*, 2500–2503.
79. McLuckey, S. A.; Glish, G. L.; Asano, K. G.; Van Berkel, G.J. *Anal. Chem.* 1988, *60*, 2312–2314.
80. Vedel, F.; Vedel, M.; March, R. E. *Int. J. Mass Spectrom. Ion Processes.* 1991, *108*, R11–R20.
81. Vedel, F.; Vedel, M.; March, R. E. *Int. J. Mass Spectrom. Ion Processes.* 1990, *99*, 125–138.
82. Eades, D. M.; Yost, R. A. *Rapid Commun. Mass Spectrom.* 1992, *6*, 573–578.

83. Kleintop, B. L.; Yost, R. A.; Abolin, C. R. *J. Am. Soc. Mass Spectrom.* 1992, *3*, 85–88.
84. Pedder, R. E.; Johnson, J. V.; Yost, R. A. *Proc. 40th ASMS Conf. Mass Spectrometry and Allied Topics.* San Francisco, 1993, p. 711a.
85. Kelley, P. E.; Hoekman, D. J.; Bradshaw, S. C.; Stiller, S. W. *Proc. Pittsburgh Conf. Analytical Chemistry and Applied Spectroscopy.* 1993, Atlanta, p. 1081.
86. Olson, E. S.; Diehl, J. W. *Anal. Chem.* 1987, *59*, 443–448.
87. Ratnayake, W. M. N.; Timmins, A.; Ohshima, T.; Ackman, R. G. *Lipids.* 1986, *21*, 518–524.
88. Watson, J. T. In *Introduction to Mass Spectrometry.* Raven Press, New York, 1985, p. 121–152.
89. McLuckey, S. A.; Goeringer, D. E.; Glish, G. L. *J. Am. Soc. Mass Spectrom.* 1993, 2, 11–21.
90. Eades, D. M.; Yates, N. A.; Yost, R. A. *Proc. 39th ASMS Conf. Mass Spectrometry and Allied Topics.* 1991, Nashville, p. 1491.
91. Julian, R. K.; Cooks, R. G. *Anal. Chem.* 1993, *65*, 1827–1833.
92. Shaffer, B. A.; Karnicky, J.; Buttrill, S. E., Jr. *Proc. 40th ASMS Conf. Mass Spectrometry and Allied Topics.* San Francisco, 1993, p. 802a.
93. Yates, N. A.; Eades, D. M.; Jones, J. A.; Booth, M. M; Stephenson, J. L. J.; Griffin, T. G.; Yost, R. A. *J. Am. Soc. Mass Spectrom.* Submitted for publication.
94. Chen, L.; Wang, T-C. L.; Ricca, T. L.; Marshall, A. G. *Anal. Chem.* 1987, *59*, 449–454.
95. Schwartz, J. C.; Jardine, I. *Rapid Commun. Mass Spectrom.* 1992, *6*, 313–317.
96. McLuckey, S. A.; Glish, G. L.; Asano, K. G. *Anal. Chim. Acta.* 1989, *225*, 25–35.
97. Bier, M. E.; Winkler, P. C.; Herron, J. R. *J. Am. Soc. Mass Spectrom.* 1992, *4*, 38–46.
98. Pinkston, J. D.; Delaney, T. E.; Morand, K. L.; Cooks, R. G. *Anal. Chem.* 1992, *62*, 1571.
99. Van Berkel, G. J.; Glish, G. L. McLuckey, S.A. *Anal. Chem.* 1990, *62*, 1284–1295.
100. Booth, M. M.; Stephenson, J. L. J.; Yost, R. A. *Proc. 41st ASMS Conf. Mass Spectrometry and Allied Topics.* San Francisco, 1993, p. 716a.
101. Todd, J. F. J.; Penman, A. D.; Thorner, D. A.; Smith, R. D. *Proc. 38th ASMS Conf. Mass Spectrometry and Allied Topics.* Tucson, 1990, p. 532.
102. Johnson, J. V.; Pedder, R. E.; Yost, R. A. *Rapid Commun. Mass Spectrom.* 1992, *6*, 760–764.
103. McLuckey, S. A.; Goeringer, D. E.; Glish, G.L. *Anal. Chem.* 1992, *64*, 1455–1460.
104. Lammert, S. A.; Cooks, R. G. *Rapid Commun. Mass Spectrom.* 1992, *6*, 528–530.
105. Wang, M.; Wells, G. *Proc. 41st ASMS Conf. Mass Spectrometry and Allied Topics.* San Francisco, 1993, p. 463a.
106. Gronowska, J.; Paradisi, C.; Traldi, P.; Vettori, U. *Rapid Commun. Mass Spectrom.* 1990, *4*, 306–313.
107. Srzic, D.; Martinovic, S.; Vujanic, P.; Meic, Z. *Rapid Commun. Mass Spectrom.* 1993, *7*, 163–166.

Chapter 5

LIQUID CHROMATOGRAPHY/MASS SPECTROMETRY

Brent L. Kleintop, Donald M. Eades, Jon A. Jones, and Richard A. Yost

CONTENTS

I. INTRODUCTION

The combination of high performance liquid chromatography (HPLC) with mass spectrometry (MS) is well recognized as possessing enormous potential for analyzing a wide variety of compounds. Although gas chromatography (GC) generally provides higher chromatographic efficiency and is easy to interface, its use is limited to compounds of sufficient volatility and thermal stability under the high temperatures normally used. Indeed, it has been estimated that only 20% of all organic compounds are amenable

0-8493-8251-3/95/$0.00+$.50

to GC and GC/MS analyses.[1] Although derivatizaton methods can expand the range of compounds amenable to GC/MS, these techniques are often time consuming and are practical only for a limited number of compounds. HPLC is capable of separating most compounds not amenable to GC, albeit not as quickly or efficiently, because analytes remain in the condensed phase during separation, so thermal decomposition is not a problem.

The major challenge to be faced when using a mass spectrometer as a liquid chromatographic (LC) detector is the design of a suitable interface which can convert the mobile phase flow into the gas phase, yet maintain typical MS operating conditions. Vaporizing this quantity of liquid generates a volume of gas that must be reduced substantially to maintain the high vacuum (typically $\sim 10^{-5}$ torr) of the mass spectrometer. Not until the emergence of membrane introduction and the aerosol-based interfaces (thermospray, particle beam and electrospray/ionspray) in the 1980s was practical LC/MS realized. Today, nearly every type of mass spectrometer has been interfaced to LC; however, a truly universal and efficient LC/MS interface remains a challenge. Nevertheless, these improved interfaces have caused a renewed interest in LC/MS for biotechnical applications and environmental analyses.

Although widely accepted as a sensitive benchtop mass spectrometric detector for GC, the quadrupole ion trap has been pursued only recently as a mass spectrometer for LC applications. Ion trap performance was expected to suffer because of the large amount of residual solvent vapor introduced by most LC interfaces. The high abundance of solvent neutrals causes high operating pressures and undesired ion/molecule reactions; the high abundance of solvent ions may cause space-charge effects which limit sensitivity and result in poor mass resolution and overall spectral quality. With recent advances in the instrumentation and operation of ion traps, these adverse solvent effects have been reduced sufficiently to the point where successful coupling with LC has been realized. The use of mass-selective ejection techniques has been demonstrated to reduce dramatically space-charge effects caused by unwanted solvent ions. For example, the RF voltage applied to the ring electrode can be elevated such that ions having a m/z value less than a chosen value are ejected from within the ion trap, that is, a low-mass cut-off is imposed.[2,3] Alternatively, a combination of RF and DC voltages (mass-selective storage) can be applied to the ring electrode during ionization to store selectively a desired range of mass/charge ratios.[4,5] Although these operational modes efficiently eject unwanted solvent ions, typically they reduce the storage efficiency of the ions of interest. The use of resonant ejection also has been shown to eject efficiently unwanted ions while efficiently storing analyte ions of interest. Resonant excitation waveforms comprised of single frequencies[6] and waveforms comprised of multiple frequencies based upon broadband patterns,[5] white noise,[7] stored-waveform inverse Fourier transform

(SWIFT),[8–10] and filtered noise fields[11] all have been implemented to eject unwanted ions from within the ion trap. The injection of ions from a differentially-pumped external ion source has been shown also to be an efficient means of preventing neutral solvent species from reaching the analyzer region. Indeed, several groups have utilized successfully and ion injection[12] techniques for coupling LC to the ion trap.[13–21]

In addition to sensitivity requirements, biotechnical applications require typically high mass analysis, better-than-unit mass resolution, and MS/MS capabilities. Recent advances in quadrupole ion trap technology have shown that this mass spectrometer has the potential of meeting these requirements. New operational modes of the quadrupole ion trap have demonstrated mass range analysis up to 70 kDa[22] and resolving power up to 1×10^6.[23] These features, along with the efficient MS^n capabilities[24,25] and true single ion monitoring,[20] have generated substantial interest in using ion traps for LC/MS analyses of biological and environmental compounds. Although all of these capabilities have yet to be implemented fully on a commercially available instrument, the potential of the quadrupole ion trap has certainly been demonstrated in several research laboratories.

Several reports involving interfacing LC to quadrupole ion traps via electrospray/ionspray interfaces and API interfaces will be the subject of other chapters of this book and, therefore, will not be discussed here. This chapter will review reports of coupling ion traps with membrane, thermospray, and particle beam interfaces.

II. MEMBRANE INTRODUCTION ION TRAP MASS SPECTROMETRY

Membranes have been used as mass spectrometer sample inlet systems since the early 1960s.[26] In general, membrane introduction of an analyte to the mass spectrometer is accomplished by selective analyte diffusion through the membrane followed by evaporation from the membrane. By separating the analyte from the aqueous solution, both EI and CI spectra can be acquired. As reported in a recent review,[27] membrane introduction mass spectrometry (MIMS) has been used for both environmental sampling[28–31] and for on-line monitoring.[32,33] Probably the most practical advantage is the ability to analyze directly ppb levels of organic compounds in a 100% aqueous solution. Indeed, with increasingly aqueous mobile phase composition, nearly every LC interface exhibits reduced efficiency (that is, solvent evaporation, analyte transport) and MS instruments exhibit reduced performance. The advantage of membranes to separate efficiently analytes from aqueous solutions has eliminated the need for many time consuming extraction and preconcentration steps

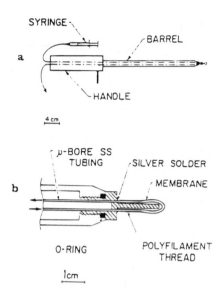

FIGURE 5.1

Schematic of the direct insertion membrane probe (DIMP) for coupling to a Finnigan ITD-700 quadrupole ion trap. (From Ref. 34: Bier, M. E.; Cooks, R. G. *Anal. Chem.* 1987, *59*, 597.)

typically associated with aqueous environmental sample analyses. The simplicity and efficiency of the MIMS technique have opened the door to continuous on-line monitoring of aqueous solutions.[32,33]

The membrane introduction system can be coupled to nearly any type of mass spectrometer. The versatility of the direct insertion membrane probe (DIMP), developed by Cooks and co-workers,[28,33–36] allowed coupling to Finnigan MAT ITD-700,[28] ITS-40,[35] and TSQ-45[28,34,36] mass spectrometers. The DIMP, shown in Fig. 5.1, is a modified ion volume insertion/removal tool. The aqueous sample (not the membrane) is heated *via* a nichrome heating element and flows through a hollow fiber capillary polymer membrane made of dimethyl vinyl silicone. Volatile analytes diffuse selectively through the membrane and into the mass spectrometer where normal mass spectral acquisition can be performed.

Lister *et al.*[28] interfaced the DIMP to an ITD-700 instrument for the purpose of developing a compact, yet sensitive analysis system for environmental water samples. The ITD was modified by the addition of an ion gauge; the GC transfer line was replaced with a gas sample line equipped with the DIMP and a variable leak valve. Analyte molecules permeating the membrane were transferred to the ion trap by molecular diffusion through the variable leak valve. Water molecules diffusing through the membrane and reaching the ion trap were ionized and used as CI reagent ions. The authors noted that a large membrane surface area generated space-charge conditions in the trap due to excessive amounts of analyte and solvent ions. Reduction of the surface area to *ca.* 16 mm^2 resulted in unit resolved mass spectra. The system allowed separation and analysis of 9 organic compounds directly from a 100% aqueous solution with detection limits varying from 1 to 100 ppb. These detection limits were typically 5 to 1000 times lower than those obtained on a TSQ-45 mass spectrometer using the same DIMP but employing isobutane as the CI reagent gas.[28] The markedly lower detection limits were reportedly due to both the ion trap's sensitivity and to the choice of CI reagent ions. A test mixture of three compounds gave a linear dynamic range of three orders of magnitude.

To characterize the DIMP/ITD system, the authors examined response times, and rise and fall times. The response time (that is, the time required to reach 50% of analyte signal) varied from 5 to 81 s. The rise and fall times characterize the peak shape by measuring the time required for the signal to go from 10 to 90% (rise) or 90 to 10% (fall). On average, the fall times were one to five times longer than the rise times. All values were reportedly similar to that reported previously for a TSQ-45 instrument.[34]

The work reported by Lister et al.[28] readily identified several methods for improving the performance of the current system. Placement of the DIMP closer to the ion trap would improve analyte ion transmission resulting in higher sensitivity. Control of the DIMP temperature was noted as another factor for sensitivity enhancement. The authors also stated that removal of unwanted solvent ions (H_3O^+) would enhance sensitivity also. Although this work utilized H_3O^+ as a CI reagent ion, removal of the solvent ions by the methods described in the introduction[2-11] would allow acquisition of EI spectra, reduce space-charge effects (with concomitant improved resolution), and allow a larger membrane surface area for increased analyte diffusion.

Bauer et al.[35] modified an ITS-40 system to allow the DIMP to be placed much closer to the ion trap analyzer than that used by Lister et al.[28] This arrangement allowed improved analyte delivery and enhanced sensitivity. This configuration also allowed the temperature of the membrane to be controlled by conduction from the ITS-40 vacuum manifold heaters. Characterization studies showed that decreasing the DIMP temperature decreased the analyte diffusion rate through the membrane, thus requiring longer analysis times. Studies of probe temperature in the ion trap system showed that the signal/noise ratio decreased with increasing ion trap and probe temperatures,[35] which is the inverse of that observed with the DIMP/TSQ-45 systems.[36] The authors concluded that this effect was due to increased water transport through the membrane at the higher temperatures. The resulting neutral water background caused the AGC[37] software of the ITS-40 to compensate automatically for the high number of ions present by selecting a shorter ionization time. Such compensation caused a decrease in the analyte ion signals and thus reduced sensitivity. Nevertheless, as shown in Table 5.1, sub-ppb detection limits were reported for several compounds analyzed from aqueous solution under

TABLE 5.1

Detection limits for aqueous solutions obtained using the modified ITS40
MIMS system

Analyte	Detection Limit (ppb)	EPQ PQL (ppb)
Benzene	0.21	2
Trichlorofluoromethane	0.60	10
Chlorobenzene	0.16	2
trans-Dichloroethene	0.81	1
Chloroform	0.58	0.5
cis-1,3-Dichloropropene	0.60	20
Hexachloroethane	0.28	0.5
Carbon tetrachloride	0.60	1

From Ref. 27.

EI-AGC conditions at flow rates of 2 mL/min. These results were quite impressive considering the very high solvent flow rates employed.

Slivon and co-workers[38] have also interfaced membrane introduction with a quadrupole ion trap for the sampling of drinking water. This interface differs from that described previously in that the aqueous sample flows over the hollow fiber membrane while the interior of the membrane is purged continuously with helium. The helium flow sweeps volatile organics which have permeated through the membrane and assists transport to the ion trap. Water flow rates were varied over the range 200 to 1000 mL/min with standards introduced over the range 4 to 120 μL/min. When coupled to a Finnigan MAT ITD-700, this configuration reportedly allowed shorter response times as well as the use of long, hollow fibers for improved sensitivity. As shown in Fig. 5.2, concentrations as low as 0.09 ppb were observed for vinyl chloride, benzene, and carbon tetrachloride. Using this same membrane coupled to a Finnigan MAT TSQ-45, comparable sensitivities were observed. The lack of increased sensitivity when coupled to the ITD was believed to be due to excessive permeant water vapor introduced to the ion trap. The authors reported a much larger membrane surface area (86 mm²)[38] than that used by Lister et al.;[28] therefore, it follows that higher partial pressures of water would be present. Solvent effects were reportedly reduced by decreasing the helium purge rate.[38]

Although receiving less attention than the spray techniques, MIMS is the easiest sample introduction method to interface with a mass spectrometer, especially for 100% aqueous solutions. Unlike other interfaces, MIMS requires no additional pumping, no use of high voltages (as in electrospray), and no nebulization or vaporization of the sample flow (as in thermospray and particle beam). This technique has been interfaced successfully with ion traps,[28,35,38] and low ppb detection limits have been

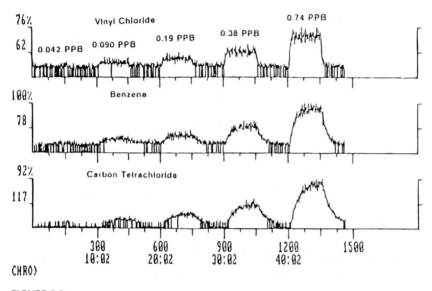

FIGURE 5.2
Mass chromatograms of sub-ppb concentrations of vinyl chloride, benzene, and carbon tetrachloride standards in reagent water obtained on a Finnigan ITD-700. (From Ref. 29: Dheandhando, S.; Dulak, J. *Rapid Commun. Mass Spectrom.* 1989, 3, 175.)

obtained routinely for low molecular weight (<300 Da) organic compounds. The MIMS technique, however, is not without disadvantages. Currently, flow injection analyses and on-line monitoring are the main applications. Due to the relatively long response times, membrane introduction has not been coupled to LC. Also, MIMS is limited to the analysis of relatively low molecular weight compounds of low polarity.

III. THERMOSPRAY/ION TRAP MASS SPECTROMETRY

Thermospray (TSP) was once the most commonly used interface for LC/MS applications. The pioneering work with TSP, directed mainly toward quadrupole mass spectrometers, led to dramatic improvements over the current techniques of the time, such as the moving belt. Although the related techniques (that is, electrospray/ionspray, and particle beam) have gained tremendous popularity, TSP remains a widely used LC/MS method, probably due to the fact that TSP was the first spray-type interface to be introduced.[39,40]

The TSP interface consists of an electrically-heated capillary vaporizing probe, a modified ion source, and a pump with a cryogenic trap. The capillary is inserted into the heated ion source with direct pumping on-line with the capillary. While under vacuum, the LC effluent (solvent,

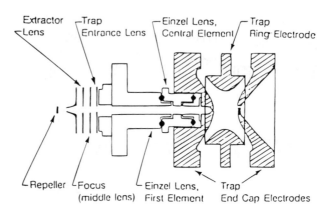

FIGURE 5.3
Schematic of TSP/ion injection system used by Kaiser et al.[43] for coupling a thermospray interface to a modified Finnigan quadrupole ion trap mass spectrometer (ITMS). (From Ref. 43: Kaiser, R. E.; Williams, J. D.; Schwartz, J. C.; Lammert, S. A.; Cooks, R. G.; Zakett, D. *Proc. 37th ASMS Conf. Mass Spectrometry and Allied Topics*, Miami Beach, 1989, p. 369.)

analyte, buffer electrolytes) exits the heated capillary, resulting in a supersonic jet expansion. The resulting fine mist of droplets and particles travels through the heated source and continues to vaporize as the rough pump and cryogenic trap remove the evaporated solvent. The mechanism for ion formation remains controversial, and interested readers are referred to the article by Arpino.[41] The resulting mass spectrum is typically CI-like. A review of general TSP LC/MS applications can be found in *Mass Spectrometry Reviews*.[42]

Although the ion trap is accepted generally as one of the most sensitive GC/MS instruments, it was not viewed as a mass spectrometer compatible with the TSP technique due to the high amounts of solvent vapor present with this interface. Kaiser et al., however, showed the first successful coupling of a TSP interface to a quadrupole ion trap.[43] They coupled a Finnigan TSP ion source to a modified ion trap mass spectrometer (ITMS) and employed an on-line ion injection system. A schematic of the design used for this work is shown in Fig. 5.3. Initial characterization studies of the TSP/ITMS using a 50:50 CH_3OH/H_2O mixture showed that low LC flow rates ~0.5 mL/min) did not affect adversely spectral acquisition, although typical TSP flow rates (~1 to 2 mL/min) did produce high pressures in the ion trap. The high neutral pressure caused broad peaks, tailing towards low mass, and loss of unit mass resolution. The authors point out that the tailing towards low mass is evidence for high pressure in the ion trap, not evidence for space charge (which would cause tailing towards high mass). Although used in these experiments, helium buffer gas was reportedly not required for efficient trapping of ions due to the high ambient pressure. This work readily identified the need to pump differentially the TSP source from the analyzer region in order to reduce the operating pressure. A spectrum of the residual solvent ions showed abundant methanol clusters which were assumed to be due to the high pressure in the TSP source and to the long ion injection times (10 to 100 ms). Their results showed that long injection times and low repeller voltages (+20 V) gave increased abundance of solvent cluster ions and increased ion/molecule reaction products. Increasing the repeller to +100 V and reducing the injection time decreased solvent effects and showed improved mass spectra.

Kaiser *et al.*[43] also observed adduct ions formed by the analyte and the solvent. Phenylalanine reportedly reacted with CH_3OH to yield a methylated adduct, while adenosine exhibited no such adducts. Flow injection analysis (FIA) of 100 ppm of phenylalanine and 100 ppm adenosine in a 80:20 CH_3OH/H_2O mixture at 0.5 mL/min was reported. Since the ion trap does not typically perform well under high pressures and high amounts of H_2O, the results were very encouraging.

To improve on the LC/ITMS interface described above, Bier *et al.*[12] made several modifications. The instrument, a modified Finnigan ITS-40, was pumped differentially by a 330 L/s turbo-molecular pump for the TSP source and 170 and 330 L/s turbo-molecular pumps for the analyzer. A DC quadrupole triplet was installed for improved ion transmission and beam shaping, and for a region of increased pumping capabilities. A schematic of the instrumentation is shown in Fig. 5.4. These improvements allowed higher solvent flow rates (1 to 1.5 mL/min) and higher percent H_2O mobile phase composition (70% H_2O), without resulting in high background pressures or adverse solvent effects. Shown in Fig. 5.5 is the mass spectrum acquired for 100 ng of reserpine injected into a 1.0 mL/min flow of 30:70 CH_3OH/H_2O. The mass spectrum exhibited unit mass resolution and no apparent solvent effects from the relatively high mobile phase flow rate. While using the same mobile phase composition, unit mass resolution was also reported for 78 ng of adenosine injected into a 1.5 mL/min flow.

As discussed above, thermospray interfaces have been coupled successfully to quadrupole ion traps. Utilizing ion injection techniques[21,43] to remove solvent neutrals has allowed the acquisition of unit resolved mass spectra at relatively high flow rates of up to 1.5 mL/min. The results shown here emphasize the potential of the quadrupole ion trap for liquid phase analyte detection.

IV. PARTICLE BEAM/ION TRAP MASS SPECTROMETRY

Particle beam (PB) LC/MS interfaces, originally known as MAGIC (monodisperse aerosol generation interface for liquid chromatogra-

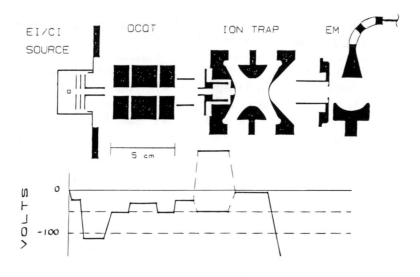

EXTERNAL IONIZATION WITH DCQT OPTICS

FIGURE 5.4
Ion injection/DC quadrupole triplet system used by Bier *et al.*[21] for coupling thermospray and particle beam interfaces to a modified Finnigan ITS-40 quadrupole ion trap. (From Ref. 21: Bier, M. E.; Hartford, R. E.; Herron, J. R.; Stafford, G. C. *Proc. 39th ASMS Conf. Mass Spectrometry and Allied Topics.* Nashville, 1991, p. 538.)

phy)[44,45] have received increased attention recently due largely to their ability to yield classical EI mass spectra for a wide variety of relatively involatile and thermally labile compounds.[46,47] The fragmentation patterns exhibited by EI spectra can provide structural information and can be compared to existing libraries for the identification and confirmation of unknown compounds, a typical goal of many LC/MS analyses. Most other LC/MS interfaces (for example, thermospray, and electrospray/ ionspray), although more applicable to the analysis of thermally labile and involatile compounds, produce spectra unique to the interface (that is, CI-like spectra). This mode of behavior typically limits their use to targeted compound analysis, although MS/MS has been employed for obtaining structural information.[14,48] Structural information from electrospray ions has been obtained also from collision-induced dissociation within the transport region of the interface, prior to the mass spectrometer.[49,50]

The PB interface works by first converting the LC eluent into an aerosol by nebulization, usually with a concentric flow of helium, typically at 0.5 to 0.7 L/min. Volatile solvent is vaporized as the aerosol travels through a heated desolvation chamber resulting in a mixture of solvent

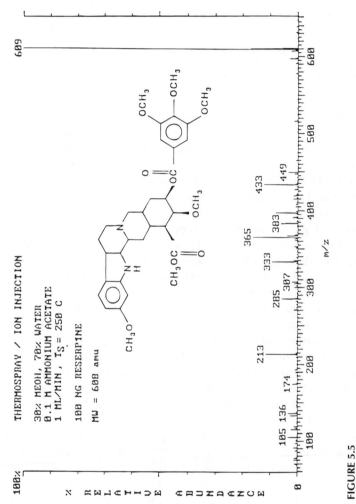

FIGURE 5.5

Mass spectrum of 100 ng reserpine acquired with thermospray/ion injection at 1.0 mL/min of 30:70 CH₃OH/H₂0. (From Ref. 21: Bier, M.E.; Hartford, R.E.; Herron, J.R.; Stafford, G.C. *Proc. 39th ASMS Conf. Mass Spectrometry and Allied Topics.* Nashville, 1991, p. 538.)

TABLE 5.2

Operating pressures for 2-stages and 3-stages of momentum separation

| | Ion Gauge Pressure (Torr)* | | | |
	100% CH$_3$CN	75/25 CH$_3$CN/H$_2$O	50/50 CH$_3$CN/H$_2$O	25/75 CH$_3$CN/H$_2$O
2-stage	1.9×10^{-4}	1.9×10^{-4}	2.8×10^{-4}	3.2×10^{-4}
3-stage	5.8×10^{-6}	5.7×10^{-6}	5.7×10^{-6}	3.5×10^{-5}

* All pressures were obtained at a flow rate of 0.3 mL/min; no supplemental helium was added to the de-solvation chamber. Pressures were read directly from the ion gauge controller and are reported with no correction factors.

vapor, helium, and largely-desolvated particles containing analyte. This mixture forms the "particle beam" as it passes through a beam collimator and into the momentum separator, which consists typically of two[44] regions but can be up to three[2,3,51] regions of pumping, each separated by a skimmer. Here, solvent and helium are pumped away while the more massive and thus higher momentum analyte particles continue in a beam towards the mass spectrometer. Typical operating pressures for 2-stages and 3-stages of momentum separation are shown in Table 5.2. Once inside the ion source, the particles are flash vaporized *via* collisions with a target heated typically to 250 to 300°C. The gas-phase analyte neutrals can then be ionized by EI, or alternately by CI, with the introduction of a suitable reagent gas.

In general, the major operating parameters of the PB interface which affect performance are the nebulizing gas (He) flow rate, desolvation chamber temperature and pressure, and source target temperature.[47,51] Variations among commercially-available PB interfaces and MS systems necessitate the need to characterize these parameters for each system in order to develop effective LC/PB/MS methods. The optimum operating conditions for most of these parameters are either largely mobile phase or analyte dependent. In general, the optimum nebulizing gas flow rate and desolvation chamber temperature and pressure increase as the aqueous content on the mobile phase increases. Optimum source target temperature is largely compound dependent; therefore, caution must be taken to avoid thermal decomposition caused by high target temperatures.

The PB interface was first coupled to an ion trap by Bier and co-workers[2] who used a prototype interface consisting of three stages of momentum separation. The third stage was required to reduce further the amount of solvent reaching the ion trap. The interface, illustrated in Fig. 5.6, was adapted with a 1/2" stainless steel probe for insertion through a probe lock assembly. Helium was added into the third stage of the momentum separator to reduce the amount of solvent passing from stage two to stage three and to prevent backstreaming of roughing pump oil

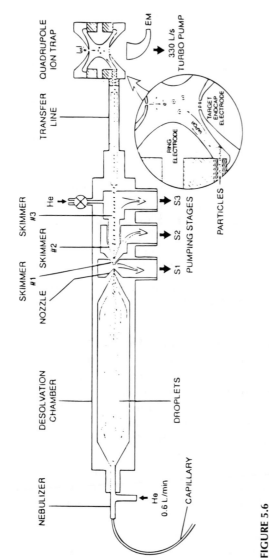

FIGURE 5.6

Schematic of prototype 3-stage PB interface used by Bier et al.[2] for coupling to a modified Finnigan ITS-40 quadrupole ion trap. (From Ref. 2: Bier, M.E.; Winkler, P.C.; Lopez, J.T. J. Am. Soc. Mass Spectrom. 1993, 4, 38.)

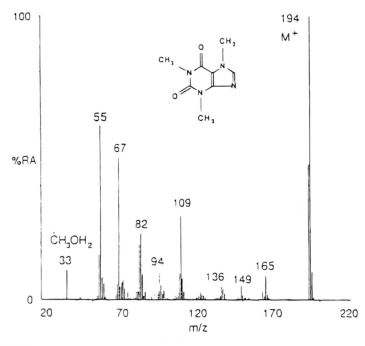

FIGURE 5.7
EI mass spectrum of 20 ng caffeine acquired with the PBI/ion trap system[2] in 100% methanol at a flow rate of 0.5 mL/min. RA, relative abundance. (From Ref. 2: Bier, M. E.; Winkler, P. C.; Lopez, J. T. *J. Am. Soc. Mass Spectrom.* 1993, 4, 38.)

caused by low pressures. The additional helium also provided the 1 mtorr of helium buffer gas used in normal ion trap operation to damp the ion trajectories towards the center of the ion trap. The ion trap was a Finnigan MAT ITS-40 prototype which was modified by mounting an ion gauge on the vacuum manifold, and was controlled by Finnigan ITMS software. The partial pressure of residual methanol solvent was measured to be about 1×10^{-6} torr at a flow rate of 0.5 mL/min. The exit end-cap electrode, heated by conduction from the vacuum manifold heater jacket, served to vaporize the particles emerging from the PB interface. The RF amplitude was set at a level such that all ions below m/z 50 were ejected during ionization to reduce the solvent ion population and to protect the electron multiplier.

Using this configuration, Bier and co-workers were able to acquire EI mass spectra from flow injection analysis of several compounds with low nanogram detection limits.[2] Fig. 5.7 illustrates the EI mass spectrum of 20 ng of caffeine obtained at a flow rate of 0.5 mL/min of 100% ethanol. The spectrum exhibited unit mass resolution and extensive EI fragmentation. Also, the relative abundance of the ^{13}C isotope agreed well with the expected value of 11%. Although the RF level was set initially

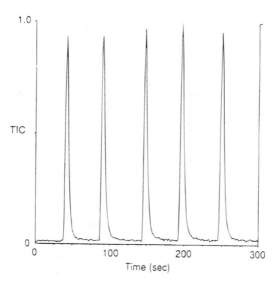

FIGURE 5.8
Sequential injections of 100 ng of carbaryl using the PB system described in Ref. 2. (From Ref. 2: Bier, M. E.; Winkler, P. C.; Lopez, J. T. *J. Am. Soc. Mass Spectrom.* 1993, *4*, 38.)

to eject all ions below m/z 50, a m/z 33 ion was observed and attributed to a protonated methanol ion formed as a result of charge exchange of methanol with the analyte ion, followed by self-CI resulting from the large population of solvent neutrals. An increase in the $(M + 1)^+/M^+$ analyte ion ratio was observed also with increasing concentration of analyte. This observation was due presumably to an increase in ion/molecule reactions such as self-CI caused by the increasing amounts of analyte ions and neutrals within the ion trap.

Although some ion/molecule reactions were evident, the PB/ion trap system exhibited excellent reproducibility. Fig. 5.8 illustrates the full-scan total ion current from five 100 ng injections of carbaryl. The chromatogram shows excellent reproducibility (RSDs of peak areas = 4.3%) and a S/N ratio of 100. RSDs of peak areas from 100 ng injections of caffeine over 2 months were reported also to be less than 40%. Nonlinear calibration curves were obtained for caffeine; however, other PB systems have been reported also to exhibit this type of response curve.[52–54]

Bier and co-workers also performed experiments using 100% water as the mobile phase. At a flow rate of 0.5 mL/min, the observed mass spectra showed space charging which was attributed to inefficient vaporization of the aqueous solvent. The EI mass spectrum of 100 ng of caffeine was observed to exhibit unit mass resolution when the de-solvation chamber temperature was increased to 80°C. Although unit resolution was observed, the total ion current was an order of magnitude less than that obtained using a 100% organic mobile phase. This decrease in sensitivity was not believed to be caused by the ion trap since similar results have been observed with other PB/MS systems.[51]

A similar PB interface has been coupled to a Finnigan ITMS by Kleintop and co-workers.[55,56] The prototype three-stage PB interface was modified by adding a pressure gauge and supplemental helium to the desolvation chamber. Addition of supplemental helium to the desolvation chamber to increase the pressure to ~400 Torr resulted in an approximate two-fold increase in analyte ion signals. The additional helium provided a higher thermally-conductive medium, which increased the rate of de-solvation. These results were consistent with those reported by Browner et al.[57] The interface was inserted into the ITMS via a 1/2" o.d. transfer line probe which allowed insertion and removal without disturbing the high vacuum of the ITMS. The probe tip was positioned 1/4" away from the ion trap entrance to prevent high solvent pressures inside the trap. Physically joining the PB to the ion trap entrance via a Rulon fitting yielded high operating pressures inside the trap, which resulted in broad mass spectral peaks which tailed to low mass. Interestingly, analyte signal did not vary significantly when the probe was positioned anywhere from 1/4" to 3" away from the analyzer. The filament end-cap electrode served as the particle beam source target; however, the end-cap was modified to allow insertion of two heater cartridges which were controlled by an external temperature controller. This procedure allowed normal PB target temperatures (250 to 300°C) to be used without heating significantly the other mass spectrometer components, for example, the electron multiplier. The higher target temperatures yielded a more efficient flash vaporization process which reduced significantly the amount of peak tailing observed in previous studies.[2,56]

Using this configuration, Kleintop and co-workers investigated several ion trap operational modes implemented with standard ITMS software which ejected unwanted ions from within the ion trap.[3] The adverse effects of residual solvent are illustrated in the profile mass spectrum from FIA of 100 ng carbaryl in Fig. 5.9. The large abundance of protonated methanol solvent molecules, m/z 33, caused space charging which resulted in poor mass resolution. The m/z 144:m/z 145 ratio illustrated in the inset indicated also that a large degree of CI had occurred. The base peaks of the EI and CI mass spectra of carbaryl are m/z 144 and m/z 145, respectively. Although some m/z 145 would be expected in the EI mass spectrum due to ^{13}C, the figure shows that the intensity of m/z 145 is twice that due to m/z 144. The relatively high intensity of m/z 145 was attributed to solvent-CI caused by the large abundance of methanol molecules within the ion trap. Since the major advantage of using PB is the ability to acquire EI mass spectra, solvent ions must be removed from the trap to prevent solvent-CI.

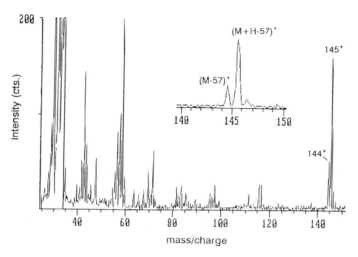

FIGURE 5.9
Profile mass spectrum of FIA of 100 ng carbaryl in 100% methanol with no solvent ion ejections. Inset is a blow-up of the region from m/z 140 to m/z 150 to better illustrate the m/z 144: m/z 145 intensity ratio. (From Ref. 3: Kleintop, B. L.; Eades, D. M.; Yost, R. A. *Anal. Chem.* 1993, *65*, 1295.)

One operational mode investigated involved elevating the RF voltage applied to the ring electrode to eject ions below a chosen m/z value, that is, to impose a low-mass cut-off.[3] This mode of operation was investigated when applied both during ionization similar to Bier *et al.*,[2] and afterward, during ion storage. Although analyte ion signal intensity was relatively unaffected by the storage RF level, signal intensity decreased significantly at increased ionization RF levels. Elevating the RF level increased the initial kinetic energy of the ions formed, which resulted in losses from large initial velocities and quasi-unstable trajectories.[58] Although elevating the RF level during ion storage also increased the ions' kinetic energies, the ions were already effectively trapped, so ion losses were minimal until they became unstable when the low-mass cut-off approached m/z 144.

Another method studied,[3] which involved resonant ejection of a single ion species of the solvent by applying a notch filter,[6] was carried out by applying a supplemental 6 v_{p-p} RF signal (resonant excitation voltage) across the end-cap electrodes at the secular frequency of the ion of interest. Using this method, the single solvent ion species was efficiently ejected (~100%) with no loss of analyte signal when resonant excitation was applied during ion storage. However, the lower efficiency (~30%) when this process was applied during the ionization event was attributed to the competing processes of ionization and ion ejection. Ions were constantly produced during the ionization event; those created towards the end of the ionization period were resonantly excited for only a very short period

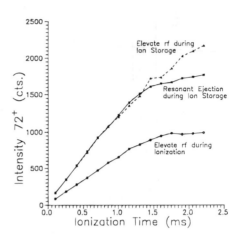

FIGURE 5.10

Plots of diuron analyte ion signal
(*m/z* 72) *versus* ionization time
comparing ion trap operational
modes which eject solvent ions from
within the ion trap prior to mass
analysis. (From Ref. 3: Kleintop, B. L.;
Eades, D. M.; Yost, R. A. *Anal. Chem.*
1993, *65*, 1295.)

of time. Also, since ionization times were typically only 1 ms, the maximum amount of time an ion could be irradiated was 1 ms, as compared to up to 50 ms of irradiation during ion storage.

These modes of operation were compared also to determine which mode would provide the highest sensitivity for LC/PB/ITMS analyses. The plots of Fig. 5.10 were obtained by varying the ionization time while flowing a constant amount of diuron through the interface. As expected, analyte signal intensity increased with increasing ionization time until the onset of space charging, which is indicated by the onset of curvature in the plots. The plots indicate also that methods which utilized postionization ejection of solvent ions provided about two times greater signal. This observation was attributed to ion losses due to quasi-unstable trajectories and smaller ionization volumes at higher ionization RF levels.[58] It should be noted that this effect may not be as significant for compounds whose EI mass spectral base peaks are of higher *m/z* values. Although elevating the RF level and resonant ejection during ion storage provided approximately the same signal intensity, Kleintop and co-workers concluded that elevating the RF level after ion storage was the best way to operate the ITMS for LC/PB/ITMS analyses of compounds whose EI base peaks are less than *m/z* 200. Elevation of the RF level ejects all ion species, including other background ions (air/water, column bleed, etc.) below the selected low-mass cut-off. Resonant ejection of multiple ions would require multiple excitation events which would allow more time for undesired ion/molecule reactions to occur. Multiple resonant ejection events would likely be necessary for LC/MS analyses which employ gradient elution or isocratic solvent mixtures.

Kleintop and co-workers also performed both isocratic[3] and gradient elution[56,59,60] LC/MS analyses of several pesticide mixtures. Mass spectra obtained from these analyses compared favorably with both library spectra and EI mass spectra obtained from solids probe/ITMS analyses (that is, a solvent-"free" method) of a pure standard. Increases in the relative abundance of $(M + 1)^+$ ions were observed at high concentrations of analyte and were attributed to self-CI caused by the increasing amount of analyte ions and neutrals within the trap. The PB/ITMS system provided typically low nanogram detection limits and excellent reproducibility between replicate injections (RSDs of peak areas typically <5%). Although linear calibrations were observed, non-linear calibrations were far more prevalent.

Several laboratories have been able to obtain linear calibrations on other PB systems by using co-eluting isotopically-labelled internal standards (IS).[52-54] The major benefit of using a co-eluting IS for PB analyses is believed to be that the co-eluting component provides for improved analyte transport through the momentum separator due to larger particle formation. However, the use of a co-eluting IS in the ion trap can cause problems by producing space-charge conditions which can limit the linear dynamic range of the analysis. A new scan method using alternating RF/DC isolations has been used to extend the dynamic range to four orders of magnitude for GC/ITMS analysis which utilize co-eluting ISs.[61] This scan method has been applied also to the LC/PB/ITMS analysis of benzidine using d_8-benzidine as the IS.[56,60] Isotope dilution calibration curves from these analyses exhibited linear behavior over two orders of magnitude. A larger dynamic range was not possible for this analysis because PB/ITMS does not provide the same low detection limits as GC/ITMS (nanograms as compared to picograms).

The performance of this PB/ITMS system was compared also to a PB/quadrupole system by Kleintop and Yost.[56,59] Mass spectral quality, limits of detection (LOD), precisions, and calibration curves were compared by performing a gradient elution analysis of a mixture of ten pesticides using the same LC conditions and PB interface on each system. The quadrupole mass filter system used was a Finnigan TSQ-70 which was operated in single MS mode. The PB interface was inserted into the TSQ through a standard probe lock which replaced the GC transfer line flange. A standard GC/EI ion volume served as the PB source target and was heated to the same temperature as the target end-cap electrode of the ion trap. In order to provide the best possible comparison, all PB and LC operating conditions were identical on both systems.

The total ion current chromatogram from this analysis obtained with the ion trap system is shown in Fig. 5.11. Also shown are peak identities and amounts of each component injected. Fig. 5.12 compares the background-subtracted mass spectra of mexacarbate (MW = 165 Da) obtained on both PB systems. The spectra exhibited the same EI fragmentation patterns in both systems; however, some minor differences in the relative intensities of the fragment ions were observed. An increase in the

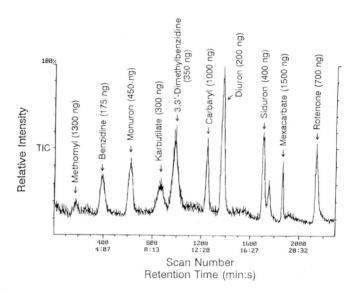

FIGURE 5.11
Total ion chromatogram (TIC) from LC/PB/ITMS of a ten component pesticide mixture. Also shown are peak identities and amounts of each component injected. (From Ref. 59: Kleintop, B. L.; Eades, D. M.; Yost, R. A. *Anal. Chem.* 1993, 65, 1295.).

$(M + 1)^+/M^+$ ratio was observed typically in the ion trap mass spectra at high analyte concentrations and was attributed to self-CI occurring within the trap. The two systems were found also to provide comparable precisions and LODs. These results were unexpected since the ion trap is known as a more sensitive mass spectrometer, particularly when compared in the full-scan mode. The LODs obtained were believed to be limited by the interface rather than the type of mass spectrometer used. Lower LODs might have been obtained on the ion trap system if the interface had been able to provide more efficient analyte transport to the mass spectrometer.

One advantage of the ion trap over other mass spectrometers is the ability to acquire different types of mass spectra (EI, CI, MS^n)[62-64] with no hardware modifications. This facility of the ion trap makes possible the acquisition of alternating EI and CI spectra over an LC peak when directly coupled to a PB interface. The ability to provide both structural information from EI spectra and molecular weight information from CI spectra may provide for more positive confirmation of unknowns. The use of sequential scans to acquire EI and CI spectra has been previously demonstrated over a GC peak[65] by Creaser and co-workers. An interesting feature of performing CI with a PB/ion trap system is that the residual solvent introduced by the interface can be used to produce the CI reagent ions. Eades and co-workers have demonstrated this ability by using pro-

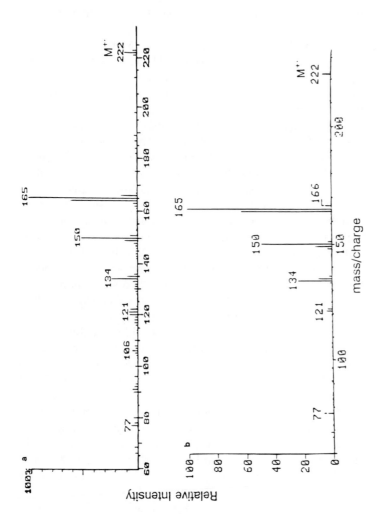

FIGURE 5.12

Comparison of background-subtracted mass spectra of mexacarbate from LC/MS analysis obtained using (a) a quadrupole ion trap, and (b) a quadrupole mass filter as the mass spectrometric detector. (From *Ref.* 59: Kleintop, B.L.; Eades, D. M.; Yost, R.A. *Anal. Chem.* 1993, 65, 1295.)

tonated methanol and acetonitrile as CI reagent ions to determine molecular weights of several environmental pesticides.[66] Isolation of a particular solvent ion species has also been shown to improve the quality of the CI spectra by ejecting analyte ions which were formed by EI. The ability to acquire alternating EI/CI scans over an LC peak has been demonstrated by Yost and co-workers.[67] Fig. 5.13 shows both the EI and CI spectra of the pesticides monuron (MW = 198 Da) and diuron (MW = 232 Da) from this type of analysis using protonated acetonitrile as the CI reagent ion. While the EI spectra exhibit fragmentation which reveals structural information about the two components, the CI spectra exhibit primarily molecular weight information.

Creaser and co-workers[62,63] have designed and constructed a PB interface which was interfaced to a Finnigan ITMS. Their design, shown in Fig. 5.14, utilized a glass de-solvation chamber, two stages of momentum separation, and allowed for interchangeable transfer lines to facilitate coupling to triple quadrupole, sector, and ion trap instruments. The PB interface was adapted with a transfer line consisting of two concentric stainless steel tubes with a heater and thermocouple assembly located at the tip. The transfer line was introduced to the ion trap *via* a solids probe lock and ionization was performed internally. The "modular design" also allowed investigation and optimization of inter-skimmer distances, nebulizer sizes, flow rates, and target temperature. The two-stage momentum separator design allowed pumping from both sides. This design resulted in an operating base pressure of approximately 1.8×10^{-5} Torr.

Using this system, Creaser and co-workers were able to acquire quality EI spectra (internal ionization) of pharmaceutical compounds by normal phase LC/PB/ITMS. The spectra reported compared favorably with probe-EI and LC/PB/sector mass spectra. The authors utilized a modified scan function composed of repeated ionization/low mass ejection steps which minimized any adverse solvent effects. This scan function allowed for sufficient accumulation of desired analyte ions while ejecting alternately unwanted low mass/charge solvent ions. Acquisition of MS/MS spectra of these compounds with the ITMS exhibited increased fragmentation as compared to the triple quadrupole MS/MS spectrum acquired with xenon collision gas at 4 eV collision energy.

Although all of the studies summarized previously in this PB section have demonstrated the abilitly to acquire quality EI spectra, all have had to employ some strategy in order to minimize problems associated with directly coupling PB with an ion trap. Space charging and undesired ion/molecule reactions (self-CI, adduct formation, etc.) caused by high ion and neutral species populations (solvent, analyte) within the trap are the two major problems associated with directly coupling a PB interface with an ion trap. To minimize these problems, it is desirable to prevent solvent ions and molecules, as well as excess analyte neutrals from reaching the

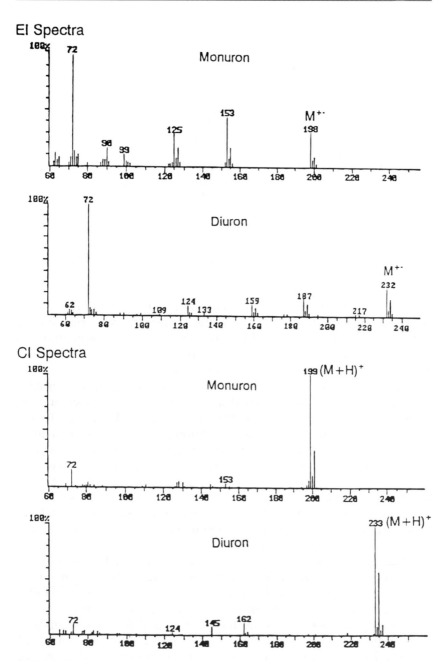

FIGURE 5.13

Background-subtracted mass spectra of monuron and diuron obtained from alternating EI/ CI scans. (From Ref. 64: Yost, R. A.; Johnson, J. V.; Yates, N. A.; Kleintop, B. L.; Coopersmith, B.; Eades, D. M. Presented at the 43rd Annual Pittsburgh Conference on Analytical Chemistry and Applied Spectroscopy, New Orleans, 1992.)

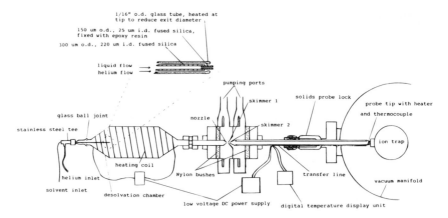

FIGURE 5.14
Schematic of the PB interface used by Creaser et al.[63] for coupling to a Finnigan ITMS. (From Ref. 63: Creaser, C. S.; Stygall, J. W. *Proc. 41st ASMS Conf. Mass Spectrometry and Allied Topics.* San Francisco, 1993, p. 470a.)

trap. The use of an external, differentially-pumped ion source and ion injection have been investigated recently as alternatives to using internal ionization *via* direct coupling. By using ion injection, excess neutrals can be pumped away prior to entering the analyzer region, which can prevent undesired ion/molecule reactions from occurring.

Bier and co-workers were the first to report using an ion injection system to couple a PB interface with an ion trap.[21] The instrument utilized Finnigan ITS-40 electronics to control the ion trap and Finnigan TSQ-70 electronics to control the ion source, vacuum pumps, and lens potentials. The instrument used was shown previously in Fig. 5.4 and is described in Section III. Using FIA, EI spectra were acquired and compared to EI spectra acquired on the PB system using internal ionization.[2] The RF amplitude applied to the ring electrode was set to eject all ions less than 34 u. Spectra obtained with the ion injection system showed no evidence of self-CI; reduced solvent effects were observed due to the removal of solvent neutrals prior to the trap.

Yost and co-workers[67] have also utilized an external ion source and ion injection for PB/ion trap analyses. A Finnigan TSQ 45 ion source, controlled by its normal electronics, was mounted on a vacuum chamber pumped by turbo pumps operating at 500 and 270 L/s. Finnigan ITS-40 electronics were used to control the ion trap and for data acquisition. Without the use of differential pumping, EI spectra were obtained by FIA of several pesticides in acetonitrile at a flow rate of 0.3 mL/min. The spectra compared favorably with library spectra and showed no evidence of self-CI or solvent-related ion/molecule reactions. Fig. 5.15 demonstrates how the $(M + 1)^+/M^+$ ratio of benzidine varied with increasing the amount of analyte injected using both direct coupling and ion injection. The

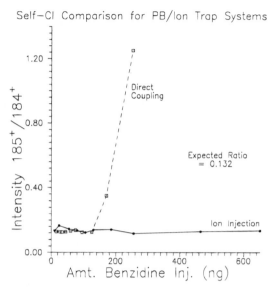

Self—CI Comparison for PB/Ion Trap Systems

FIGURE 5.15
Plots of (M + 1)⁺/M⁺ ratio for m/z 185 and m/z 184 *versus* amount of benzidine injected from PB/ITMS analyses using both (a) direct coupling and (b) external ion source and ion injection. (From Ref. 67: Kleintop, B. L.; Yost, R. A. Presented at the 44th Annual Pittsburgh Conference on Analytical Chemistry and Applied Spectroscopy, Atlanta, 1993.)

increase in the m/z 185/m/z 184 intensity ratio when directly coupling the interface with the ion trap was attributed to self-CI caused by a large neutral analyte population within the trap. This same behavior was not exhibited in the ion injection system. The use of ion injection prevented self-CI from occurring by pumping away the neutral analyte species prior to the ion trap.

V. CONCLUSIONS

Although currently seeing widespread use as sensitive benchtop mass spectrometric detectors for GC, the use of ion traps has been pursued only recently for LC/MS applications. The major challenge with interfacing LC with ion traps is the abundance of solvent neutrals which need to be pumped away prior to the mass analyzer. Residual solvent, inherent to most LC/MS interfaces, can cause space charging and undesired ion/molecule reactions which result, typically, in both poor mass resolution and poor spectral quality. Indeed, all of the techniques described above have utilized some strategy to reduce adverse solvent effects. Ion injection techniques have been implemented to reduce the amount of solvent neutrals which reached the trap. Ion ejection techniques have been employed to rid the ion trap of abundant solvent ions. Clearly, the ideal solution would be to employ an ion injection system coupled with efficient ejection of unwanted ions, such as SWIFT or filtered noise fields. Indeed, several researchers have shown dramatic improvements in analyte ion signal intensity when ejecting selectively unwanted ions during the ion accumulation process. With continued advances in ion trap technology and instrumentation, practical LC on ion trap mass spectrometers will be realized.

References

1. Brown, R. *Chem. Eng. News.* 1989, *31*, 89.
2. Bier, M. E.; Winkler, P. Ć.; Lopez, J. T. *J. Am. Soc. Mass Spectrom.* 1993, *4*, 38.
3. Kleintop, B. L.; Eades, D. M.; Yost, R. A. *Anal. Chem.* 1993, *65*, 1295.
4. Creaser, C. S.; Mitchell, D. S.; O'Neil, K. E. *Int. J. Mass Spectrom. Ion Processes.* 1991, *106*, 21.
5. Eades, D. M.; Yates, N. A.; Yost, R. A. *Proc. 39th ASMS Conf. Mass Spectrometry and Allied Topics.* Nashville, 1991, p. 1491.
6. Kelley, P. E.; Stafford, G. C.; Syka, J. E. P.; Reynolds, W. E.; Louris, J. N.; Todd, J. F. J. *Proc. 33rd ASMS Conf. Mass Spectrometry and Allied Topics.* San Diego, 1985, p. 707.
7. McLuckey, S. A.; Goeringer, D. E.; Glish, G. L. *Anal. Chem.* 1992, *64*, 1455.
8. Marshal, A. G.; Wang, T.-C. L.; Ricca, T. L. *J. Am. Chem. Soc.* 1985, *107*, 7893.
9. Julian, R. K.; Cox, K.; Cooks, R. G. *Proc. 40th ASMS Conf. Mass Spectrometry and Allied Topics.* Washington, DC, 1992, p. 943.
10. Julian, R. K.; Cooks, R. G. *Anal. Chem.* 1993, *65*, 1827.
11. Kelley, P.; Hoekman, D.; Bradshaw, S. *Proc. 41st ASMS Conf. Mass Spectrometry and Allied Topics.* San Francisco, 1993, p. 453a.
12. Louris, J. N.; Amy, J. W.; Ridley, T. Y.; Cooks, R. G. *Int. J. Mass Spectrom. Ion Processes.* 1989, *88*, 97.
13. Kaiser, R. E. Jr.; Williams, J. D.; Lammert, S. A.; Cooks, R. G. *J. Chromatogr.* 1991, *562*, 3.
14. Van Berkel, G. J.; Glish, G. L.; McLuckey, S. A. *Anal. Chem.* 1990, *62*, 1284.
15. McLuckey, S. A.; Glish, G. L.; Asano, K. G. *Anal. Chim. Acta.* 1989, *225*, 25.
16. McLuckey, S. A.; Van Berkel, G. J.; Glish, G. L.; Huang, E. C.; Henion, J. D. *Anal. Chem.* 1991, *63*, 375.
17. McLuckey, S. A.; Glish, G. L.; Asano, K. G.; Bartmess, J. E. *Int. J. Mass Spectrom. Ion Processes.* 1991, *109*, 171.
18. McLuckey, S. A.; Glish, G. L.; Van Berkel, G. J. *Anal. Chem.* 1991, *63*, 1971.
19. Van Berkel, G. J.; McLuckey, S. A.; Glish, G. L. *Anal. Chem.* 1991, *63*, 1098.
20. Mordehai, A. V.; Henion, J. D. *Rapid Commun. Mass Spectrom.* 1993, *7*, 205.
21. Bier, M. E.; Hartford, R. E.; Herron, J. R.; Stafford, G. C. *Proc. 39th ASMS Conf. Mass Spectrometry and Allied Topics.* Nashville, 1991, p. 538.
22. Kaiser, R. E.; Cooks, R. G.; Stafford, G. C.; Syka, J. E. P.; Hemberger, P. H. *Int. J. Mass Spectrom. Ion Processes.* 1991, *106*, 79.
23. Williams, J. D.; Cox, K.; Morand, K. L.; Cooks, R. G.; Julian, R. K.; Kaiser, R. E. *Proc. 39th ASMS Conf. Mass Spectrometry and Allied Topics.* Nashville, 1991, p. 1481.
24. Louris, J. N.; Brodbelt-Lustig, J. S.; Cooks, R. G.; Glish, G. L.; Van Berkel, G. J.; McLuckey, S. A. *Int. J. Mass Spectrom. Ion Processes.* 1990, *96*, 117.
25. McLuckey, S. A.; Glish, G. L.; Van Berkel, G. J. *Int. J. Mass Spectrom. Ion Processes.* 1991, *106*, 213.

26. Hoch, G.; Kok, B. *Arch. Biochem. Biophys.* 1963, *101,* 160.
27. Kotiaho, T.; Lauritsen, F. R.; Choudhury, T. K.; Cooks, R. G.; Tsao, G. T. *Anal. Chem.* 1991, *63,* 875A.
28. Lister, A. K.; Wood, K. V.; Cooks, R. G.; Noon, K. R. *Biomed. Environ. Mass Spectrom.* 1989, *18,* 1063.
29. Dheandhando, S.; Dulak, J. *Rapid Commun. Mass Spectrom.* 1989, *3,* 175.
30. Sturaro, A.; Doretti, C.; Parvoli, G.; Lecchinato, F.; Frison, G.; Traldi, P. *Biomed. Environ. Mass Spectrom.* 1989, *18,* 707.
31. LaPack, M. A.; Tou, J. C.; Enke, C. G. *Anal. Chem.* 1990, *62,* 1265.
32. Degn, H.; Cox, R. P.; Lloyd, D. *Methods Biochem. Anal.* 1985, *31,* 165.
33. Kotiaho, T.; Hayward, M. J.; Cooks, R. G. *Anal. Chem.* 1991, *63,* 1794.
34. Bier, M. E.; Cooks, R. G. *Anal. Chem.* 1987, *59,* 597.
35. Bauer, S. J.; Cooks, R. G. *American Lab.* 1993, 25(8), 36.
36. Bier, M. E.; Cooks; R. G. *Anal. Chim. Acta.* 1990, *231,* 175.
37. Stafford, G. C.; Taylor, D. M.; Bradshaw, S. C.; Syka, J. E. P. *Proc. 35th ASMS Conf. Mass Spectrometry and Allied Topics.* Denver, 1987, p. 775.
38. Slivon, L. E.; Bauer, M. R.; Ho, J. S.; Budde, W. L. *Anal. Chem.* 1991, *63,* 1335.
39. Blakely, C. R.; Vestal, M. L. *Anal. Chem.* 1983, *55,* 750.
40. Garteiz, D. A.; Vestal, M. L. *LC Magazine.* 1985, *3,* 334.
41. Arpino, P. *Mass Spec. Reviews.* 1990, *9,* 631.
42. Arpino, P. *Mass Spec. Reviews.* 1992, *11,* 3.
43. Kaiser, R. E.; Williams, J. D.; Schwartz, J. C.; Lammert, S. A.; Cooks, R. G.; Zakett, D. *Proc. 37th ASMS Conf. Mass Spectrometry and Allied Topics,* Miami Beach, 1989, p. 369.
44. Willoughby, R. C.; Browner, R. F. *Anal. Chem.* 1984, *56,* 2626.
45. Winkler, P. C.; Perkins, D. D.; Williams, W. K.; Browner, R. F. *Anal. Chem.* 1988, *60,* 489.
46. Behymer, T. D.; Bellar, T. A.; Budde, W. L. *Anal. Chem.* 1990, *62,* 1686.
47. Voyksner, R. D.; Smith, C. S.; Knox, P. C. *Biomed. Environ. Mass Spectrom.* 1990, *19,* 523.
48. Smith, R. D.; Loo, J. A.; Edmonds, C. G.; Barinaga, C. J.; Udseth, H. R. *Anal. Chem.* 1990, *62,* 882.
49. Loo, J. A.; Edmonds, C. G.; Smith, R. D. *Anal. Chem.* 1991, *63,* 2488.
50. Katta, V.; Chowdhury, S. K.; Chait, B. T. *Anal. Chem.* 1991, *63,* 174.
51. Ligon, W. V.; Dorn, S. B. *Anal. Chem.* 1990, *62,* 2573.
52. Doerge, D. R.; Miles, C. J. *Anal. Chem.* 1991, *63,* 1999.
53. Brown, F. R.; Draper, W. M. *Biol. Mass Spectrom.* 1991, *20,* 515.
54. Ho, J. S.; Behymer, T. D.; Budde, W. L.; Bellar, T. A. *J. Am. Soc. Mass Spectrom.* 1992, *3,* 662.
55. Kleintop, B. L.; Eades, D. M.; Yost, R. A. *Proc. 40th ASMS Conf. Mass Spectrometry and Allied Topics.* Washington, D.C. 1992, p. 1308.
56. Kleintop, B. L.; Ph.D. Dissertation, University of Florida, 1992.
57. Browner, R. F. *J. Microchem.* 1989, *40,* 4.
58. Dawson, P. H. *Quadrupole Mass Spectrometry and Its Applications.* Elsevier, Amsterdam, 1976.
59. Kleintop, B. L.; Yost, R. A. *Anal. Chem.* Eades, D. M. 1993, 65, 1295.
60. Kleintop, B. L.; Eades, D. M.; Yost, R. A.; Behymer, T. D.; Budde, W. L. Presented at the 19th Annual Meeting of the Federation of Analytical Chemistry and Spectroscopy Societies. Philadelphia, 1992.
61. Kleintop, B. L.; Yost, R. A.; Abolin, C. R. *J. Am. Soc. Mass Spectrom.* 1992, *3,* 85.
62. Creaser, C. S.; Stygall, J. W. *Proc. 22nd Annual Meeting British Mass Spectrometry Society.* St. Andrews, UK, 1992, p. 8.
63. Creaser, C. S.; Stygall, J. W. *Proc. 41st ASMS Conf. Mass Spectrometry and Allied Topics.* San Francisco, 1993, p. 470a.

64. Yost, R. A.; Johnson, J. V.; Yates, N. A.; Kleintop, B. L.; Coopersmith, B.; Eades, D. M. Presented at the 43rd Annual Pittsburgh Conference on Analytical Chemistry and Applied Spectroscopy. New Orleans, 1992.
65. Creaser, C. S.; Mitchell, D. S.; O'Neil, K. E.; Trier, K. J. *Rapid Commun. Mass Spectrom.* 1990, *4*, 217.
66. Eades, D. M.; Kleintop, B. L.; Yost, R. A. *Proc. 40th ASMS Conf. Mass Spectrometry and Allied Topics.* Washington, DC, 1992, p. 1290.
67. Kleintop, B. L.; Yost, R. A. Presented at the 44th Annual Pittsburgh Conference on Analytical Chemistry and Applied Spectroscopy. Atlanta, 1993.

Chapter 6

Ion Spray Liquid Chromatography/ Mass Spectrometry and Capillary Electrophoresis/Mass Spectrometry on a Modified Benchtop Ion Trap Mass Spectrometer

A. Mordehai, H.K. Lim, and J.D. Henion

CONTENTS

0-8493-8251-3/95/$0.00+$.50

I. INTRODUCTION

We, in this laboratory, have been interested in coupling the modern separation sciences on-line with mass spectrometry for 15 years.[1] The consistent goal remains high sensitivity and ease of use along with the specificity for which mass spectrometry is well known. In the early eighties, we explored the analytical potential of atmospheric pressure ionization (API) combined with triple-stage quadrupole mass spectrometric techniques to accomplish these goals.[2] The potential of this technology for accomplishing liquid chromatography/mass spectrometry (LC/MS), ion chromatography/mass spectrometry (IC/MS), capillary electrophoresis/mass spectrometry (CE/MS) and supercritical fluid chromatography/mass spectrometry (SFC/MS) has surpassed our expectations.[3] This capability is now available commercially from several mass spectrometer maufacturers and reports of real-world applications are beginning to appear more regularly.

Two very useful API techniques include atmospheric pressure chemical ionization (APCI) using, for example, corona discharge ionization with the heated pneumatic nebulizer interface and electrospray. It is the combination of these two different API modes that can be used on one mass spectrometer that makes coupling to condensed-phase separation sciences so practical. Both of these ionization techniques preclude direct liquid introduction of chromatographic effluent into the mass spectrometer vacuum system, so we can avoid some of the problems associated with removing excess solvent or the thermal effects that can occur with particle beam and thermospray LC/MS.

There have been significant improvements with API techniques on both quadrupole, magnetic sector mass spectrometers[4] as well as recent developments with the implementation of electrospray on Fourier transform mass spectrometers (FTMS).[5] The tremendous success of electrospray mass spectrometry for the determination of high molecular weight biomolecules has increased the need and interest in this technology. In particular, there is increased need for accurate mass determination such as is available from magnetic sector[6] and FTMS[7] systems as well as the intrinsic analytical utility of tandem mass spectrometry (MS/MS).[8] However, quadrupole-based systems lack the high mass resolution capability for accurate mass measurement and cannot readily provide multiple stages of mass analysis that may be of interest in certain instances. There is a need for a system that combines the relatively low cost and ease of use common to single-stage quadrupole systems with the unique analytical capabilities of FTMS.[7]

The ion trap mass spectrometer would appear to be the ultimate system for combining all the needs of the modern laboratory interested in routine, high performance LC/MS and related techniques. In our opinion, the most significant breakthrough for demonstrating the potential of the ion trap coupled with the electrospray technique was reported by McLuckey et al. in 1990.[9]

Following earlier reports that described the implementation of an atmospheric sampling glow discharge source on an ion trap,[10] the Oak Ridge group simply installed a sprayer in an electrospray format on this same device and demonstrated the feasibility and potential of an electrospray/ion trap combination.[9] Following the initial report of this work, we collaborated with the Oak Ridge group to explore the analytical potential of pneumatically-assisted electrospray (ion spray) LC/MS on an ion trap mass spectrometer (ITMS) system.[11] This and related work has shown clearly that the ion trap offers exciting possibilities for an electrospray-based system that can provide routine high sensitivity LC/MS, as well as accurate mass and high molecular weight determination.[12,13]

Following our preliminary collaboration with McLuckey et al.,[11] it was clear that the ion trap offered analytical capabilities that we needed for our continuing LC/MS studies and related work. However, the relatively high cost of a commercial ITMS system precluded the early installation of an instrument in our laboratory. Not to be thwarted, we decided to implement an API interface on a relatively inexpensive commercial benchtop ion trap, because a successful outcome of the experiments to be carried out would demonstrate the feasibility of this analytical capability on a much less expensive system. In this report, we summarize our experience to date with this project and focus on the hardware development and applications of this model system.

The goals of this project focused on achieving routine LC/MS and CE/MS capability without losing the benefits of the ion trap that have been well documented in the gas chromatography/mass spectrometry (GC/MS) mode. In addition, we planned to address some of the problems common to GC/MS on the ion trap, such as reduction of background chemical noise, and space-charge effects that contribute to ion peak shape distortion.

A. Sensitivity

The most important goal following successful implementation of ion spray LC/MS capability on the benchtop ion trap was to achieve high sensitivity. The success of this effort is a function of how efficiently the ions are sampled from atmospheric pressure into the trap as well as how efficiently these ions are trapped once inside the trapping region. High analyte sensitivity is also critically dependent upon the level of background ion current in the system which can come from chemical as well as electrical noise. In ion trap GC/MS experiments, considerable chemical noise can originate from capillary GC column bleed and the sample matrix. We anticipated that contributions from both of these sources would be significantly reduced in LC/MS experiments. Interference from high performance liquid chromatography (HPLC) column bleed is likely to be reduced because the very nonpolar material that may elute under LC/MS conditions is not likely to ionize under ion spray conditions. Similarly, matrix interferences should be reduced because ionization occurs external to the ion trap such that only the ionized matrix components enter the ion trap. In addition, the adsorption–desorption of sample matrix components to the ion trap surfaces that occur under GC/MS conditions are not likely to occur in LC/MS experiments. Of course, when high levels of ionized matrix components are introduced, or co-elute with components of interest, there are likely to be unavoidable matrix interferences. This problem can often be addressed by changing the chromatographic conditions.

The other potential source of chemical interference is cluster ion formation between the HPLC eluent components under atmospheric pressure ionization conditions.[3,14] These cluster ions may consist of water plus a variety of organic solvents such as methanol and acetonitrile, along with a myriad of commonly used buffer additives and impurities that are often in these buffers. The latter may include a wide variety of low molecular weight acids and bases as well as ion-pair reagents and other additives. Since the concentration of these additives is often much higher than that of the analytes of interest, interference from these components is likely to be considerable. Therefore, an important goal of this project was to implement adequate "declustering" conditions in the API interface using the ion optics of the ion trap to minimize these interferences.

B. Trapping Externally Produced Ions

Effective introduction into and cumulative trapping of externally produced ions in a physical trap is an intrinsic problem of any ion trapping technology such as FT-ICR (Fourier transform-ion cyclotron resonance, an electromagnetic static trap) or the QUISTOR (radio-frequency dynamic trap).[15] The problem is intimately related to the basic law of the conservation of energy. According to this law, any particle having kinetic energy and entering the trap from an external source must be ejected eventually from the trap; thus, where there are no energy losses in the trap, the resulting trapping efficiency is zero. To overcome the problem, it has been suggested that the mechanism of collisional relaxation be used *via* introduction of a momentum-dampening gas to achieve effective accumulation of externally produced ions in a QUISTOR device.[16] By this mechanism, each externally produced ion must experience at least one collision with a damping gas particle as it passes through the trap to be trapped successfully and stored.

The trapping efficiency is defined usually as the percentage of the accumulated ions relative to the total number of ions that passed through the trap from the external source. This definition of trapping efficiency is the analog of the transmission efficiency definition used for beam-type MS systems such as quadrupole and magnetic sector instruments. Therefore, these definitions describe the amount of mass-analyzed ions as a percentage of the total number of ions exiting the ion source. These important characteristics allow one to estimate and compare the performance of beam *versus* trapping-type mass spectrometers.

The reported theoretical transmission efficiency for a quadrupole instrument ranges from 5 to 40% (depending upon mass resolution) although the experimental data show much lower values of 1 to 10%.[17] For the QUISTOR-type instrument to be useful for analytical applications, it should have a trapping efficiency comparable to the transmission efficiency in quadrupole instruments. Actually, in the case of the QUISTOR instrument, all ions in the mass range of interest are accumulated in the trap at the same time (the Felgett advantage),[18] thereby allowing very fast scan rates. In contrast, the quadrupole instrument scan rate is somewhat less. One advantage of the ion trap is that its very fast scan rates make it especially beneficial as a mass analyzer for on-line coupling with separation techniques that afford very narrow profile peaks.

Optimization of the trapping efficiency for the QUISTOR instrument has been studied in detail elsewhere.[19] It has been shown that by introducing helium damping gas at unusually high pressures (higher than 1×10^{-3} Torr), the experimental trapping efficiency continues to increase with increasing helium pressure. Optimization of the damping gas pressure to achieve higher trapping efficiency effects a compromise between the trapping efficiency and mass resolution, because mass resolution tends to decrease as the pressure increases.[20] It has been established that the optimum helium pressure in the QUISTOR is 6×10^{-3} Torr (absolute reading) which provides unit mass resolution across the mass region of the instrument with a trapping efficiency in the range of 5 to 10%. With these factors in mind we set out to design, construct, and implement an API interface on a benchtop Varian Saturn II ion trap.

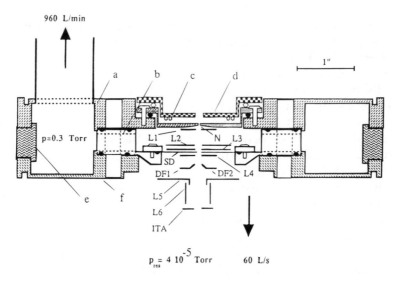

FIGURE 6.1

API/ion trap interface: (a) top flange with pump-out port; (b) ion trap mounting flange; (c) heater; (d) liquid shield lens; (e) outer flange; (f) bottom flange. Ion optical system: N, nozzle plate; L1, L2, L3, electrostatic lenses in first vacuum region; SD, separating diaphragm equipped with a 400 µm orifice; DF1 and DF2, deflector electrodes; L4, L5, L6 electrostatic lenses in second vacuum region; ITA, ion trap aperture. (Reprinted with permission from John Wiley & Sons)

C. The Phase I System and Its API Interface

The details of our first benchtop ion trap LC/MS system have been described elsewhere.[21] The system was designed with a minimum alteration of the standard Saturn II benchtop ion trap vacuum system provided by Varian. The gas chromatograph was removed while the filament and its mounting flange were replaced by the API interface shown in Fig. 6.1. The standard 60 L/s turbo-molecular pump was used to evacuate the high vacuum region and the first stage of the API interface region was evacuated by a 960 L/min two-stage rotary pump. The standard software was modified by Varian to increase the accumulation time from the standard 25 ms to 250 ms.

The most significant design feature of the API/ion trap system is the API interface shown in Fig. 6.1. In brief, this device incorporates a heated liquid shield (Fig. 6.1, d) closely coupled to a nozzle plate (Fig. 6.1, N). Ions produced under API conditions pass through the heated atmospheric pressure liquid shield and enter the first vacuum region through the nozzle plate (N) *via* an orifice of diameter 75 µm. The thickness of the nozzle plate is 150 µm which leads to an aspect ratio of 0.5. The first vacuum region contains specially designed ion optics which include lens electrodes (L1, L2, L3, and SD) that are constructed from pure nickel. Ion sampling in the second vacuum region occurs through a 400 μm diameter orifice in the diaphragm separating the two vacuum regions. Collision-induced dissociation processes may be invoked in this first vacuum region by imposing a voltage difference between N and SD (Fig. 6.1). By varying the potential difference in this region we may either increase the declustering of cluster ions or, when greater potential differences are used, dissociate covalent bonds. The latter higher energy conditions are referred to as "up-front" CID,[22] and may be used as an analytical tool to produce reproducible CID mass spectra for interpretive or identification purposes.

Under ion spray conditions,[23] the pressure in the first vacuum region is 0.3 Torr while the base pressure in the high vacuum region is typically 4×10^{-5} Torr. Pressures are measured using a Bayard–Alpert ion gauge tube installed directly in the mass analyzer vacuum chamber *via* the GC/MS transfer line port of the standard SATURN II system. During operation, helium damping gas is introduced into the high vacuum region to produce a total pressure of 1.3×10^{-3} Torr (corrected reading).

The ion optics in the second vacuum region consist of an electrostatic lens assembly (SD, L4, L5, L6) and pulsed ion beam deflector electrodes (DF1, DF2). The ion trap is housed in its original vacuum chamber within the Saturn II system. The nozzle plate, ion sampling diaphragm, and all ion optics electrodes are controlled by eight independent voltages taken from a single DC power supply ($\pm$ 130 V). After the accumulation step (0.25 s) the ion beam is deflected from the entrance of the ion trap and MS detection followed. Deflection of the ion beam occurs by applying a 0 to 100 V pulse to one of the deflector electrodes.

The atmospheric pressure liquid shield lens (d) is maintained at the same potential as the nozzle plate (N). Between the liquid shield lens and nozzle plate a coiled resistive wire electrical heater (c) is installed which provides heat to both the nozzle plate and atmospheric pressure liquid shield lens. The temperature of the nozzle plate during these experiments could be maintained between 100 to 250°C.

II. RESULTS AND DISCUSSION

A. Applications of LC/MS and CE/MS on the Phase I System

To test the ruggedness and utility of this instrument as an on-line detector, a series of experiments was conducted coupling HPLC and CE to the API/ion trap system. The compounds chosen for these studies were small quaternary ammonium drugs. These charged compounds were selected because they are not amenable to gas chromatographic analysis without pyrolysis, so the combination of liquid chromatography or capillary electrophoresis with mass spectrometry may provide a preferred tool

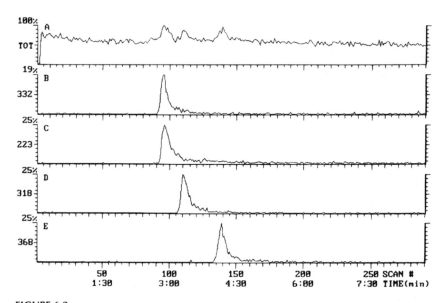

FIGURE 6.2

Ion spray HPLC/ITMS analysis of a synthetic mixture containing 5 ng each of four quaternary ammonium drugs. Conditions: 1 mm i.d. × 150 mm C-8 HPLC column operated in the isocratic mode with 40 mL/min of 80/20 acetonitrile/water containing 20 mM ammonium acetate and 0.1% trifluoroacetic acid (TFA). Potential difference across first vacuum region was 70 V. (A) TIC from m/z 100 to 400, (B) EICP for m/z 332, ipratropium, (C) EICP for m/z 223, neostigmine; (D) EICP for m/z 318, glycopyrrolate; (E) EICP for m/z 368, propantheline. (Reprinted with permission from John Wiley & Sons)

for their separation with on-line identification at the low nanogram level. Results of the on-line coupling of HPLC/ITMS are shown in Fig. 6.2. A synthetic mixture containing 5 ng of each drug was injected on-column under micro HPLC conditions. The total column effluent (40 mL/min) was directed to the mass spectrometer in this experiment. The potential drop across the first vacuum region was 70 V. Panel A (Fig. 6.2) is the total ion current (TIC) using a scan range from 100 to 400 amu, while panels B–E are the extracted ion current profiles (EICP) for the molecular ions of ipratropium, neostigmine, glycopyrrolate and propantheline, respectively. Fig. 6.3 shows the full-scan ion spray LC/MS mass spectrum for glycopyrrolate taken from scan #112 in Fig. 6.2 as an example of a typical mass spectrum obtained from 5 ng of drug using this system. The base peak is the molecular ion at m/z 318, and the mild de-clustering conditions used in this experiment lead to essentially no fragmentation.

The analysis of a synthetic mixture containing four quaternary ammonium compounds by CE/ITMS is shown in Fig. 6.4. The coupling of CE to the ITMS was accomplished using a coaxial (sheath) flow incorporated in the ion spray interface. The sheath flow was maintained at 8 μL/min

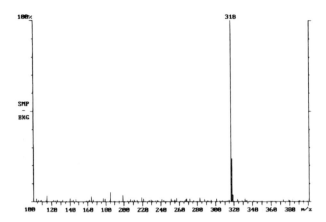

FIGURE 6.3
Background-substracted ion spray LC/MS mass spectrum for glycopyrrolate taken from scan #112 in Fig. 6.2.

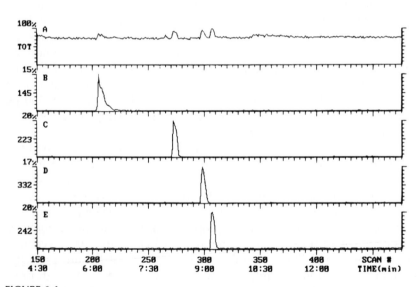

FIGURE 6.4
Ion spray CE/ITMS analysis of a synthetic mixture containing 150 pg each of four quaternary ammonium compounds. Conditions: 50 μm i.d. × 90 cm bare fused-silica capillary with a 30 kV potential applied to the anode. The CE running buffer contained 30 mM ammonium acetate and acetic acid, pH 4. The sheath was 3 mM ammonium acetate in methanol, pH 6.7, and was pumped at a rate of 8 μL/min. (A) TIC from m/z 54 to 400, (B) EICP for m/z 145, succinylcholine (doubly charged), (C) EICP for m/z 223, neostigmine, (D) EICP for m/z 332, ipratropium, (E), EICP for m/z 242, tetrabutylammonium cation. (Reprinted with permission from John Wiley & Sons)

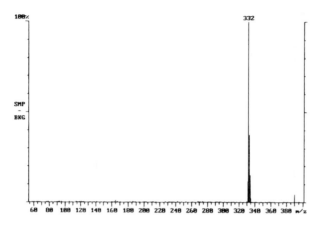

FIGURE 6.5
Background-subtracted ion spray CE/MS mass spectrum for ipratropium taken from scan #295 in Fig. 6.4.

and was helpful to maintain a stable spray because the effluent originating from the CE column is much less than 1 μL/min. Panel A of Fig. 6.4 shows the total ion electropherogram (scan range m/z 54 to 400) for 150 pg each of the four drugs. The extracted ion current profiles of the molecular ions for each compound are illustrated in panels B–E of Fig. 6.4. As was observed for the LC/MS results, the base peak for each compound is the molecular ion. A representative full-scan mass spectrum of the data obtained by CE/ITMS is shown in Fig. 6.5 for 150 pg of ipratropium taken from scan #297 (Fig. 6.4D). These full-scan CE/MS results from 150 pg per component suggest that the ion trap may provide improved detection limits over our previous approach.[24]

The phase I system described above demonstrated the feasibility of implementing an API interface on a commercial benchtop ion trap system. It also proved to be a relatively easy system to use and demonstrated sensitivity performance comparable to our established API/quadrupole-based system. However, the standard 60 L/s turbo-molecular pump was not adequate for maintaining an appropriate base vacuum in the analyzer housing where the electron multiplier resides. In addition, the ion current response remained linear to only 1.0×10^{-3} Torr of helium pressure, wherein the electron multipler would fail due to excessively high pressures. Therefore, the system was redesigned to include significantly improved high vacuum pumping speed in addition to other improvements.

B. The Phase II Differentially Pumped System

The goal of this design phase was to develop a more sensitive ion trap mass spectrometer for effective trapping and mass analysis of ions

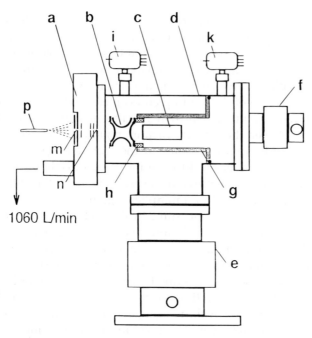

FIGURE 6.6

API/ion trap mass spectrometer: (a) API interface; (b) ion trap; (c) electron multiplier; (d) vacuum housing; (e) ion trap turbo pump; (520 L/s); (f) electron multiplier turbo pump (60 L/s); (g) baffle; (h) Teflon end-cap extending unit; (i,k) ion gauges; (m) ion sampling nozzle; (n) separating diaphragm; (p) ion spray interface. (Reprinted with permission from John Wiley & Sons)

produced by API techniques for on-line LC/MS and related techniques. The ion trap vacuum chamber was redesigned to implement differential pumping of the ion trap and electron multiplier regions, thus allowing ion trap operation at pressures up to 1×10^{-2} Torr of helium while maintaining high vacuum in the detector region. Optimizing trapping efficiency for externally produced ions as well as studying sensitivity and mass resolution in the medium vacuum pressure range was of particular interest. In this section are presented our initial results from the implementation of this new differentially-pumped API/ITMS system.[25]

A schematic diagram of the API ion trap system is shown in Fig. 6.6. It consists of the previously reported API interface (a), the differentially-pumped ion trap (b), and electron multiplier (c) vacuum regions. The API interface (a) is backed by a 1060 L/min rotary pump *via* two 1.25" i.d. vacuum hoses. The housing of the mass spectrometer (d) is specially designed to implement differential pumping between the ion trap and electron multiplier regions. The housing is a 6" vacuum "T" modified for the introduction of high voltage radiofrequency (RF) driving power, as well as for accommodating electrical feedthroughs for the electron multiplier. The ion trap vacuum pump is a 520 L/s turbo-molecular pump (e) while the electron multiplier utilizes a 60 L/s turbomolecular pump (f).

The ion trap vacuum region is separated from the electron multiplier region by a baffle (g) which is mounted inside the vacuum "T" and sealed with an O-ring. One of the ion trap end-cap electrodes was coupled to a cylindrical Teflon unit (h) which fits tightly into the baffle. The only connections between the ion trap vacuum region and the electron multiplier region are the ion exit holes in the exit end-cap electrode. Two Bayard–Alpert ion gauges (i, k) were mounted directly on the vacuum housing for independent pressure measurements in the differentially-pumped ion trap and electron multiplier regions. All ion trap pressures referred to in this work were measured by the analyzer housing ion gauge.

During operation, helium is introduced into the analyzer housing region of the ion trap through a variable leak valve, up to a pressure of 6×10^{-3} Torr. The ion gauges are installed directly on the ion trap vacuum chamber housing and it is assumed that the pressures inside the ion trap itself are comparable to those in the surrounding analyzer housing. Pressures were measured by the ion gauge controller which was calibrated in the factory, with nitrogen. The pressure in the API interface region was measured by a thermocouple manometer. The base pressures during normal API operating conditions were 0.3 Torr in the API interface region (a), 5×10^{-5} Torr in the ion trap region (b), and 4×10^{-6} Torr in the electron multiplier region (c). The differential pressure ratio in the ion trap region and electron multiplier detector region was at least 100:1.

The API/ITMS interface used in this work is the same as that which was used in the Phase I system described previously. In general, ions are sampled from atmospheric pressure through a nozzle (m) heated to 200°C and protected by a liquid shield lens. In the API vacuum region, ions are extracted from the supersonic jet by a three-electrode lens assembly and are guided through an orifice in the flat diaphragm (n, Fig. 6.6) separating the API interface from the ion trap vacuum region. In the ion trap region, the ion beam is focused toward the end-cap entrance orifice by a three-electrode lens assembly. The only differences between the present API interface and the one described previously are the increased diameters of the orifices in the ion sampling nozzle (130 μm) (m) and the separating diaphragm (600 μm) (n). System operation was controlled by the Saturn II electronics; however, the original single cone electron multiplier, suitable only for detecting positive ions, has been replaced with a high energy dynode (HED) electron multiplier. The HED of the new multiplier is maintained at $\pm$ 6 kV depending on the polarity of the ions under investigation. The HED was powered by a separate power supply of reversible polarity.

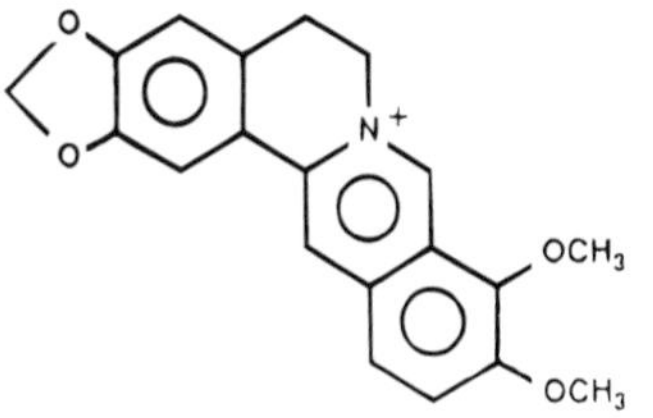

Berberine, MW 336

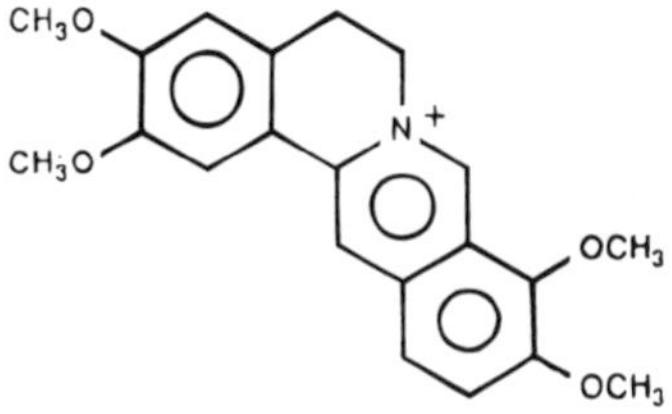

Palmatine, MW 352

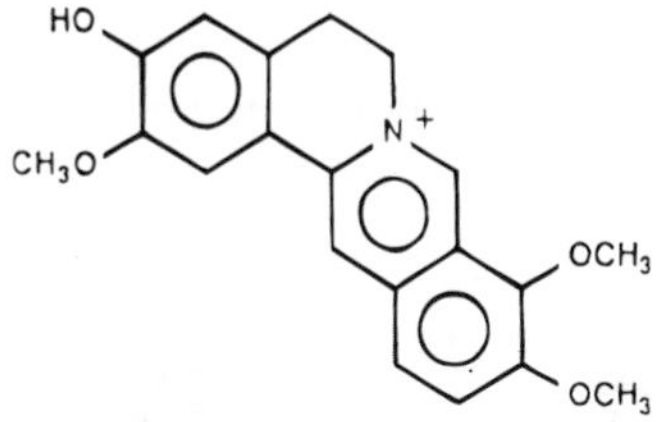

Jatrorrhizine, MW 338

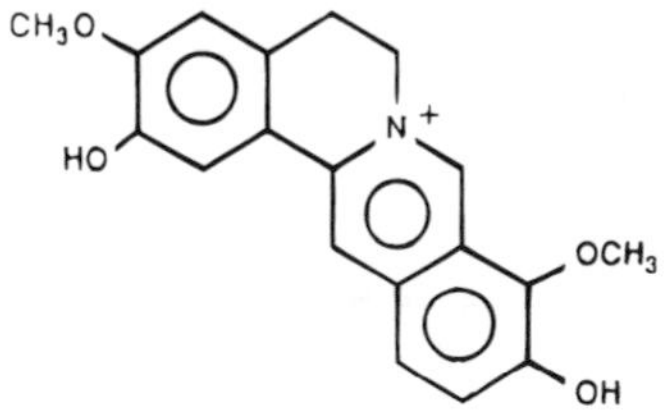

Phellodendrine, MW 342

FIGURE 6.7
Chemical structures for the isoquinoline quaternary ammonium natural product compounds studied by ion spray/ion trap LC/MS.

C. Applications of LC/MS on the Phase II API/Ion Trap Mass Spectrometer System

Among the design goals of this system were improved analytical ruggedness for routine LC/MS applications. In addition, improved full-scan MS sensitivity is needed for the increasingly challenging analytical problems encountered in today's real-world trace analysis problems. With improved high vacuum pumping along with optimized ion optics, we are hopeful that this benchtop API/ion trap concept will be applicable to many challenges that are currently problematic using existing commercial equipment. Initial LC/MS results from this system are quite encouraging; some of these results are described below.

Fig. 6.7 shows the chemical structures for four representative quaternary isoquinoline alkaloids that are of considerable therapeutic interest. These compounds have been of interest for many years and have challenged previous attempts to characterize their structures by electron ionization (EI), chemical ionization (CI), and fast atom bombardment (FAB) mass spectrometry.[26] Their highly polar nature and thermal instability preclude practical chemical characterization by these ionization techniques. However, their presence in solution as quaternary ammonium ions affords highly sensitive detection using ion spray without any exposure to elevated temperatures or other detrimental conditions. Their fused-ring

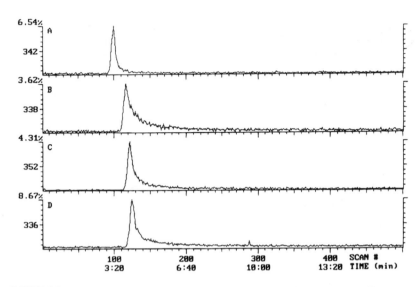

FIGURE 6.8

Ion spray/ion trap micro-LC/MS analysis of a synthetic mixture containing 1.25 ng each of four quaternary isoquinoline alkaloids. Extracted ion current profiles show: (A) m/z 342 for phellodendrine; (B) m/z 338 for jatrorrhizine; (C) m/z 352 for palmatine; and (D) m/z 336 for berberine. Chromatographic conditions were: BDS-cyanopropyl microbore column, 1 mm i.d. × 150 mm o.d. maintained at a flow of 60 mL/min of 20% 5 mM ammonium formate, pH 3.5, in acetonitrile.

structures make them quite stable to fragmentation or disruption of the aromatic ring system, but the varied substituents on the rings afford predictable fragment losses in the up-front higher pressure CID region of the API/ion trap mass spectrometer interface. Therefore, preliminary investigations were initiated to determine the relative sensitivity of the API/ion trap mass spectrometer system as well as the potential for extracting structural information from CID mass spectra of these compounds.

Fig. 6.8 shows the extracted ion current profiles for the LC/MS analysis of a synthetic mixture containing 1.25 ng of each of the four quaternary ammonium isoquinoline alkaloids shown in Fig. 6.7. The HPLC separation was carried out isocratically using a 1 mm i.d. × 150 mm BDS-cyanopropyl column with a flow of 20% 5 mM ammonium formate solution in acetonitrile (pH 3.5) maintained at 60 mL/min. The total effluent was directed unsplit to the ion spray LC/MS interface which was positioned on-axis and 1 cm away from the ion sampling orifice. The ion trap was scanned from m/z 50 to 500 with an accumulation time of 250 ms and up-front CID voltage difference of 40 V. The ion current profiles shown in Fig. 6.8 reveal the stability of the signal and the signal/noise ratio obtained from the analysis of a sample containing 1.25 ng per component. This full-scan ion current response is comparable to or better than those typically obtained from modern GC/MS systems, but this API/ion trap mass spectrometer system full-scan ion current response is for analytes that are not amenable to GC/MS characterization.

One of the problems associated with API techniques is elevated chemical noise originating from the cluster ion formation between the HPLC solvent and buffers at atmospheric pressure.[3,14] All ions formed at atmospheric pressure tend to form clusters and multiple adducts with ambient water and solvent gases, so effective means must be implemented for de-clustering these unwanted interferences. The high chemical noise problem is especially bothersome in the lower mass region extending from m/z 30 to 200, and sometimes higher. This problem is caused by multiple interactions between ions and neutral molecules due to the very short mean-free paths that exist between these species at atmospheric pressure. When ion/molecule reactions take place, the major products are of higher molecular weight that can extend considerably into the higher mass region where the analytes of interest are expected. This problem is particularly bothersome in the LC/MS determination of small molecules whose molecular weights are less than, for example, 500 Da.

In order to deal with the problem of solvent-mediated, lower mass chemical noise, effective de-clustering mechanisms must be invoked. The previous ploys to accomplish this task include, but are not limited to, the nitrogen curtain gas of the SCIEX system,[3] the bath gas of the electrospray system developed by Fenn and Whitehouse,[27] the heated capillary transfer device described by Chait et al.,[28] and the electrospray/ion trap system described by McLuckey et al.[9] Each of these devices deals with the de-clustering problems to varying degrees, but none of them completely eliminates the low-mass solvent-mediated clusters. Ultimately, the optimum sensitivity of any API LC/MS system will be determined by the detection efficiency of the low ion current signal from the analyte in the absence of the baseline chemical noise.

The API/ion trap MS interface shown in Figs. 6.1 and 6.6 must deal also with the low-mass chemical noise problem. Although it does not use a bath gas or curtain gas, it does use some heat in the liquid shield (Fig. 6.1(d)) and it does expose the solvent-mediated clusters to multiple collisions across an electrical potential in the de-clustering region (Fig. 6.1, N–SD). In addition, when the ions are stored in the trap, their lifetimes are extended such that they are exposed to a "heating" effect that elevates their temperature to some extent above room temperature.[29,30] The effect of this heating is to de-cluster the adducts that have formed during the formation of ions at atmospheric pressure.

An indication is given in Fig. 6.9 of the operational effectiveness of this API/ion trap interface in the on-line LC/MS mode for the analysis of the synthetic mixture described for Fig. 6.8. Fig. 6.9 shows a background

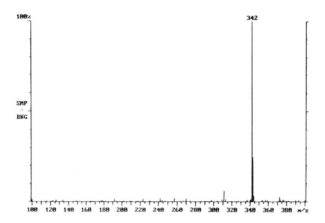

FIGURE 6.9
Ion spray/ion trap LC/MS mass spectrum for 1.25 ng of phellodendrine (scan #93 shown in Fig. 6.8A) in the positive ion mode of detection.

subtracted, full-scan (m/z 50 to 500) mass spectrum for phellodendrine (1.25 ng on-column) obtained from the 3.2 min retention-time component observed in Fig. 6.8. The unsubtracted mass spectrum for this component (not shown here) was relatively free of low-mass chemical noise with the exception of an abundant m/z 102 ion that appeared to be due to an unknown impurity in the HPLC eluent. Although this impurity could be removed from the spectrum by background subtraction, as shown in Fig. 6.9, efforts to remove it from the eluent were unsuccessful.

This problem highlights the importance of using very clean HPLC solvents and buffers that are free of trace impurities. Often, this type of problem arises because some solvents contain non-UV absorbing stabilizers and antioxidants that do not interfere with conventional LC/UV applications, but do interfere with LC/MS studies. We recommend using HPLC solvents and analytical reagent grade buffer additives that are known to be free of unwanted materials. The full-scan mass spectrum shown in Fig. 6.9 is, however, relatively free of other interfering ions, thus making the determination of the molecular weight for this compound straightforward.

The full-scan ion spray mass spectrum shown in Fig. 6.10 was obtained by increasing the up-front CID voltage potential to 68 V to induce fragmentation of the molecular ion of berberine. Thus, the m/z 336 molecular ion appears to fragment by losing several substituents attached to the aromatic ring system. The CID mass spectra obtained in this manner from the ion trap system agree well with both up-front CID and MS/MS mass spectra of this compound obtained from our commercial SCIEX API tandem triple-stage quadrupole mass spectrometer (unpublished results). Reference to previously-reported EI as well as FAB mass spectra for this compound[26] show very similar fragmentation behavior for this compound, but several

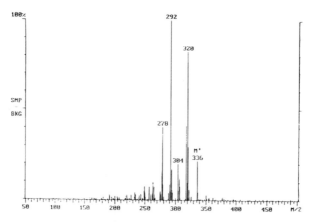

FIGURE 6.10

Ion spray/ion trap CID mass spectrum of 5.0 ng berberine. The fragmentation observed is due to the application of a 68 V potential difference in the high pressure nozzle-skimmer region (Fig. 6.1, N-SD) of the API/ion trap interface.

hundred micrograms of material were required in order to obtain a mass spectrum. Using the ion spray/API/ion trap MS system described here, low nanogram levels of this and related compounds readily afford interpretable mass spectra from this interesting class of compounds.

Fig. 6.11 shows the extracted ion current profiles for the LC/MS analysis of a synthetic mixture containing 1.25 ng each of lysergic acid diethylamide (LSD), its N-demethylated metabolite (nor-LSD), as well as impurities such as iso-LSD and iso-nor-LSD. The HPLC separation was carried out isocratically using a 1 mm i.d. × 150 mm C-8 column with a flow of 20% 5 mM ammonium formate solution in acetonitrile (pH 3.5) maintained at 40 µL/min. The total effluent was directed unsplit to the ion spray LC/MS interface which was positioned on-axis and 1 cm away from the ion sampling orifice. The ion trap was scanned in this case from m/z 250 to 350 to avoid scanning through lower mass impurity components present in the HPLC eluent. The accumulation time was 250 ms and the up-front CID voltage difference was 50 V.

The ion current profiles shown in Fig. 6.11 reveal the stability of the signal and the signal/noise ratio obtained from the analysis of a sample containing 1.25 ng per component. This full-scan ion current response is comparable to or better than that typically obtained from modern GC/MS systems, but this system does so without the need for chemical derivatization and elevated temperatures that can be conducive to thermal degradation when done by capillary GC/MS.[31] It is important, however, to remove the interfering components present in the HPLC eluent so that the full-scan mass acquisition mode may be used in these experiments.

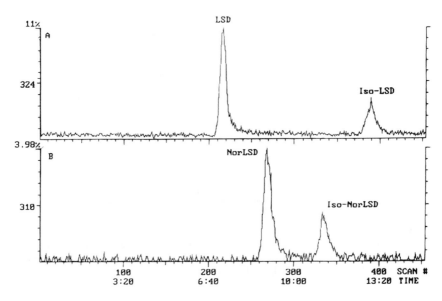

FIGURE 6.11

Ion spray/ion trap analysis of a synthetic mixture containing 1.25 ng each of LSD and three potential metabolites. Extracted ion current profiles show: (A) m/z 324 for LSD and iso-LSD; (B) m/z 310 for nor-LSD and iso-Nor-LSD. Chromatographic conditions were: C-8 microbore column, 1 mm i.d. × 150 mm o.d. maintained at a flow of 40 µL/min of 20% 5 mM ammonium formate, pH 3.5, in acetonitrile.

Fig. 6.12 shows the LC/MS ion current profile obtained from the differentially-pumped API/ion trap MS system following the on-column injection of 250 pg of LSD using the same experimental conditions as described above for Fig. 6.11. Since we have reported previously preliminary efforts to develop LC/MS/MS trace analysis techniques for the determination of LSD in human urine,[32] we have initiated efforts to evaluate the analytical potential of the API/ion trap MS system for the determination of this drug and its metabolites in urine. For forensic purposes, it is desirable to have the maximum specificity possible for the confirmation of a positive drug finding. Therefore, it is often necessary to use full-scan mass spectrometric acquisition procedures to obtain structurally important fragment ions for compounds such as LSD. This goal is compromised in the single MS mode, however, when interference ions are present in the mass spectrum. We are exploring various approaches to remove this problem.

Fig. 6.12 shows a limited mass scan range (m/z 300 to 350) for the LC/MS analysis of a synthetic sample containing 250 pg LSD. The abbreviated scan acquisition range was necessary here in order to avoid scanning through interfering chemical constituents present in the HPLC eluent. Clearly, improvements must be made in the availability of these solvents

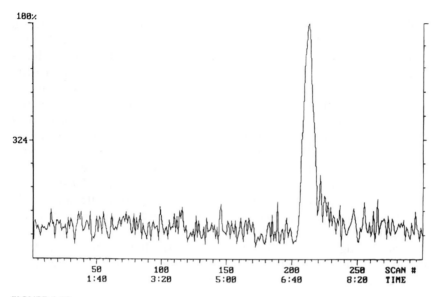

FIGURE 6.12

Ion spray/ion trap micro-LC/MS analysis of a synthetic standard containing 250 pg of LSD using the same LC/MS conditions as described for Fig. 6.11. An abbreviated scan acquisition range of m/z 300 to 350 was used in this experiment to reduce the chemical noise resulting from scanning through interfering ions in the lower mass range.

and the overall cleanliness of the system. We continue to explore possibilities for improvements here.

As a final example, Fig. 6.13 shows a full-scan mass spectrum of cholesterol sulfate obtained in the negative ion detection mode *via* infusion of a solution of the analyte at a concentration of 5 ng/mL. The delivery of the sample solution was maintained at 20 μL/min with an accumulation time of 250 ms and an up-front CID voltage of 40 V. These mild ionization

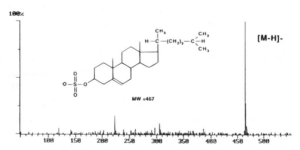

FIGURE 6.13

Ion spray/ion trap mass spectrum of cholesterol sulfate obtained in the negative ion mode of detection. The data were obtained from the accumulation of one scan of an infused solution containing 5 ng/mL of cholesterol sulfate dissolved in 50% methanol/5 mM ammonium acetate delivered at a flow of 20 μL/min.

and ion sampling conditions produce the mass spectrum shown in Fig. 6.13 with an abundant deprotonated molecule ion at m/z 467 with little or no fragmentation. This class of compounds is important for the detection of steroid abuse in athletes since sulfate conjugation of the parent drug is a major route of excretion.[33] Sulfoconjugates have been characteristically difficult to determine in biological samples due to their extreme polarity and thermal instability. Ion spray LC/MS ion trap characterization of these potentially important metabolites could preclude time-consuming hydrolysis, sample cleanup, and derivatization procedures that are typically required for their identification by GC/MS techniques. The mass spectrum shown in Fig. 6.13, as well as related up-front CID studies of this class of compounds (unpublished results), appear to offer a uniquely useful approach to the characterization of these negatively-charged sulfate conjugates. The glucuronide conjugates common to steroids and other classes of drugs may be determined similarly, but we have found the positive ion mode of detection to be the preferred detection mode for these compounds (unpublished results).

III. SUMMARY AND CONCLUSIONS

The seminal work of McLuckey *et al.*[9] that demonstrated the feasibility of implementing electrospray ionization on the ion trap was an exciting glimpse of the potential for electrospray in combination with an ion trap mass spectrometer. Their work, as well as that of others[34] and our own,[21,25] convinces us that the development of API techniques in conjunction with the ion trap mass spectrometer is of great importance for the future of analytical chemistry. It is intuitively obvious that just as GC/MS has matured and become an essential component of many laboratories, LC/MS and other condensed phase separation/MS techniques (IC/MS, CE/MS, SFC/MS, etc.) will also become essential, routine tools for analytical chemists.

The benchtop GC/MS concept, that was first introduced by Hewlett-Packard in the late seventies, has provided access to mass spectrometry for many gas chromatographers who accept the mass spectrometer as a "detector." There is no doubt in our mind that general acceptance of LC/MS is possible and, in fact, is likely. If one accepts the postulate that GC techniques are amenable to perhaps 20% of the known organic compounds and that HPLC techniques are amenable to most of the remaining organic compounds, then it appears obvious why HPLC equipment has been implemented in multiplets in laboratories worldwide. If one accepts also that mass spectrometry has proven to be a tremendous means of detection for GC, then it is also likely that coupling mass spectrometry to HPLC

(LC/MS) to give an analytically-rugged combination should have far-reaching appeal and acceptance.

Of course the "appeal" will survive only so long as the equipment is reliable, easy to use, trouble-free, and relatively inexpensive. Skeptical chromatographers have come to learn that the first three factors are nonissues for the current benchtop GC/MS systems. The economic factor is less of a concern now because one usually gets what one pays for, and it is now agreed that a modern commercial benchtop GC/MS system is a "good buy" given the quality, quantity, and value of the information produced.

Perhaps the key element for growing acceptance of a benchtop LC/MS system can be provided by a benchtop ion trap MS system similar to the one described in this report. The future needs of many analytical chemists may be met by an ion trap-based LC/MS system. The modern, next generation ion trap of the future should be able to incorporate most of the needs of the practical user as well as the researcher. Its inherently high sensitivity would appeal to the trace analysis person while its apparent operational simplicity would appeal to the routine operator. The ultimate in sensitivity might be provided by the equivalent of selected ion monitoring that has been reported recently.[35] More advanced options that have been reported recently include extended mass range, increased mass resolution, and MS^n which will provide structural analysis capability as well as the possibility for increased sensitivity by reducing chemical noise.[36]

For the benchtop ion trap LC/MS system of the future to be truly successful, some existing problems must be solved. Certainly, there is a need for improved de-clustering of ion/molecule clusters, and removal of interfering solvent impurities in condensed-phase mobile phases must be achieved. The system must be automated in addition to incorporating most of the developments mentioned above. The LC/MS system of the future most certainly must include implementation of an API interface on an ion trap system such as has been described in this report, yet such an instrument is not commercially available as this report is being written. The ion trap hardware and software used so far is out-of-date, and a new generation of ion trap MS system is needed that incorporates many of the improvements that have been reported or learned about ion traps since their commercial introduction.

When all the available knowledge is incorporated into the next generation of ion trap and coupled to a simple, reliable API interface in a benchtop format, the ultimate LC/MS system will have arrived. When this has become commercially available for less than $100,000 per system, the analytical and financial opportunities for this system will foreshadow the present benchtop GC/MS market.

ACKNOWLEDGMENTS

We thank Dr. W.N. Wu for providing samples of the quaternary isoquinoline drugs, and Ms. Karen Bean for providing the cholesterol sulfate sample. The technical expertise and able collaboration of Drs. T. Wachs and G. Hopfgartner are gratefully appreciated. We also thank the Eastman Kodak Company for substantial financial support of this project as well as the Xerox Corporation, Balzers High Vacuum Products, and Galileo Electro-Optics Corporation for supplying key components of the ion trap system described here. H.K.L. and J.D.H. thank the Wyeth-Aeyrst Research Inc. for financial support of a sojourn at Cornell University.

REFERENCES

1. Henion, J. D. *Anal. Chem.* 1978, *50*, 1687–1703.
2. Henion, J. D.; Thomson, B. A.; Dawson, P. H. *Anal. Chem.* 1982, *54*, 451–456.
3. Huang, E. D; Wachs, T.; Conboy, J. J; Henion, J. D. *Anal. Chem.* 1990, *62*, 713A–725A.
4. Allen, M. H.; Lewis, I. A. S. *Rapid Commun. Mass Spectrom.* 1989, *3*, 255–258.
5. Henry, K. D.; Williams, E. R.; Wang, B. H.; McLafferty, F. W.; Shabanowitz, J.; Hunt, D. F. *Proc. Natl. Acad. Sci.* 1989, *86*, 9075–9078.
6. Busch, K. L.; Glish, G. L.; McLuckey, S. A. *Mass Spectrometry/Mass Spectrometry Techniques and Applications of Tandem Mass Spectrometry*, VCH: New York, 1988, Chapter 2.3.
7. Cody, R, B.; Kinsinger, J. A.; Ghaderi, S.; Amster, I. J.; McLafferty, F. W.; Brown, C. E. *Anal. Chim. Acta.* 1985, *178*, 43–66.
8. Busch, K. L.; Glish, G. L.; McLuckey, S. A. *Mass Spectrometry/Mass Spectrometry Techniques and Applications of Tandem Mass Spectrometry*, VCH: New York, 1988, Chapter 2.2.2.
9. Van Berkel, G. J.; Glish, G. L.; McLuckey, S. A. *Anal. Chem.* 1990, *62*, 1284–1295.
10. McLuckey, S. A.; Glish, G. L.; Asano, K. G.; Grant, B. C. *Anal. Chem.* 1988, *60*, 2220–2227.
11. McLuckey, S. A.; Van Berkel, G. J.; Glish, G. L.; Huang, E. C.; Henion, J. D. *Anal. Chem.* 1991, *63*, 375–383.
12. March, R. E. *Org. Mass Spectrom.* 1991, *26*, 627–632.
13. Todd, J. F. J. *Mass Spectrom. Rev.* 1991, *10*, 3–52.
14. French, J. B.; Reid, N. M.; Buckley, J. A. *Method and Apparatus for Focussing and Declustering Trace Ions.* U.S. Patent No. 4,121,099, Oct. 17, 1978.
15. Lichtenberg, A. J. *Phase-Space Dynamics of Particles.* John Wiley S. Sons, NY, 1969. Chapter 5.5.
16. Louris, J. N.; Amy, J. W.; Ridley T. Y.; Cooks, R. G. *Int. J. Mass Spectrom. Ion Processes.* 1989, *88*, 97.
17. Dawson, P. H. (Ed), *Quadrupole Mass Spectrometry and Its Applications.* Elsevier, Amsterdam, 1976, Chapter 1B, 6E.
18. March, R. E.; Hughes, R. J. *Quadrupole Storage Mass Spectrometry.* Wiley Interscience: New York, 1989.
19. Asano, K. G.; Glish, G. L.; McLuckey, S. A. *Proc. 38th ASMS Conf. Mass Spectrometry and Allied Topics.* San Francisco, 1988, p. 636–637.

20. Goeringer, D. E.; Whitten W. B.; Ramsey, M. J.; McLuckey S. A.; Glish G. L. *Anal. Chem.* 1992, *64*, 1434–1439.
21. Mordehai, A. V.; Hopfgartner, G.; Huggins, T. G.; Henion, J. D. *Rapid Commun. Mass Spectrom.* 1992, *6*, 508–516.
22. Hopfgartner, G.; Wachs, T.; Bean, K.; Henion, J. D. *Anal. Chem.* 1993, *65*, 439–446.
23. Bruins, A. P.; Covey, T. R.; Henion, J. D. *Anal. Chem.* 1987, *59*, 2642–2646.
24. Johansson, M.; Pavelka, R.; Henion, J. D. *J. Chromatogr.* 1991, *559*, 515–528.
25. Mordehai, A. V.; Henion, J. D. *Rapid Commun. Mass Spectrom.* 1993, *7*, 205, 209.
26. Masucci, J. A.; Caldwell, G. W.; Wu, W. N.; Isensee, R. K.; Slayback, J. R. B. *Proc. 35th ASMS Conf. Mass Spectrometry and Allied Topics.* Denver, May 24–29, 1987, p. 514–515.
27. Fenn, J. B.; Mann, M.; Meng, C. K.; Wong, S. J.; Whitehouse, C. M. *Science.* 1989, *246*, 64–71.
28. Chowdhury, S. K.; Katta, V.; Chait, B. T. *Rapid Commun. Mass Spectrom.* 1990, *4*, 81–87.
29. Basic, C.; Eyler, J. R.; Yost R. A.; *J. Am. Soc. Mass Spectrom.* 1992, *3*, 716–726.
30. Nourse, B. D.; Kenttamaa, H. I. *J. Phys. Chem.* 1990, *94*, 5809–5812.
31. Nelson, C. C.; Foltz, R. L. *Anal. Chem.* 1992, *64*, 1578–1585.
32. Henion, J. D.; Wachs, T.; Foltz, R. L. *Proc. 39th ASMS Conf. Mass Spectrometry and Allied Topics,* Nashville, May 19–24, 1991, p. 1653–1654.
33. Weidolf, L. O. G.; Lee, E. D.; Henion, J. D. *Biomed. Environ. Mass Spectrom.* 1988, *15*, 283–290.
34. Lin, H. Y.; Vooyksner, R. D. *Proc. 40th ASMS Conf. Mass Spectrometry and Allied Topics.* Washington, D.C. May 31–June 5, 1992, p. 703–704.
35. Buttrill, S. E.; Shaffer, B.; Karnicky, J.; Arnold, J. T. *Proc. 40th ASMS Conf. Mass Spectrometry and Allied Topics.* Washington, D.C. May 31–June 5, 1992, p. 1015–1016.
36. March R. E. *Int. J. Mass Spectrom. Ion Processes.* 1992, *118/119*, 71–135.

Chapter 7

CHEMICAL IONIZATION IN ION TRAP MASS SPECTROMETRY

Colin S. Creaser

CONTENTS

I. INTRODUCTION

The ion trap has been likened to a test tube in which ion/molecule reactions, such as those associated with chemical ionization (CI), can be performed in the absence of solvation effects. The development of CI procedures, particularly those which have high selectivity for a chosen analyte is potentially, therefore, one of the most promising areas of ion trap development.

II. PRINCIPLES OF CHEMICAL IONIZATION

In chemical ionization, a reaction occurs between a reagent ion and an analyte molecule which leads to the formation of a charged molecular species. The reagent ion is derived typically from a reagent gas, such as methane, ammonia or isobutane, in a multi-stage process involving primary electron ionization followed by reagent ion formation. This multi-stage process is illustrated for the formation of the NH_4^+ reagent ion from ammonia *via* the following reactions:

1. Primary ion formation

$$NH_3 + e^- \rightarrow NH_3^{+\bullet} + 2e^- \tag{7.1}$$

2. Reagent ion formation

$$NH_3^{+\bullet} + NH_3 \rightarrow NH_4^+ + NH_2^\bullet \tag{7.2}$$

Other common reagent ions are CH_5^+ from methane, and $C_3H_7^+$ and $C_4H_9^+$ from isobutane.

The introduction of an analyte into the trap results in reactions between the reagent ions and the sample molecules. The most important of these reactions are proton transfer from the reagent ion to the sample molecule and electrophilic addition of the reagent ion. These reactions are indicated below for the NH_4^+ reagent ion:

1. Proton transfer

$$NH_4^+ + M \rightarrow MH^+ + NH_3 \tag{7.3}$$

where M is the sample molecule and MH^+ is the protonated molecular ion, and

2. Electrophilic addition

$$NH_4^+ + M \rightarrow [M + NH_4]^+ \tag{7.4}$$

where $[M + NH_4]^+$ is an adduct ion.

For proton transfer to occur, the proton affinity (PA) of the sample molecule must be greater than that of the reagent gas, that is, the reaction must be exothermic. The energy released by the reaction appears as internal energy in the product ion, but since the exothermicity is usually

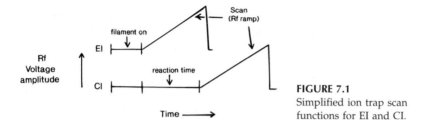

FIGURE 7.1
Simplified ion trap scan
functions for EI and CI.

small, CI spectra are characterized by strong protonated molecular ions, or adduct ions, and few fragment ions.

The primary ions formed during the initial electron ionization stage may undergo charge exchange reactions with sample molecules, when the ionization energy of the sample molecule is less than the recombination energy of the reagent ion, for example:

$$NH_3^{+\cdot} + M \rightarrow M^{+\cdot} + NH_3 \qquad (7.5)$$

Negative chemical ionization may also be carried out by production of reagent ions such as NH_2^- and OH^-, from ammonia and water, respectively. These ions can react with sample molecules by proton abstraction and nucleophilic addition, in a manner analagous to positive ion CI.

III. CHEMICAL IONIZATION IN THE ION TRAP

The chemical processes leading to positive ion CI in the ion trap are the same as in the source of a sector mass spectrometer or quadrupole mass filter, but the experimental procedures are quite distinct for the two types of instrument. The difference lies in the pressure regimes and reaction times in the ion trap and in conventional CI sources, since the formation of the products of chemical ionization is dependent upon the product of reagent gas pressure and reaction time. Thus, in a conventional CI source, a high reagent gas pressure of 0.1 to 1 Torr is required to produce CI product ions, because the ion residence time in the source is very short, typically a few microseconds. However, CI is a low pressure process in the ion trap, where the typical reagent gas pressure is 10^{-5} Torr and the helium buffer gas pressure is 1 mTorr; therefore, the same extent of product formation requires much longer reaction times.[1-4]

Simplified ion trap scan functions for electron ionization (EI) and CI are shown in Fig. 7.1.[3] In both experiments, electrons are gated into the trap for periods up to a few milliseconds to ionize the sample or, in the case of CI, sample and reagent gas molecules present in the trap. In EI, this primary ion formation step is followed immediately by a radio frequency (RF) mass-selective instability scan of the resulting mass spectrum, whereas in CI, the primary ion formation step is followed by a reaction period in which primary ions react to form reagent ions and reagent ions react with the sample to produce quasi-molecular ions or protonated molecular ions. Clearly, if the low-mass cut-off during ionization is set above the mass of the reagent ion, then reagent ion/sample molecule reactions will not occur because the primary reagent ions are ejected immediately from the trap as they are formed. Variation of the low-mass cut-off has been used as the basis of an alternating EI/CI procedure by rapid switching between scan routines with and without reagent ion storage.

The temporal changes in reagent and sample product ion intensities are shown in Fig. 7.2 for the formation of NH_4^+ reagent ions from ammonia (Eq. (7.2)) and their subsequent reaction with the sample molecule M, as in Eq. (7.3); the data shown in Fig. 7.2 are for the case when M is pyrrole.[5] The pyrrole $[M + H]^+$ ion intensity rises rapidly with increasing time before leveling off as the reaction proceeds to completion. The reaction time required for CI in the ion trap depends on the nature of the reagent and sample molecules, and on the trap pressure but is, typically, a few tens to hundreds of milliseconds. Because of the differing nature of EI and CI experiments in the ion trap, discussions of relative sensitivity are complicated.[3] Experience indicates that it is possible to obtain CI spectra on 50 or less pg of standards by ion trap CI. The EI sensitivity in terms of practical signal-to-background noise observed in gas chromatography/ mass spectrometry, may be better by a factor of two to three, or more in favorable cases.

In addition to the differences in pressure regime and reaction time-scale for the CI process in the ion trap and in a conventional source, there is also a difference between the reagent ion kinetic energies. The kinetic energy of an ion in a conventional CI source is usually less than 1 eV, and is determined largely by the source temperature. However, in the ion trap, ions may gain additional kinetic energy from the applied RF voltage. The effect of enhanced ion kinetic energy in ion trap CI may be to impart additional internal energy to the sample molecule during the ionization process, leading to increased fragmentation in the ion trap CI mass spectrum.[6,7] The extent of this kinetic energy transfer depends on the low-mass RF cut-off level, or "reaction level," selected for the CI scan. When this level is close to the reagent ion mass, then the working point of the ion is close to the boundary of the stability diagram, resulting in a higher kinetic energy collision and greater fragmentation.

Adduct ions resulting from electrophilic (or nucleophilic) addition reactions have been reported in ion trap CI mass spectra,[8,9] but are less commonly observed than in conventional CI mass spectra. The reason for this lies in the pressures used in the ion trap experiment. In an ion

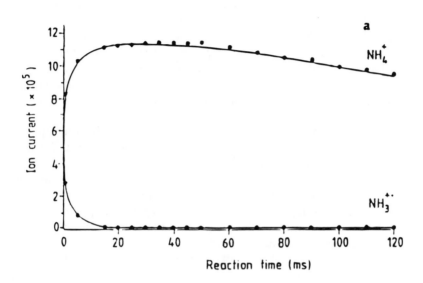

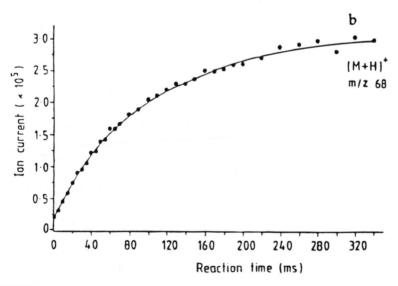

FIGURE 7.2

Variation of ion current with time during (a) ammonium reagent ion formation, (b) pyrrole $[M + H]^+$ ion formation by NH_4^+. (From Ref. 5)

trap operated under normal conditions, an analyte ion will undergo many collisions with helium, but only a small number with the neutral reagent gas,[8] resulting in a less efficient collisional stabilization of loosely-bound adducts than is achieved as a result of collisions with the reagent gas in a conventional source. In consequence, ion trap CI mass spectra are characterized by intense quasi-molecular ions with few adducts. Fig. 7.3 compares the quadrupole and ion trap CI mass spectra of perfume components, separated by gas chromatography (GC), showing the effect on fragment and adduct ion formation.[10]

The readiness with which CI reactions can occur in the ion trap often leads to enhanced [M + H]$^+$ ion intensities in EI mass spectra, a phenomenon termed "self-CI."[11,12] Proton transfer arises in these cases from the reaction of a sample fragment ion and a neutral sample molecule when there is a high analyte concentration in the trap. In the commercial ion trap detector, the effects of self-CI are minimized by adjustment of the electron ionization time which controls the number of ions in the trap. However, reactions may also be induced by matrix species (matrix-CI) arising from co-eluting components of complex samples following chromatographic separation.[13]

The concentration of sample ions in the trap is determined by the length of the reaction time in which reagent and sample ions are formed following primary ionization of the reagent gas. Variation of this time allows the population of product ions to be controlled and space-charge effects minimized. The consequences of trap overloading may be avoided by a short pre-scan which is used to measure the sample concentration in the trap and to determine the reaction time required to yield an optimum population of ions prior to the main scan.[9,10,14] This pre-scanning technique allows a wide dynamic range of analyte concentrations to be detected in a single chromatographic run. For example, the GC analysis of a urine sample containing morphine at a concentration of 4000 ng/mL and 6-monoacetylmorphine at a concentration of 10 ng/mL has been shown to give high quality positive ion CI mass spectra of both components (Fig. 7.4).[14]

Negative CI has been performed on research ion trap mass spectrometers, but not on routine instruments. Reactant ions such as OH$^-$ may be produced in the ion trap from water, by dissociative electron capture, and these ions will react with sample molecules by proton abstraction to yield characteristic [M–H]$^-$ ions with little or no fragmentation.[15,16] However, even with long ionization times (20 to 100 ms), the number of negative reagent ions produced during the ionization step is, typically, 10^3 lower than the number of positive ions, and negative ions may be quenched further by reaction with positive ions stored in the ion trap. This loss process in addition to the low cross-section for formation of negative ions, leads to very poor sensitivities in negative ion CI unless steps are taken

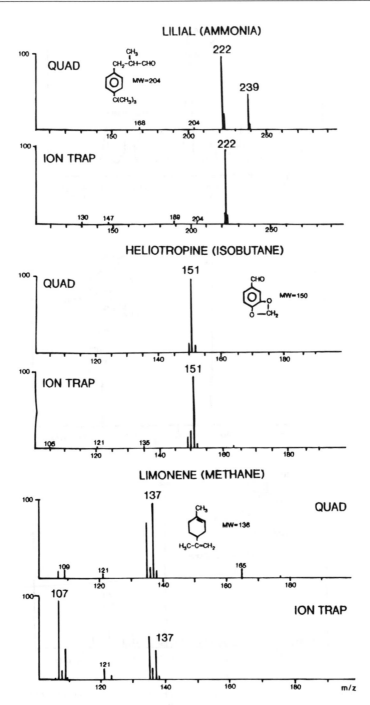

FIGURE 7.3
Comparison of quadrupole and ITD CI spectra. (From Ref. 10)

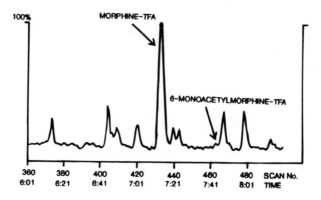

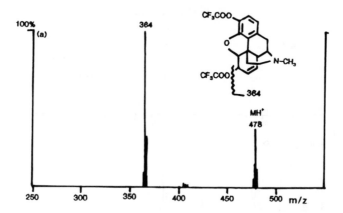

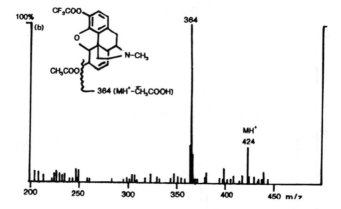

FIGURE 7.4

GC/Ion trap analysis of the trifluoroacetyl derivatives of (a) morphine and (b) 6-monoace-tylmorphine from urine sample containing 4000 ng/mL of morphine and 10 ng/mL 6-monoacetylmorphine. (From Ref. 7)

to enhance the population of negative ions.[7] The ion trap is not amenable to resonance electron capture negative ion CI, unless an external ion source is used, a technique used widely in high pressure CI, because thermal electrons produced by primary ionization are not retained in the trap.

IV. SELECTIVE CHEMICAL IONIZATION PROCEDURES

The performance of CI in the ion trap may be enhanced significantly, relative both to the simple CI procedures described above and a conventional CI source, by the adoption of selective procedures. Such procedures fall into two categories: methods which remove residual EI components from CI spectra, and those in which selectivity is improved by mass-selection of reagent ions.

The presence of EI spectral characteristics in CI ion trap mass spectra may arise from the ionization of matrix and sample molecules present in the ion trap during the primary ionization stage. These ions will remain in the trap during the subsequent chemical ionization reaction period and will be detected when all ions are ejected during the mass-instability scan. A convenient way to exclude these EI-type ions is to modify the scan function so as to introduce an isolation step after reagent ion formation. In the simplest case, this isolation step acts as a low-mass pass filter which retains the reagent ion and ejects all high mass ions. Alternatively, RF and DC voltages may be applied to the ring electrode which isolate only the reagent ion and reject all others.[17] The effect of reagent ion isolation is illustrated in Fig. 7.5 for a sample consisting of a mixture of pyrrole and toluene.[5] The normal ammonia CI spectrum (Fig. 7.5, middle) shows ions at m/z 67 ($C_4H_5N^{+\cdot}$) and m/z 91 ($C_7H_7^+$) in addition to the $[M + H]^+$ ion of pyrrole [m/z 68]. These ions are observed in the EI spectrum of the pyrrole/toluene sample (Fig. 7.5, top). Modifying the routine to isolate the NH_4^+ reagent ion and to eject other ions from the trap prior to the ionization of sample molecules, results in the spectrum shown in Fig. 7.5, bottom, in which EI-type ions are absent. This improved CI scan routine, therefore, yields ion trap CI spectra which can be compared directly with CI spectra produced on conventional sector and quadrupole instruments.[18]

A natural extension of this approach is the mass-selection of other reactant ions for selective CI processes using an ion isolation procedure.[1,3,8,19] The reactant ion need not be derived from a conventional CI reagent gas, but may be any ion generated in the trap by any means. Variation of the reactant ion can radically alter the CI behavior of a sample molecule. For example, the selective CI spectra of phenatole generated using reactant ions derived from isobutane reagent gas are shown in Fig. 7.6.[17] The selectivity of these ions towards the formation of $M^{+\cdot}$ (m/z 122), [M + H]$^+$, or the adduct ions at 13 and 41 u above the molecular weight is evident.

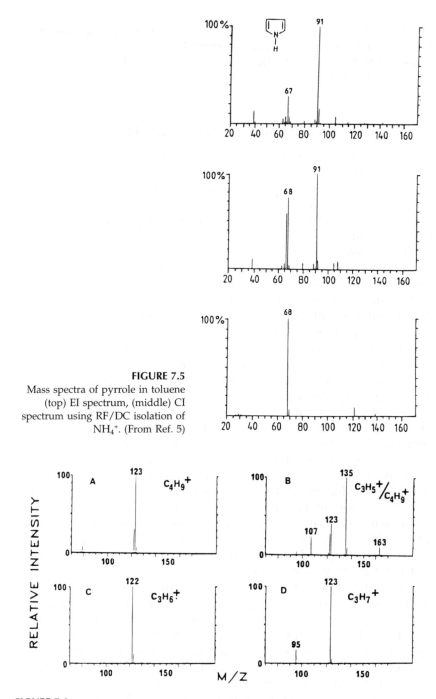

FIGURE 7.5
Mass spectra of pyrrole in toluene (top) EI spectrum, (middle) CI spectrum using RF/DC isolation of NH_4^+. (From Ref. 5)

FIGURE 7.6
CI mass spectra of phenatole using various reagent ions. (From Ref. 25)

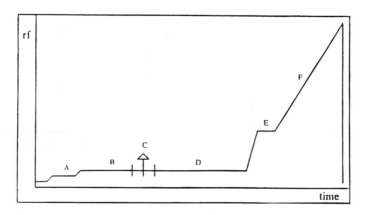

FIGURE 7.7
ITMS scan routine for the selective ammonia CI of nitrogen heterocycles (not to scale): A = ionization (0.2 ms); B = formation of reagent ions (20 ms); C = isolation of reagent ions with DC applied voltage (1 ms); D = formation of sample ions (180 ms); E = selection of low-mass cut-off for start of acquisition (50 amu); F = analytical scan (50 to 300 amu). (From Ref. 5)

Several mass-selected ion/molecule reagent ions for charge exchange, proton transfer, and adduct formation have been reported,[20] and the scope for future development in this area is clearly very great.

An example of the application of a selective CI procedure is the ionization of nitrogen heterocycles in a petroleum fraction.[5] The scan routine used for the analysis, shown in Fig. 7.7, was designed with the objective of creating a large population of NH_4^+ reagent ions while minimizing the abundance of $NH_3^{+\bullet}$ and EI products. This objective was achieved by an ion isolation step (C) following reagent ion formation (steps A and B) and prior to sample ionization (step D). In addition to the removal of the unwanted EI ions, the effect of the routine was to promote proton transfer reactions of the NH_4^+ ion with nitrogen heterocycles and to suppress competing charge exchange reactions of $NH_3^{+\bullet}$ with aromatic components of the sample. The advantages of this ion trap procedure are demonstrated by the selective ionization of the four methylcarbazole isomers following GC separation. The single ion chromatogram for the [M + H]$^+$ ion (m/z 182) and the selective proton transfer spectrum of the 3-methylcarbazole peak (I) are shown in Fig. 7.8(a) and (b). These data show significantly higher selectivity than spectra obtained by using either the standard ITD-type CI scan routine with-

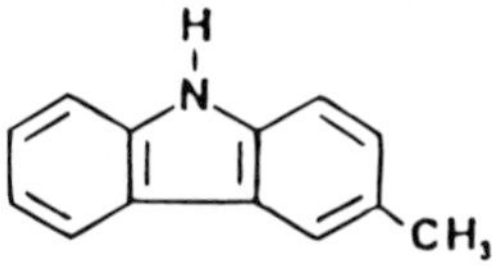

SCHEME 7.1

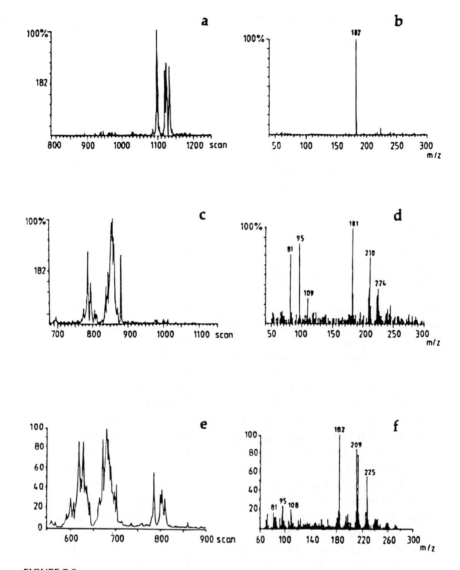

FIGURE 7.8
GC/CIMS analysis of a gas oil in toluene. Single-ion chromatogram of m/z 182 for (a) ITMS, (c) ITD, and (e) quadrupole. Mass spectra of 3-methylcarbazole peak (b) ITMS (scan 1123), (d) ITD (scan 1003), and (f) quadrupole (scan 803). (From Ref. 5)

out an RF/DC isolation step (Fig. 7.8(c) and (d)), or by ammonia CI on a conventional quadrupole spectrometer (Fig. 7.8(e) and (f)). In the latter cases, charge exchange and EI products of the oil matrix interfere extensively with the detection of the nitrogen heterocycles.

V. APPLICATIONS

Ion trap procedures have been reported for a variety of chemical, biological, and environmental samples including, pesticides,[18,21,22] herbicides,[19] amphetamines,[23] volatile organics,[24] fuels,[5,25] perfumes,[12,26] veterinary and clinical drugs, and their metabolites.[9,14,27] In all these applications, sample extracts were separated by GC prior to ion trap chemical ionization.

The application of chemical ionization/ion trap mass spectrometry for the analysis of pesticides in complex samples, in particular the triazines, tolclofos-Me and parathion-Et, has been reviewed recently.[21] CI procedures appear to show promise and, for example, Mattern et al.[22,28,29] have reported the multi-residue determination of several pesticides and their metabolites in a variety of fruit, vegetable and surface water samples after separation on a nonpolar capillary GC column. Methane and isobutane reagent gases were investigated. Fewer fragment ions and enhanced $[M + H]^+$ intensities were observed for isobutane CI. Residues were detected in several samples at levels below legal tolerances and the authors concluded that the procedure should be explored by regulatory authorities for the analysis of pesticide residues. The use of ion trap detection for environmental samples has also been discussed by Huebschmann,[19] with special reference to the analysis of herbicides, including the relative merits of EI and CI ionization.

An examination of the methane-CI spectra of a variety of pesticides obtained using ion trap detection showed that some spectra consisted of superimposed EI and CI mass spectral features.[18] These spectra were found to be useful for confirmation of identity and did not affect quantitative precision and accuracy. This phenomenon has been reported by other authors[18] and arises from the presence of EI products formed during the primary ionization step and charge exchange processes. Procedures for the removal of these ions are discussed in Section IV (Selective Chemical Ionization Procedures).

A novel approach to the determination of polar volatile organic compounds in air used the water present in the samples as the CI reagent gas.[24] Proton transfer from the $[H_3O]^+$ reagent ions produced intense quasi-molecular ions and characteristic fragments for these low molecular weight compounds. Water mediated CI mass spectra have also been reported for volatile organics in aqueous samples introduced into the trap using a membrane probe.

Several clinical assays have been reported which successfully employed CI on the ion trap. A CI/ion trap detector procedure was used to determine the cholinergenic drug, arecoline, in plasma extracts.[9] The CI reaction was carried out with acetonitrile. Satisfactory precision (relative standard deviation <20%) and accuracy (relative error <15%) were achieved using homoarecholine as the internal standard. The performance of EI and methane-CI was compared by Wu *et al.* for the identification of methamphetamines and sympathometric amines in urine.[23] CI spectra were obtained for the underivatized analytes and EI spectra of the heptafluorobutyric anhydride and 4-carbethoxyhexafluorobutryl chloride derivates. The EI spectra of these analytes were similar for both derivatizing reagents, while the CI spectra showed discernible differences. The within-run precision was slightly worse for CI (7 to 9%) than EI (5 to 6%) and the limits of detection were, as expected, higher.

Improvements in the quantitation of methylmalonic acid (MMA) in urine, a standard test for vitamin B_{12} deficiency, were achieved using isobutane-CI.[27] Quantitation of the trimethylsilyl derivatives of dicarboxylic acids by EI ionization is difficult because of the low intensity of the molecular ion formed, whereas the enhanced $[M + H]^+$ ion obtained by CI assisted the detection of MMA.

The use of the ion trap CI technique is also illustrated by the determination of the anthelmintic drug, levamisole, in milk[13] by on-line LC-GC-ITMS. Even with the highly specific multidimensional LC-GC cleanup procedure employed, the presence of co-eluting matrix components resulted in EI spectra which contained unwanted interferences and an enhanced $[M + H]^+$ ion intensity arising from matrix-induced CI processes. Thus, poor spectral quality and quantitative precision prevented the mass spectrometric confirmation of levamisole at sub-pbb levels. This is frequently observed for analytes with high proton affinites in complex samples, where it is difficult to reduce trap overloading without losing the required sensitivity. In this case, the problem of competing matrix-CI reactions was overcome by using ammonia-CI to give quantitative conversion of levamisole to the CI product ion. Structural confirmation was achieved by isolation of the $[M + H]^+$ ion (m/z 205) and subjecting it to collisionally-activated dissociation MS/MS. As a general rule, if matrix-CI processes result in poor quantitative performance in an EI trace analysis, then improvements in precision and dynamic range can usually be obtained by employing CI.

REFERENCES

1. Brodbelt, J. S.; Louris, J. N.; Cooks, R. G. *Anal. Chem.* 1987, *59(9)*, 1278–1285.
2. Louris, J. N.; Syka, J. E. P.; Kelley, P. E. *Eur. Pat. Appl.* 1987, 23pp. Method of operating a quadrupole ion trap. European patent 0,215,615,1987, Japanese patent 62,115,641, 1987; U.S. Patent 4, 686, 367, 1987.
3. Eckenrode, B. A.; McLuckey, S. A.; Glish, G. L. *Int. J. Mass Spectrom. Ion Processes.* 1991, *106*, 137–157.
4. Boswell, S. M.; Mather, R. E.; Todd, J. F. J. *Int. J. Mass Spectrom. Ion Processes.* 1990, *99(1–2)*, 139–149.
5. Creaser, C. S.; Krokos, F.; O'Neill, K. E.; Smith, M. J. C.; McDowell, P. G. *J. Am. Soc. Mass Spectrom.* 1993, *4*, 322–326.
6. Dorey, R. C.; Freeman, J. P. *Proc. 40th ASMS Conf. Mass Spectrometry and Allied Topics.* Washington, D.C., June 1992, p. 1763.
7. Dorey, R. C. *Org. Mass Spectrom.* 1989, *24(10)*, 973–975.
8. Brodbelt, J.; Liou, C. C.; Donovan, T. *Anal. Chem.* 1991, *63(13)*, 1205–1209.
9. Shetty, H. U.; Daly, E. M.; Greig, N. H.; Rapoport, S. I.; Soncrant, T. *J. Am. Soc. Mass Spectrom.* 1991, Vol. 2, 168–173.
10. Keller, P. R.; Harvey, G. J.; Foltz, D. J. *Proc. 36th ASMS Conf. Mass Spectrometry and Allied Topics.* San Francisco, June 1988, p. 643.
11. McLuckey, S. A.; Glish, G. L.; Asano, K. G.; Van Berkel, G. J. *Anal. Chem.* 1988, *60(20)*, 2312–2314.
12. Eichelberger, J. W.; Budde, W. L.; Slivon, L. E. *Anal. Chem.* 1987, *59(22)*, 2730–2732.
13. Chappell, C. G.; Creaser, C. S.; Stygall, J. W.; Shepherd, M. J. *Biol. Mass Spectrom.* 1992, *21*, 688–692.
14. Lim, H. K.; Sahashita, C. O.; Foltz, R. L. *Rapid Commun. Mass Spectrom.* 1988, *2(7)*, 129–131.
15. McLuckey, S. A.; Glish, G. L.; Kelley, P. E. *Anal. Chem.* 1987, *59*, 1670.
16. Eckenrode, B. A.; Glish, G. L.; McLuckey, S. A. *Int. J. Mass Spectrom. Ion Processes.* 1990, *99(1–2)*, 151–167.
17. Strife, R. J.; Keller, P. R. *Org. Mass Spectrom.* 1989, *24(3)*, 201–204.
18. Cairns, T.; Chiu, K. S.; Siegmund, E. G. *Rapid Commun. Mass Spectrom.* 1992, *6*, 331–338.
19. Huebschmann, H. J. *LaborPraxis.* 1990, *14(10)*, 810–812.
20. Berberich, D. W.; Hail, M. E.; Johnson, J. V.; Yost, R. A. *Int. J. Mass Spectrom. Ion Processes.* 1989, *94(1–2)*, 115–147.
21. Huebschmann, H. J. *LaborPraxis.* 1990, *14(12)*, 1014–1017.
22. Mattern, G. C.; Singer, G. M.; Louis, J.; Robson, M.; Rosen, J. D. *J. Agric. Food Chem.* 1990, *38(2)*, 402–407.
23. Wu, A. H. B.; Onigbinde, T. A.; Wong, S. S.; Johnson, K. G. *J. Anal. Toxicol.* 1992, *16(2)*, 137–141.
24. Gordon, S. M.; Miller, M. *EPA Report.* EPA/600/3-89/070; 1989, Order No. PB 90-106451, 37 pp.
25. Black, B. H.; Hardy, D. R.; Morris, R. E.; Mushrush, G. W. *Fuel Sci. Technol. Int.* 1990, *8(9)*, 935–945.
26. Keeler, P. R.; Harvey, G.; Foltz, D. J. *Lab 2000.* 1990, *4(6)*, 51–55.
27. Geypens, B.; Ghoos, Y.; Hiele, M.; Rutgeerts, P.; Vantrappen, G.; Joosten, E.; Pelemans, W. *Anal. Chim. Acta.* 1991, *247(2)*, 243–248.
28. Mattern, G. C.; Louis, J. B.; Judith, B.; Rosen, J. D. *J. Assoc. Off. Anal. Chem.* 1991, 982–986.
29. Mattern, G. C. *Diss. Abstr. Int. B.* 1991, *52(3)*, 1155, 82 pp. Univ. Microfilms Int., Order No. DA9125388.

Part 3

APPLICATIONS INVOLVING SMALL MOLECULES

Chapter 8

ENERGETICS AND COLLISION-INDUCED DISSOCIATION EFFICIENCIES IN THE QUADRUPOLE ION TRAP MASS SPECTROMETER

Jodie V. Johnson, Cecilia Basic, Randall E. Pedder, Brent L. Kleintop, and Richard A. Yost

CONTENTS

0-8493-8251-3/95/$0.00+$.50

I. INTRODUCTION

The energy of an ion stored in a quadrupole ion trap mass spectrometer (QITMS) is a function of both the heating effects induced by the radio frequency (RF) trapping field and the cooling effects due to collisions with any gases present in the trap. Characterization of the energy of the ions is critical to a complete understanding of the ion/molecule reactions occurring in the QITMS, and to the role which the QITMS can assume in the determination of the fundamental kinetic and thermochemical properties of ion/molecule reactions. Knowledge of the amount of internal energy deposited into the ion and of the dissociation efficiencies obtained under a given set of activation conditions provides insight into the ion dissociation process and allows the practical comparison of tandem mass spectrometric (MS/MS) analyses obtained on a QITMS with those obtained using conventional beam mass spectrometers. Moreover, an understanding of the ion energies defines the limits of a given ion activation method, and in so doing defines the range of studies amenable to a given activation technique.

To this end, numerous theoretical and experimental studies aimed at the determination of trapped ion energies have been presented. This chapter will begin with a brief review of ion motion in RF ion traps, followed by a survey of the various ion activation methods currently employed in MS/MS studies in the QITMS. Summaries of a selection of the theoretical models and computer programs developed to simulate ion motion and to calculate ion energies will be presented, as will a survey of the experimental studies designed to measure effective ion energies. The remainder of this chapter will be devoted to a detailed examination of the various experimental parameters which define the energetics and collision-induced dissociation (CID) efficiencies obtained upon resonant excitation in the QITMS.

II. ION MOTION IN A QUADRUPOLE ION TRAP

The motion of an ion in a QITMS is described by a second-order linear differential equation of the Mathieu form,[1,2]

$$\frac{d^2u}{d\xi^2} + (a_u - 2q_u\cos2\xi)u = 0 \tag{8.1}$$

where u represents one of the cylindrical coordinate axes, r in the radial direction or z in the axial direction, and ξ is equal to $\Omega t/2$ where Ω is the angular frequency of the RF potential (rad s^{-1}) and t is time (s). The Mathieu stability parameters a_u and q_u are given by,

$$a_u = a_z = -2a_r = \frac{-16eU}{m(r_0^2 + 2z_0^2)\Omega^2} \tag{8.2}$$

and

$$q_u = q_z = -2q_r = \frac{-8eV}{m(r_0^2 + 2z_0^2)\Omega^2} \tag{8.3}$$

where U is the DC potential applied to the ring electrode, V is the zero-to-peak amplitude of the RF potential (V_{0-p}), m is the mass of the ion (kg), e is the charge (C) of the ion, r_0 is the radius of the ring electrode (m), z_0 is the radius of the ion trap in the axial direction where, ideally, $r_0^2 = 2z_0^2$.

If an ion of a given mass/charge ratio, m/z, has a_u and q_u parameters which give rise to stable solutions to the Mathieu equation (Eq. (8.1)) in both the r- and z-directions, then the ion will follow a stable trajectory and will be confined within the volume of the ion trap provided that its excursions from the center of the ion trap do not exceed the physical limitations of the device. Stable solutions are ones in which u remains finite as ξ increases and are of the form,

$$u(\xi) = A \sum_{-n=\infty}^{\infty} C_{2n}\cos(2n + \beta)\xi + B \sum_{-n=\infty}^{\infty} C_{2n}\sin(2n + \beta)\xi. \tag{8.4}$$

The range of a_u and q_u values which give rise to stable solutions in both

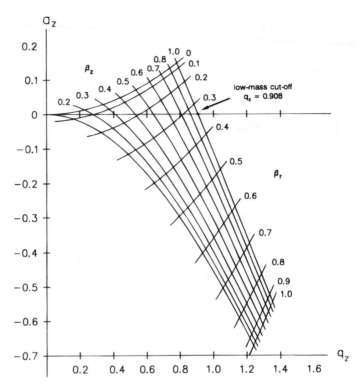

FIGURE 8.1

Stability diagram for a QITMS. Ions with a m/z ratio greater than the low-mass cut-off at $q_z = 0.908$ (for $a_z = 0$) follow stable trajectories. The β_r and β_z lines define the fundamental secular frequencies of the ions in the r- and z- directions, respectively. Adapted from Ref. 4 with permission.

the r- and z-directions can be plotted in the form of a stability diagram (Fig. 8.1).

Eq. (8.4) describes effectively the frequency spectrum of the ion motion, $u(\xi)$. Each term corresponds to an oscillation with an amplitude given by the coefficient, C_{2n}. The coefficients A and B are real and imaginary constants of integration, respectively, related to the initial position and velocity of the ion. The frequencies of the ion motion are represented by the $(2n + \beta)$ terms, where, transforming into real time,

$$\cos(2n + \beta)\xi \equiv \cos\omega_n t \tag{8.5}$$

and recalling $\xi = \Omega t/2$,

$$\omega_n = \frac{(2n + \beta)\Omega}{2} \qquad n = -\infty, \ldots -1, 0, 1 \ldots, \infty. \tag{8.6}$$

Defining ω_n as the frequency of the nth term, the fundamental angular frequency, ω_0 in either the r or the z plane is given by $\beta\Omega/2$. Values for β_z for the stability region $q_z \leq 0.4$ ($q_r \leq 0.2$, where $0 \leq \beta_z \leq 1$) can be approximated by,[3]

$$\beta_z = \left[a_z + \frac{q_z^2}{2} \right]^{1/2} \tag{8.7}$$

Complete solutions for β_u have been presented elsewhere.[4] Thus, ions with the same β_u values will have trajectories of the same secular frequency; however, the amplitude of their trajectories will be governed by their a_u and q_u values and further modified by the A and B values defined by the initial position and velocity of the ions at the time they are formed in the field.

III. ION ACTIVATION IN A QITMS

Ion activation to induce dissociation for MS/MS and MS^n analyses is typically performed in the QITMS by first mass-selecting the parent ion of interest and then applying a supplemental RF potential (or "tickle" voltage) to the end-cap electrodes.[5–8] The frequency of the supplemental potential is generally chosen to be resonant with the fundamental secular frequency of the ion's motion in the z-direction, that is, $\omega_{0,z} = \beta_z\Omega/2$ from Eq. (8.6). For a QITMS with an RF drive frequency of 1.1 MHz, the practical range of secular frequencies within the stability envelope in Fig. 8.1 varies from 55 to 550 kHz. Application of this resonant excitation potential results in an increase in the ion's axial kinetic energy, and the ion undergoes increasingly energetic collisions with any gases present, typically the He buffer gas present at $\approx 1 \times 10^{-3}$ Torr.

Collisional activation *via* resonant excitation in the QITMS differs from that in other instruments employed for CID experiments in that the parent ions are excited between collisions; as such, the ions undergo stepwise activation coupled with collisional cooling by the He buffer gas. Further, only ions of a single m/z (or a small range of m/z) are in resonance with the excitation potential; thus, once formed, the daughter ions will not be resonantly excited, but rather their motion will be damped through collisions with the helium buffer gas. The resonant excitation potential is applied typically at amplitudes of 50 mV to 5 V for 2 to 20 ms, although excitation times as long as 100 ms have been employed.[9,10] This excitation duration can be compared to the 10 to 50 μs residence times of the ions in the center quadrupole of a triple quadrupole mass spectrometer.[11]

The amount of internal energy deposited into the parent ion during the CID process can be varied by changing either the amplitude or the

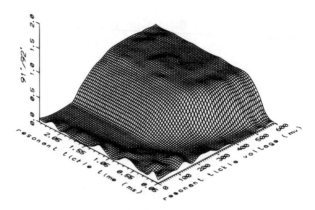

FIGURE 8.2
Collision-induced dissociation breakdown surface of the $M^{+\bullet}$ ion of n-butylbenzene showing the $91^+/92^+$ daughter ion ratio as a function of resonant excitation time (or duration) and voltage amplitude. From Ref. 13.

duration of the resonant excitation potential, the amplitude of the RF trapping field, as reflected by the Mathieu parameter q_z of the parent ion, or the pressure and nature of the buffer gas. The effects of these parameters on the energetics and CID efficiencies are discussed in the final section of this chapter.

CID breakdown graphs have been constructed by plotting the relative intensities of the parent and daughter ions as a function of either the amplitude (voltage-resolved) or duration of the resonant excitation potential (time-resolved).[7,9,12] CID breakdown surfaces have also been constructed.[13,14] Fig. 8.2 shows the ratio of the m/z 91 and m/z 92 daughter ions from the fragmentation of the molecular ion ($M^{+\bullet}$, m/z 134) of n-butylbenzene as a function of both resonant excitation time and voltage.[13] The $91^+/92^+$ ratio arising from the $M^{+\bullet}$ ion of n-butylbenzene has also been presented as a function of both the resonant excitation voltage and the q_z of the $M^{+\bullet}$ ion.[14]

Recently, several new approaches to ion activation in the QITMS for CID studies have been reported: (1) broadband excitation; (2) boundary-effect activated dissociation (BAD); (3) random noise activation; and (4) pulsed radial and axial activation.

Broad-band excitation[15–17] involves mass-selecting the parent ion and then applying a wide band of excitation frequencies to the end-cap electrodes. In so doing, excitation of the parent ion remains independent of slight shifts in the ion's secular frequency due to space-charge effects. Bandwidths of 12.5[15] and 10 kHz[16,17] have been employed. Ion excitation in the region of the axial secular frequency was found to be enhanced by a factor of greater than ten compared to single-frequency excitation.[15] Moreover, the approach does not involve the often lengthy procedure of optimizing single-frequency resonant excitation conditions. Yates *et al.*[17] have reported recently the use of stored waveform inverse Fourier transform (SWIFT) broadband excitation for automated GC/MS and GC/MS/MS analyses on a benchtop QITMS. A more detailed discussion of broadband excitation is presented in Chapter 8 of Volume 1 of this series.[18]

Boundary-effect activated dissociation was first reported by Curcuruto *et al.*[19] This technique involves the application of suitable RF and direct current (DC) potentials to the ion trap electrodes to position the ion of interest at a_z and q_z values close to the theoretical boundaries $\beta_r = 0$ or $\beta_z = 0$.[19,20] The amplitudes of the trajectories of ions positioned at the stability edges will be greater than those positioned well within the stability region, resulting in an increase in the ion kinetic energies. Thus, CID can be effected without the application of any supplemental excitation potentials. Studies have shown that energy deposition using BAD is generally less than or equal to that obtained from resonant excitation CID.[14,21] However, while it is possible to increase internal energy deposition with resonant excitation methods,[5,10] with BAD maximum ion internal energy coincides with optimum CID efficiency. Any attempt to further increase the ion internal energy results in ion ejection.[14] The BAD approach to ion activation is discussed in further detail in Chapter 7 of Volume 1 of this series.[18]

Ion activation through the application of a random noise signal has recently been presented by McLuckey *et al.*[22] In this approach, white noise is applied to the end-cap electrodes to effect non-mass-selective CID. Because it is non-mass-selective, this method of activation can result in more extensive fragmentation since the daughter ions formed also undergo resonant excitation CID. However, the approach leads to abundant daughter ion information without the need for MS^n studies (that is, MS^2 to form daughter ions, MS^3 to form granddaughter ions, etc.), lengthy tuning is not required and the method is less sensitive to the number of ions in the trap. CID efficiencies with random noise activation ranged from being equivalent to a factor of three less than those found with single-frequency resonant excitation.[22]

Pulsed radial and axial activation for CID and surface-induced dissociation (SID) studies has also been performed in the QITMS.[23,24] In this approach, ions are activated in the radial direction by applying a short (<5 μs), fast-rising (<20 ns rise-time), high voltage DC pulse symmetrically to the end-cap electrodes.[23] Alternatively, the ions can be activated in the axial direction by applying the pulse asymmetrically to one end-cap electrode.[24] Application of a very short pulse changes the initial position of the ions without changing the a_z, q_z operating conditions of the trap. In so doing, the kinetic energy of the ions increases due to the stronger RF trapping potential present at their new positions. The amount of internal energy deposited into the ions can be controlled by varying the amplitude of the DC pulse. The CID efficiencies are less than those observed with resonant excitation and range from 5% at maximum internal energy deposition (where SID is considered to contribute significantly to the fragmentation of the parent ion) to 50% at lower internal energies.[24]

While collisional activation is the most widely used method for inducing ion fragmentation for MS/MS studies in the QITMS, the use of a low-epower infrared laser for photodissociation (PD) studies have also been reported.[25–32] The experimental procedures and applications of PD in the QITMS are presented in Chapter 5 of Volume 2 of this series.[18]

IV. CALCULATIONS OF TRAPPED ION ENERGIES

Models of trapped ion trajectories provide a means of estimating ion kinetic energies under a variety of experimental conditions. Several approaches have been taken to the modelling of trapped ion trajectories both prior to and upon application of an excitation potential. These models include: (1) calculations of individual ion trajectories; (2) the pseudo-potential well model; (3) the phase space dynamical model; and (4) the temporal invariance model; and (5) expanded analytical solution method for calculation of ion velocity and kinetic energy. These simulation methods are designed for ideal quadrupole ion traps where the ion motion is described by equations of the Mathieu form and where the fundamental axial and radial frequencies are given by $\omega_{0,z} = \beta_z\Omega/2$ and $\omega_{0,r} = \beta_r\Omega/2$, respectively. Simulations of the trapping fields and the subsequent effects on ion motion due to nonlinear resonances have been presented elsewhere[33,34] and are discussed in Chapter 3 of Volume 1 of this series.[18]

A. Calculation of Individual Ion Trajectories

Several approaches have been taken to the calculation of individual ion trajectories. The position, u, and velocity, $\dot{u}$, of an ion in a three-dimensional quadrupolar field can be calculated by numerically integrating the Mathieu equation (Eq. (8.1)) using a fourth-order Runge–Kutta step-wise integration method.[35] Pedder and Yost[36,37] have used this approach in the development of a simulation program, HYPERION, which allows calculation of ion trajectories and kinetic energies under a variety of experimental conditions using a personal computer. HYPERION is a BASIC computer program compiled using Borland's TurboBASIC on an IBM PC-compatible and has an interactive graphical interface, allowing the user to vary any parameter and to monitor visually effects of the parameter change on the ion trajectory. The simulation program models dipolar resonant excitation (that is, application of an excitation potential

180° out-of-phase to both end-cap electrodes) by considering the field contributions of this dipolar resonant excitation as the electrical equivalent of a parallel plate capacitor. Thus, the ions will experience the superimposition of the quadrupole trapping field created by the RF and DC voltages applied to the ring electrode and the dipolar resonant excitation field. Such a plate capacitor model can only approximate the real field since the end-caps are not actually parallel plates; however, the model is reasonably accurate for ions having little radial energy, that is, for ions whose motion has been dampened to the center of the trap by collisions with a buffer gas.

The equation for dipolar resonant excitation employed in HYPERION is a modified form of the Mathieu equation of motion in the z-direction:[37]

$$\frac{d^2z}{d\xi^2} + (a_z - 2q_z\cos2\xi)z + b_z\cos\Omega_{res}t = 0 \qquad (8.8)$$

The $b_z \cos\Omega_{res}t$ term describes the dipolar excitation and is defined as:

$$b_z = \frac{-2eV_{res,p-p}}{mz_0\Omega^2} \qquad (8.9)$$

where $V_{res,p-p}$ is the voltage of the applied resonant excitation potential, and Ω_{res} is the resonant excitation frequency. Eq. (8.8) is essentially the same as the Mathieu equation (Eq. (8.1)) but with an added perturbation term. Note that in the plate capacitor model of resonant excitation, the field caused by the perturbation term is assumed to be zero in the r-direction. Thus, motion in the r-direction can be calculated by solving the unperturbed Mathieu equation. The perturbed Mathieu equation can be numerically evaluated using a fourth-order Runge–Kutta algorithm similar to that for the unperturbed equation. HYPERION calculates the average and maximum velocity and kinetic energies of an ion in the z-direction. A comparison of HYPERION simulations and experimentally-determined CID results for the fragmentation of the $M^{+\cdot}$ ion of n-butylbenzene is presented in detail in this chapter.

Estimates of the ion position and velocity can also be obtained from the first derivative of the generalized solutions to the Mathieu equation (Eq. (8.4)) for specific initial positions, velocities, and phases of the RF drive potential. When it is assumed that the initial velocity of the ion is zero, then the velocities for successive values of ξ can be calculated and averaged. This "complete general solution" requires lengthy calculations and can lead to large errors if the ξ increments are too large.[39,40] Another approach to estimating ion energies based on the first derivative of Eq. (8.4) is that of the "smoothed general solution."[39-41] In this approach, it is assumed that for $0 < \beta_z < 0.4$, motion due to the higher harmonics in

Eq. (8.6) can be ignored. For all values of q this assumption results in an underestimation of ion velocities by a factor of three leading to calculated kinetic energies that are low by an order of magnitude.[37] Both the average and maximum velocities can be calculated for each of the three coordinate directions, x, y, z. Both the complete general solution and the smoothed general solution result in single ion energies under collision-free conditions.[39-41]

Statistical Monte Carlo methods have been used to calculate the trajectories for up to 150 ions present in the trap.[42-44] In this approach, the initial position, u_0, the initial RF phase, ξ_0, and the RF phase at the moment of collision are varied randomly. For each set of randomly-generated parameters, a_u, q_u, β_u, and A, B, and C_{2n} in Eq. (8.4), and the maximum ion excursion, u_{max}, are calculated in both the r- and z-directions. Changes in ion position and energy are then determined following a series (up to 15) charge-exchange collisions using a simplified model of the collision process.[43,44] These calculations estimate an average kinetic energy of ≈ 0.5 eV for Ar^+ ions following 15 collisions over the q_z range 0.0782 to 0.8595. Similar energies were found for Ne^+ and CO_2^+ following 15 collisions.[44] A comparison of the average kinetic energies calculated for Ar^+ ions using the smoothed general solution approach, the pseudo-potential well model (described below), and Monte Carlo methods has been presented.[43]

Reiser and co-workers[45,46] have developed an ion trap simulation program, ITSIM, in which numerical solutions to the Mathieu equation (Eq. (8.1)) are found using a Taylor series expansion. ITSIM simulates the mass-selected instability mode of operation[47] by incorporating an RF voltage scan speed term in the equation which describes the RF potential applied to the ring electrode. RF and DC potentials used for ion isolation are treated in a stepwise fashion. Resonant excitation, including mass-range extension,[48,49] and He buffer gas and space-charge effects are modelled by adding appropriate perturbation terms to the Mathieu equation. The simulation program provides kinetic energy distributions for collections of ions as well as calculated mass spectra. ITSIM has been used to compare simulated and experimental results in a wide range of QITMS applications including resonant ion ejection[46] and ion injection from an external source.[50] The program has been extended recently to calculate the trajectories of up to 10^4 ions and also to approximate the quadrupole field created by nonideal QITMS electrodes.[51]

March et al.[38,52-56] have presented a series of computational programs: the Specific Program for Quadrupolar Resonance (SPQR); the field interpolation method (FIM); and, recently, the Integrated System for Ion Simulation (ISIS). Briefly, the programs are based on numerical integration of the equations of motion arising from the equation describing the field formed within the volume of the ion trap under a variety of experimental conditions. Variations of the ion positions and kinetic energies in both the r- and z-directions as well as the direction of ion ejection were determined as a function of ion mass, a_u, q_u, initial ion position and velocity, drive and supplemental potential phase angles, supplemental potential voltage and frequency (in monopolar, dipolar and quadrupolar modes), and the mass of the inert buffer gas.[38,52,53] Recent improvements allow continuous variation of the RF and DC potentials, thus permitting calculation of the ion trajectories while the a_z and q_z values are moved both within and outside the stability diagram.[54–56] RF/DC apex ion isolation[57–59] and consecutive ("two-step") isolation[60–62] under collision and collision-free conditions have been simulated.[54–56] Program details as well as the fundamental results of these simulations can be found in Chapter 6 of Vol. 1 of this series.[18]

B. Pseudo-Potential Well Model

The pseudo-potential well model was first presented by Dehmelt[63,64] and developed subsequently by Todd *et al.*[39–41,65] This model also assumes that motion due to the higher harmonics can be ignored, and as such provides another means for estimating the average kinetic energy of the ions due to their secular motion. In this approach, the ion's motion is considered to exhibit simple oscillatory motion at the secular frequencies ω_{0z} and ω_{0r} in pseudo-potential wells of depth $\overline{D}_z$ and $\overline{D}_r$. The model is valid for $q_z < 0.4$.[41] The average displacement of the ion over one period of the RF cycle corresponds to simple harmonic motion of the z component of the secular motion, with a frequency equal to the fundamental secular frequency, ω_{0z}. The depth of the potential well in the z-direction, $\overline{D}_z$, is given by:

$$\overline{D}_z = -\frac{eV^2}{4mz_0^2\Omega^2} = \frac{q_z V}{8} \tag{8.10}$$

The well depth in the r-direction is derived similarly and, since $r_0^2 = 2z_0^2$, $\overline{D}_z = 2\overline{D}_r$. The average kinetic energy of an ion, $K.E._{ave,z}$, over one cycle of the secular motion, and the maximum kinetic energy $K.E._{max,z}$ in the z-directions are:

$$K.E._{ave,z} = \frac{4}{\pi^2}\,e\overline{D}_z; \quad K.E._{max,z} = e\overline{D}_z \tag{8.11}$$

while the average total kinetic energy, $K.E._{ave,total}$ and the maximum kinetic energy, $K.E._{max,total}$ due to motion in both z and r are given by:

TABLE 8.1

Pseudo-potential well depth and kinetic energy calculations for Ar^+ ions

q_z	RF Voltage (V_{0-p})	$e\overline{D}_z$ (eV)	$K.E._{max,z}$ (eV)	$K.E._{ave,z}$ (eV)
0.200	99	2.48	2.48	1.00
0.295	146	5.39	5.39	2.19
0.386	191	9.22	9.22	3.74
0.454	225	12.7	12.7	5.17

$a_z = 0$; $\Omega/2\pi = 1.1$ MHz; $z_o = 0.707$ cm.

$$K.E._{ave,total} = \frac{8}{\pi^2} e\overline{D}_z; \quad K.E._{max,total} = 2e\overline{D}_z \qquad (8.12)$$

Calculated $\overline{D}_z$, $K.E._{max,z}$, and $K.E._{ave,z}$ values at a variety of q_z values for Ar^+ ions (m/z 40) are presented in Table 8.1.

C. Phase-Space Dynamics

In contrast to the models presented above, the phase-space dynamics model calculates changes in the position and velocity distribution of an ion "cloud" as a function of time.[39,40,65–69] The phase-space approach considers a group of noninteracting particles whose motions in the three orthogonal directions, x, y, z, are considered to be independent of each other. Changes in the ion cloud are mapped by considering variations in the initial position and velocity with time. A plot of position and velocity in Cartesian co-ordinates forms a two-dimensional "phase-space." The properties of phase-space state that if a certain group of particles has a boundary, C_1, at time t_1, then at time t_2, C_1 will have transformed into C_2 which will bound the same group of particles. Thus, a large group of particles undergoing a transition in time can be accounted for, or mapped, by following the changes in the phase-space boundaries with time. The average and maximum ion kinetic energies at a given RF phase angle are described by the eccentricity of the ellipse defined by a given set of experimental conditions.

One of the advantages of the phase-space model over that of the pseudo-potential well is that it can give rise to energies over the entire range of q_z, as opposed to being limited to q_z values less than 0.4. Moreover, phase-space dynamical calculations reflect the interrelationship between the a_u,q_u pairs and β_u. A comparison of the average energies obtained from the "smoothed general solution," pseudo-potential well, and phase-space dynamics models for ions of m/z 16 stored at a variety of q_z values has been presented.[40]

D. Temporal Invariance

The temporal invariance model for ions stored in an RF ion trap was first presented by André and co-workers[70,71] and is discussed in detail elsewhere.[4] As in the phase-space dynamics model, the temporal invariance approach considers changes in the properties of the ion cloud. The model calculates the spatial and energy probability densities of the ion cloud based on the assumption that after a fixed reorganization time following ionization, collisions with residual gases and the buffer gas will lead to an equilibrium repartition of the ion spatial and velocity components. The probability densities of these components are then described by a Gaussian repartition function. This function is not stationary due to the periodicity of the RF field; it is, however, invariant to the temporal translation of the RF drive field. Since the model is based on a Gaussian distribution, the ion population is described in terms of an ion "temperature" as opposed to a kinetic energy. Because the ion temperature changes as a function of time, it is referred to as a "pseudo-temperature." An average ion pseudo-temperature in either the x-, y-, or z-directions, T (K), can be obtained by averaging the time-dependent temperature over one cycle of the RF potential.

The temporal invariance model has been used to calculate the spatial and energy probability densities of the stored ions as a function of the a_u, q_u parameters,[72] the pressure and nature of buffer gas species[71,72] and the number of stored ions, n.[72–76] Calculated ion pseudo-temperatures in the z-direction for a population of Cs^+ ions cooled by He buffer gas range from 550 to 5000 K as a function of a_z over the q_z range of 0 to 0.908.[72]

V. EXPERIMENTAL DETERMINATION
OF TRAPPED ION ENERGIES

Theoretical calculations of the ion motion in an RF ion trap provide estimates of ion energies under ideal conditions. Models which incorporate any of the number of possible deviations from ideal conditions, such as field inhomogeneities due to holes in the end-cap electrodes or the temporal variations of space-charge effects encountered over the timeframe of a chromatographic peak, can be complex and time consuming to simulate. Thus, a number of experimental determinations of the energies of the ions stored in quadrupole ion traps both prior to and upon resonant excitation have been presented. Because the trapped ions do not define simple, linear trajectories in an ion trap and because the ions experience multiple collisions with neutrals during the course of an experiment, indirect methods of measurement have been used to determine the ion kinetic energies, the ion/molecule center-of-mass collision energies, and the internal energies deposited upon ion activation. Along with the following presentation, the reader is referred to an excellent review of the experimental and theoretical determinations of the energetics and dynamics of trapped ions presented by Vedel.[77]

A. Ion Energies Prior to Activation

A number of experiments designed to measure the average energies of trapped monatomic species have been reported. Bolometric methods,[78,79] optical measures,[80–86] and a time-of-flight method for profiling the extracted ion cloud[87] have been used to probe ion energies in the absence and presence of He buffer gas. In the absence of He buffer, the ion energies were found to be approximately 10% of the pseudo-potential well depth, $e\overline{D}$ (eV),[78,80–82] although a lower average energy equal to 2% $e\overline{D}$ has also been reported.[86] Ion energies measured in the presence of He buffer vary from 10% $e\overline{D}$,[85] to energies between 0.2% and 5% $e\overline{D}$.[82,83,87]

A limited number of experimental determinations of the energies of the stored ions using chemical "thermometer" reactions have also been reported. A chemical thermometer reaction is defined as an ion/molecule reaction possessing a physiochemical property which changes in a well-defined manner as a function of ion energy and which can be measured easily.[77,88] Early reports by Lawson et al.[89] estimated the average ion energy to be on the order of 1 to 3 eV with no He buffer gas present in the trap. These energies were based on the relative abundances and known appearance potentials of the m/z 15 (CH_3^+) and m/z 27 ($C_2H_3^+$) ions from ionized methane and on the appearance of the m/z 16 (NH_2^+) and m/z 18 (NH_4^+) ions from ionized ammonia. Brodbelt-Lustig and Cooks[90] used the "kinetic method"[91] to determine the effective ion temperature T = 335 K for the proton-bound dimers of a series of substituted pyridines.

Basic et al.[92] have presented a series of studies in which center-of-mass kinetic energies ($K.E._{cm}$) of Ar^+ ions stored in a QITMS were assigned based on measurements of the rate constants, k, for the well-characterized charge exchange reaction of Ar^+ with N_2.[93–95] The rate constants and assigned $K.E._{cm}$ values for the Ar^+ ions as a function of He buffer gas pressures are presented in Table 8.2. The $K.E._{cm}$ of the Ar^+ ions at a variety of q_z values at two Ar/N_2 pressures both with and without He buffer gas are presented in Figs. 8.3(a) and (b). While the sometimes large experimental error arising from the use of the Ar^+/N_2 chemical "thermometer" reaction made assignment of absolute $K.E._{cm}$ under any set of experimental conditions difficult, certain trends in the $K.E._{cm}$ values are noteworthy. (1) There is a consistent decrease in $K.E._{cm}$ upon addition of He buffer gas to the reaction system; however, increasing the He buffer pressure above 1.0 $\times\ 10^{-4}$ Torr (uncorrected ion gauge reading) appears to have little addi-

TABLE 8.2

Rate constants, k, and K.E.$_{cm}$ for the reaction of Ar$^+$ with N$_2$ at a variety of He buffer gas pressures

Total pressure with He	q_z (Ar$^+$) = 0.295		q_z (Ar$^+$) = 0.454	
(/10^{-5} Torr)	k (/10^{-11} cm^{3}s^{-1})	K.E.$_{cm}$ (eV)	k (/10^{-11} cm^{3}s^{-1})	K.E.$_{cm}$ (eV)
No He	5.4 ± 1.5	0.14 ± 0.04	6.4 ± 2.4	0.16 ± 0.06
2.6	5.5 ± 0.89	0.14 ± 0.02	5.82 ± 0.20	0.15 ± 0.01
5.1	5.4 ± 1.4	0.14 ± 0.03	5.2 ± 1.7	0.13 ± 0.04
10.0	5.5 ± 2.6	0.14 ± 0.06	5.1 ± 2.3	0.13 ± 0.05
25.0	5.0 ± 0.7	0.13 ± 0.02	5.3 ± 1.5	0.13 ± 0.04
51.0	4.8 ± 1.0	0.12 ± 0.03	4.4 ± 1.3	0.11 ± 0.03
Average with He	5.2 ± 0.4	0.13 ± 0.01	5.2 ± 0.7	0.13 ± 0.02

All K.E.$_{cm}$ values are for a 1:1 Ar:N$_2$ pressure ratio at a total pressure of 1.0×10^{-5} Torr prior to addition of He buffer and a 100 ms reaction time.

tional cooling effect. (2) An increase in K.E.$_{cm}$ is seen with increasing q_z of the Ar$^+$ ion both with and without He buffer gas. This increase in energy is more marked at lower pressures of Ar and N$_2$, presumably due to the decreased buffering abilities of these gases at lower total pressures.

All of the experimentally-determined K.E.$_{cm}$ values arising from the reaction of Ar$^+$ with N$_2$ under a variety of experimental conditions were found to lie within the range 0.11 to 0.34 eV. These values indicate that the ions are a factor of 2 to 7 times above thermal energy (0.05 eV). "Effective" ion temperature for He buffered Ar$^+$ ions were presented based on the assigned K.E.$_{cm}$ and on the assumption that the ions are described by a Maxwell–Boltzmann distribution of kinetic energies. The resulting ion temperatures lie within the range ≈1700 to 3300 K. This temperature range can be compared to the effective ion temperatures of between 600 to 700 K measured for O$_2$$^+$ in He buffer gas prior to resonant excitation[96] using a similar approach for assigning ion energies.

B. Ion Energies Upon Activation

Quantitative determinations of the average internal energy deposited into the ions in a QITMS upon resonant excitation have been limited. Louris et al.[5] used the fragmentation of the M$^{+\bullet}$ ion of n-butylbenzene (m/z 134) as a thermometer to measure the internal energy deposited as a function of the q_z value and the voltage of the resonant excitation potential. (Use of the M$^{+\bullet}$ ion of n-butylbenzene as a chemical thermometer is discussed below.) In general, the ion internal energies were found to increase as the q_z value of the M$^{+\bullet}$ ion was increased and ranged from ≈1 to 5 eV, over the q_z values (0.36 and 0.60) and voltages (2.7 to 14.4 V$_{p-p}$) employed. It was also noted that at higher resonant excitation volt-

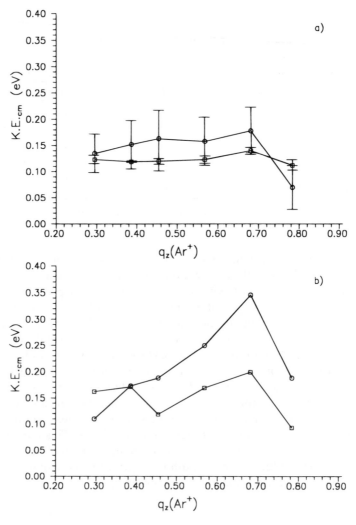

FIGURE 8.3

Center-of-mass kinetic energies (K.E._{cm}) *versus* q_z value of Ar^+ ions determined from the rate constants for the reaction $Ar^+ + N_2 \rightarrow N_2^+ + Ar$. (a) ($\circ$) 1:1 $Ar:N_2$ at 1.0×10^{-5} Torr; ($\square$) He buffer gas added to 1.0×10^{-4} Torr; (b) ($\circ$) 1:1 $Ar:N_2$ at 2.2×10^{-6} Torr; ($\square$) He buffer gas added to 1.0×10^{-4} Torr. From Ref. 92 with permission.

ages, loss of the $M^{+\bullet}$ ions competes with dissociation; thus, while the fragmentation efficiency increased, the overall CID efficiency decreased. Louris *et al.*[5] also estimated the distribution of ion internal energies by measuring the fragment ion intensities arising from CID of the $M^{+\bullet}$ ion of tetraethylsilane which fragments *via* the sequential loss of ethyl groups. Since the activation energy, ϵ_0, for each step is known, the internal energy distribution was estimated from the relative intensities of the fragment ions. The fragmentation patterns indicate that as the resonant excitation voltage is increased, the distribution of internal energies broadens, thus allowing access to higher energy pathways, that is, the loss of three C_2H_5 groups with an $\epsilon_0 = 3.5$ eV.

Brodbelt et al.[9] have presented estimated ion internal energies following resonant excitation based on the CID spectra of the $M^{+\cdot}$ ions of dimethyl phosphonate and dimethyl phosphite. The CID spectra reflect the isomerization of the phosphate ion to the phosphite tautomer, which is strongly dependent on the collision conditions employed.[97] The activation energies for the various fragmentation pathways of the two isomers are also well known.[98] The authors suggest that the absence of the m/z 47 fragment ion ($\epsilon_0 = 5.8$ eV) in the CID spectrum of the $M^{+\cdot}$ ion (m/z 110) of dimethyl phosphonate indicates that the average internal energy of the activated ions is less than 5.8 eV.

A recent study by Nourse et al.[10] reports that a total of 29 eV of internal energy can be deposited into the $M^{+\cdot}$ ion of pyrene using 6 sequential stages of resonant excitation and mass-analysis (that is, MS^7) each of which can deposit ≈ 5 eV of energy. They also report that under "extreme" activation conditions (6 V_{0-p} resonant excitation voltage applied for 100 ms) fragment ions with estimated activation energies of 17 eV are formed from the $M^{+\cdot}$ ion of pyrene. Lammert and Cooks[24] report that in excess of 20 eV of energy can be deposited into the $M^{+\cdot}$ ion of n-butylbenzene (m/z 134) using pulsed axial activation. Thus, while the majority of studies of the ion internal energies deposited upon resonant excitation in the QITMS indicate that the energies are in the low 2 to 4 eV range, substantially higher energies have also been reported.

VI. RESONANT EXCITATION IN THE QITMS: CID ENERGETICS AND EFFICIENCIES

As discussed above, the energetics and efficiencies of the resonant excitation CID process are dependent upon a number of interrelated parameters which include the frequency, voltage, and duration of the resonant excitation potential, the Mathieu parameters a_z and q_z of the parent ion, and the pressure and nature of the buffer gas. The remainder of this chapter is devoted to a discussion of the effects of these different parameters on the CID energetics and efficiencies observed in the QITMS. The effects will be monitored using the fragmentation of the $M^{+\cdot}$ ion (m/z 134) of n-butylbenzene as a chemical "thermometer." In addition, the computer simulation program HYPERION will be used to explain the experimentally-observed ion behavior during resonant excitation. Along with the studies presented below, the reader is also referred to several other studies on the effects of these different parameters on the CID process in quadrupole ion traps[5,7,14,99] as well as a study of negative ion CID in a QITMS.[100]

All the studies which follow were performed using electron ionization (EI) on a Finnigan MAT ITMS ion trap mass spectrometer which has been described in detail elsewhere.[7] n-Butylbenzene was introduced to the ion trap *via* a variable leak valve to an indicated pressure of 0.6 to 2.0 × 10^{-6} mTorr. Unless indicated otherwise, the m/z 134 $M^{+\cdot}$ ion of n-butylbenzene was resonantly excited for 5 ms with a ring RF voltage level yielding a $q_z(m/z\ 134)$ of 0.30. The resonant excitation frequency and potential were varied to maximize the yield of daughter ions. For most of the studies, helium was introduced directly into the ion trap to yield an indicated vacuum chamber pressure of 0.8 to 1.0 × 10^{-5} Torr (as measured by a Bayard–Alpert ion gauge on the vacuum chamber). Calibration of the ion gauge showed that these pressures resulted in an absolute pressure of ≈0.5 mTorr helium.[13] For some studies, helium (and other buffer gases) were introduced into the vacuum chamber to yield pressures indicated in the text. Calibration of the ion gauge readings was performed with a capacitance Baratron gauge for both introduction methods.[13] When helium was introduced directly into the ion trap volume (with the electrode spacers in place), the pressure inside the trap was approximately ten times higher than the pressure outside of the ion trap. Of course, when helium is introduced into the QITMS chamber, the pressure inside and outside the ion trap volume is the same.

Fragmentation of the $M^{+\cdot}$ ion of n-butylbenzene has been used extensively as a thermometer to measure the average internal energy deposited into an ion upon activation. The $M^{+\cdot}$ ion of n-butylbenzene fragments *via* two principal pathways:

$$C_6H_5(CH_2)_3CH_3^{+\cdot} \rightarrow C_6H_5CH_3^{+\cdot}\ (m/z\ 92) + C_3H_6 \qquad (8.13)$$

$$\rightarrow C_6H_5CH_2^+\ (m/z\ 91) + C_3H_7^{\cdot}$$

Detailed studies of the relative rates of formation of the m/z 91 and m/z 92 ions[101–103] reveal that the m/z 91 ion forms *via* direct cleavage of C_3H_7. with a high activation energy, $\epsilon_0 \approx 1.7$ eV, while the m/z 92 ion forms *via* a rearrangement and loss of C_3H_6 with a low $\epsilon_0 \approx 1$ eV.[103] Photodissociation,[102,104,105] charge exchange mass spectrometry (CEMS),[106] and photoelectron-photoion coincidence (PEPICO) studies[102] have revealed that the branching ratio of the abundances of m/z 91 and m/z 92 ions, 91$^+$/92$^+$, increases with increasing internal energy of the $M^{+\cdot}$ ion. The PEPICO[102] and CEMS[106] breakdown graphs for the $M^{+\cdot}$ ion of n-butylbenzene are presented in Fig. 8.4. A detailed comparison of the photodissociation and CEMS breakdown and the variety of internal energy effects encountered in studies of

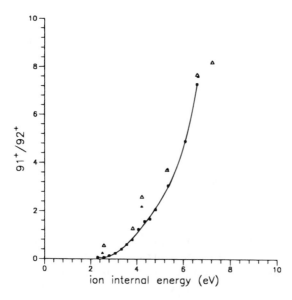

FIGURE 8.4

$91^+/92^+$ ratio as a function of internal energy of the $M^{+\bullet}$ ion of n-butylbenzene ($\bullet$) PEPICO[103] technique at 298 K and ($\triangle$) charge exchange mass spectrometry[106] at 373 K.

the $91^+/92^+$ ratio of the $M^{+\bullet}$ ion of n-butylbenzene has been presented by Boyd *et al.*[107]

Measurements of the $91^+/92^+$ ratio of the $M^{+\bullet}$ ion of n-butylbenzene have been used to assign ion internal energies upon high-energy CID,[108] low-energy CID in a triple quadrupole mass spectrometer,[109] and on both ion cyclotron resonance and hybrid magnetic sector/quadrupole, BQQ, mass spectrometers.[110] Fragmentation of the $M^{+\bullet}$ ion of n-butylbenzene has also been used as a chemical thermometer in a variety of QITMS studies, including the characterization of resonant excitation CID,[5,14,111] He buffer gas studies,[99] photodissociation studies,[30,31] boundary-effect dissociation,[14,21] and pulsed axial activation.[24] Berberich *et al.*[112] have also constructed the charge exchange breakdown graph for the $M^{+\bullet}$ ion of n-butylbenzene observed with a QITMS.

It should be noted that assigning absolute ion internal energies from the breakdown graphs in Fig. 8.4 using experimentally-measured $91^+/92^+$ ratios is difficult since assignments can vary based on the type of instrument, the activation method and the time-scales employed. However, it is generally agreed that an increase in the observed $91^+/92^+$ ratio corresponds to an increase in the average internal energy deposited in the ion. As such, throughout the discussion below, the energetics of CID upon resonant excitation will be discussed in terms of the observed $91^+/92^+$ ratio as opposed to the assignment of absolute ion internal energies from the breakdown graphs.

The CID efficiencies obtained under a given set of resonant excitation conditions will be calculated from the well-known equations:[113]

$$fragmentation\ efficiency = E_F = \frac{\Sigma\ F_i}{(P + \Sigma\ F_i)} \tag{8.14}$$

$$collection\ efficiency = E_C = \frac{(P + \Sigma\ F_i)}{P_0} \tag{8.15}$$

$$CID\ efficiency = E_{CID} = E_F \times E_C = \frac{\Sigma\ F_i}{P_0} \tag{8.16}$$

where P_0 is the initial parent ion intensity, and P and F_i are the parent and daughter ion intensities following CID. The ion intensities of P_0, P, and F_i were all obtained with the same scan function with P_0 being measured in the absence of resonant excitation and P and F_i being measured after resonant excitation. The F_i intensities were corrected for any ion intensities or noise which may have been present in the P_0 spectra (that is, often the m/z 92 fragment ion was formed following mass-isolation of the m/z 134 but in the absence of resonant excitation).

A.   HYPERION Simulations of Resonant Excitation

An ion trapped in the quadrupole field of a QITMS will undergo a periodic trajectory in both the xz and xy planes as illustrated by the HYPERION[36,37] simulations of an ion of m/z 134 presented in Figs. 8.5(a) and (b). The secular frequencies of ion motion in the axial and radial directions are a function of the Mathieu a_u and q_u parameters (Eqs. (8.2) and (8.3)), that is, for a set electrode geometry and size and a set RF drive frequency, the axial and radial frequencies of an ion of a given m/z ratio are a function of the RF and DC voltages applied to the ring electrode as described by Eqs. (8.4) and (8.6). Every ion of differing m/z ratio has a unique set of axial frequencies for a given set of RF and DC voltages.

Fig. 8.6 presents the trajectory of a m/z 134 ion in the xz plane (inset) along with the corresponding Fourier transform of z *versus* time of the ion trajectory. The transformation results in not only a fundamental axial frequency component at 118.8 kHz, but two lower amplitude harmonic overtones at 981.2 and 1218.8 kHz, respectively. The fundamental and harmonic overtones of a given ion always fall within discrete ranges as described by Eq. 8.6. For the Finnigan ITMS with an RF drive frequency of 1.1 MHz, the fundamental frequencies fall within the range of 0 to 550 kHz, where 550 kHz corresponds to one-half of the RF drive frequency. The higher frequency harmonic overtones fall within the ranges of 550 to 1100 kHz and

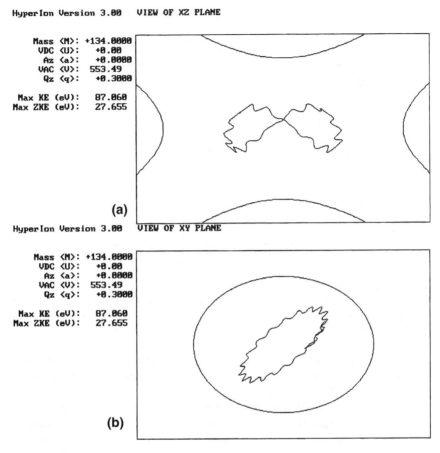

FIGURE 8.5
HYPERION simulation of the trajectory of an ion of m/z 134 at $q_z = 0.300$ in the (a) xz plane and (b) xy plane.

1100 to 1650 kHz, respectively. When these frequencies are summed, the result is a beat pattern consisting of the principal frequency overlaid with a small ripple as seen in the inset in Fig. 8.6. Note that this contradicts the interpretation of Wuerker *et al.* of the ripple seen in photomicrographs of trapped alumina microparticles, namely that such ripples are at the RF drive frequency.[3]

Figs. 8.7(a)–(d) present the results of a series of HYPERION simulations of an ion of m/z 134 stored in a QITMS with no He buffer gas. Fig. 8.7(a) shows the simulated ion trajectory both before and after application of a 118.8 kHz resonant excitation potential. The voltage of the supplemental potential is 400 mV_{p-p}. The small Lissajous figure in the center of the ion trap is the trajectory for the ion before excitation, the larger Lissajous, after excitation. Note that because the excitation potential is resonant with the

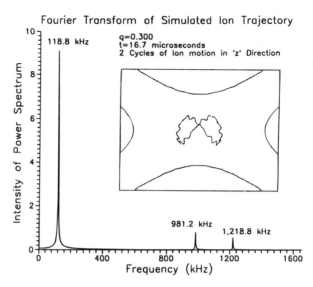

FIGURE 8.6

Fourier transform of the HYPERION simulated trajectory of a m/z 134 ion at $q_z = 0.300$ (inset) showing the frequency components of the ion's axial motion. The secular frequency component is at 118.8 kHz.

fundamental secular component of the ion's axial frequency spectrum, the amplitude of the ion's trajectory increases in the axial direction, with no increase in its radial amplitude.

Fig. 8.7(b) presents the corresponding change in z-coordinate of the m/z 134 ion as a function of time upon resonant excitation. No resonant excitation is applied for the first 200 μs, at which time the resonant excitation voltage is changed from 0 to 400 mV_{p-p}. After 1000 μs of irradiation, the ion is seen to be ejected from the bounded limits of the ion trap in the z-direction. Note that the increase in amplitude of the oscillatory motion in the z-direction is linear with time. The beating pattern evident in the boundaries of the trajectory envelope is attributed to the higher-order harmonic frequencies beating against themselves and the fundamental frequency. Fig. 8.7(c) is a plot of the calculated maximum axial ion kinetic energy as a function of time for the simulation performed in Fig. 8.7(a), while Fig. 8.7(d) shows the corresponding plot of the maximum axial kinetic energy *versus* the z-excursion of the ion from the center of the trap in the z-direction. It can be seen that the maximum axial kinetic energy of the ion increases with time until a maximum energy of ≈225 eV is attained, at which point the ion is ejected from the trap. Moreover, the highest axial kinetic energy corresponds to an ion trajectory with maximum excursion of the ion in the z-direction. These simulated results can be understood in light of the fact that the strength of the RF trapping field increases with increasing distance from the center of the trap. Thus, as the distance the ion travels from the center of the trap increases upon resonant excitation, the ion's kinetic energy increases. Note that the ion kinetic energy does not increase linearly as the excursion of the ion's trajectory increases (Fig. 8.7(d)), since kinetic energy is proportional to velocity squared and velocity is proportional to the distance from the center of the trap.

B.   Effects of the Resonant Excitation Frequency

Theoretically, strong axial resonant excitation should occur only upon application of a supplemental frequency corresponding to the fundamental axial secular frequency. However, simulation (Fig. 8.8(a)) and experiment (Fig. 8.8(b)) both reveal that a range of supplemental frequencies can effect resonant excitation. In Fig. 8.8(a), the final maximum kinetic energy of an ion of m/z 134 calculated in HYPERION is plotted *versus* the frequency of a 600 mV supplemental excitation potential applied for 0.50 ms. Fig. 8.8(b) presents the experimentally-determined loss in intensity of the m/z 134 $M^{+\bullet}$ ion of n-butylbenzene as a function of the frequency of the applied excitation potential under the same conditions. No buffer gas was used in either the simulation or the experimental study. It can be seen that the simulated frequency yielding the maximum ion kinetic energy (118.8 kHz) is almost identical with that corresponding to the maximum loss of the parent ion observed experimentally (118.7 kHz). It is interesting to note that the bandwidth at 15% height of the experimental curve (Fig. 8.8(b)) is 2.2 kHz, which is less than the 3.3 kHz simulated bandwidth at 15% height (Fig. 8.8(a)). The simulated curve is, in fact, 2.2 kHz wide at a peak height corresponding to a calculated kinetic energy of approximately 36 eV. This difference in simulated *versus* experimental bandwidths may be attributed to the fact that the experimentally-monitored decrease in the $M^{+\bullet}$ ion intensity only occurs once the threshold for either CID or ejection has been reached. Since these thresholds will be attained only after a sufficient increase in kinetic energy, the experimentally-determined bandwidth at a given peak height will be less than that of the simulated frequency bandwidth at the same peak height.

Fig. 8.9(a) presents the change in intensity of the m/z 91, m/z 92 and m/z 134 ions of ionized n-butylbenzene as a function of the frequency of the applied resonant excitation potential. The excitation voltage of 500 mV was applied for 10 ms with He buffer gas present in the trap. It can be seen that as the applied frequency approaches that of the $M^{+\bullet}$ ion's theoretical axial secular frequency (118.8 kHz), CID occurs, resulting in a decrease in the intensity of the m/z 134 ion and a subsequent increase in m/z 91 and 92 daughter ion intensities. (A number of other less intense daughter ions at m/z 39, 65, 105, and 119 are also observed but are not shown in Fig. 8.9(a).)

Hyperlon Version 3.00   UIEW OF XZ PLANE

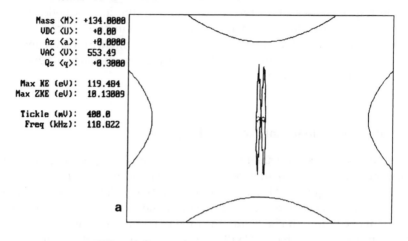

Mass ⟨M⟩: +134.0000
VDC ⟨U⟩:   +0.00
Az ⟨a⟩:   +0.0000
VAC ⟨V⟩:   553.49
Qz ⟨q⟩:   +0.3000

Max KE (eV): 119.484
Max ZKE (eV): 10.13009

Tickle (mV): 400.0
Freq (kHz): 118.822

a

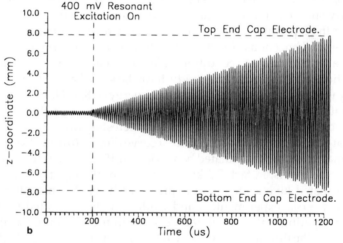

b

FIGURE 8.7

This frequency optimization curve indicates that CID is occurring over an effective bandwidth of 1400 Hz (at 10% loss of m/z 134) and not at a single excitation frequency. Note that this frequency bandwidth is approximately one-half of that observed for the experimental loss of m/z 134 in the absence of buffer gas (Fig. 8.8(b)). As will be seen below, in the presence of buffer gas, the frequency bandwidths are substantially narrower than those in the absence of buffer gas.

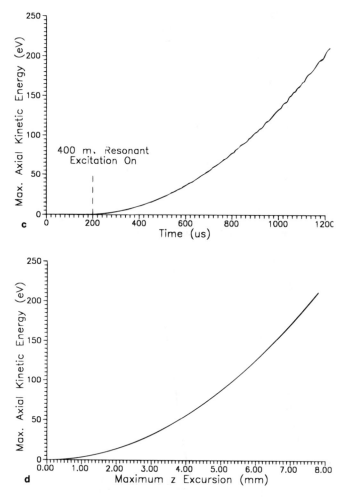

FIGURE 8.7

(a) HYPERION simulation of the motion of a m/z 134 ion at $q_z = 0.300$ before (smaller orbit) and after resonant excitation (larger orbit) in the xz plane using a 400 mV, 118.8 kHz resonant excitation potential; (b) HYPERION calculation of the motion of an ion of m/z 134 at $q_z = 0.300$ in the z-direction before and after application of a 400 mV, 118.8 kHz resonant excitation potential; (c) Calculated maximum axial kinetic energy of an ion of m/z 134 as a function of resonant excitation time; $q_z = 0.300$; 400 mV, 118.8 kHz excitation potential; (d) Calculated maximum axial kinetic energy of m/z 134 as a function of the ion's maximum excursion in the z-direction; $q_z = 0.300$; 400 mV, 118.8 kHz excitation potential.

The CID efficiencies (E_F, E_C, E_{CID}, which consider all observed daughter ions) and the $91^+/92^+$ ratio of the data in Fig. 8.9(a) are presented in Fig.

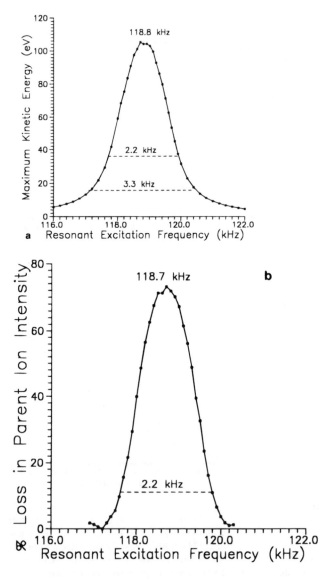

FIGURE 8.8

(a) Calculated kinetic energy of an ion of m/z 134 at $q_z = 0.300$ and (b) the experimentally determined loss of the m/z 134 ion $M^{+\cdot}$ ion of n-butylbenzene as a function of resonant excitation frequency. Conditions: $q_z(m/z\ 134) = 0.300$; resonant excitation amplitude and duration are 600 mV_{p-p} and 0.50 ms, respectively; no buffer gas.

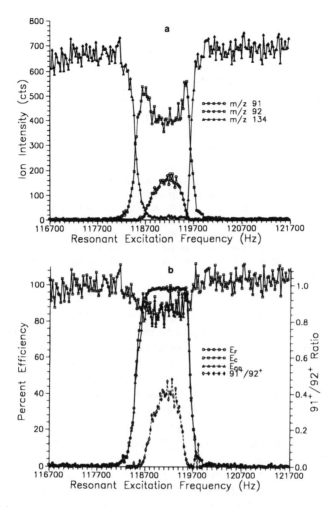

FIGURE 8.9
Optimization of the resonant excitation frequency for the CID of the $M^{+\cdot}$ ion (m/z 134) of
n-butylbenzene in terms of (a) the intensities of the m/z 91, m/z 92, and m/z 134 ions and
(b) the corresponding CID efficiencies (E_F, E_C, E_{CID}) and $91^+/92^+$ ratio. Conditions: $q_z(m/z$
134) = 0.30; resonant excitation amplitude and duration are 500 mV_{p-p} and 10 ms, respectively; helium was admitted into the ion trap to yield an indicated vacuum chamber pressure
of 1.0×10^{-5} Torr.

8.9(b). An increase in the E_F and E_{CID} are observed as the frequency of the
excitation potential approaches the resonant frequency of the m/z 134 ion.
Often the E_F can reach 95 to 100%, that is, little or no parent ion survives the
CID process. Although the E_{CID} is often very high, it is less than the E_F due
to the observed decrease in the E_C. A decrease in the E_C can result from at
least three processes: (1) formation of daughter ions which are below the

TABLE 8.3

Secular frequencies and coupling coefficients for resonant excitation CID of the $M^{+\cdot}$ ion of n-butylbenzene at a q_z ($M^{+\cdot}$) = 0.3

Type Frequency	Secular Frequency[a] (Hz)		Coupling Coefficients[b]	
	Experimental	Simulation	Experimental	Simulation
Fundamental	120,847	121,877	100%	100%
First harmonic	980,453	979,080	6.9%	7.6%
Second harmonic	1,221,960	1,221,968	3.1%	3.6%

[a] The experimental and simulation frequencies were determined from the resonant excitation frequency yielding the highest E_{CID} and the Fourier transform of the z-position of the ion *versus* time from ion trajectory calculations, respectively.
[b] The experimental coupling efficiencies were calculated as the ratio of the appearance potential (that is, the resonant excitation voltage required to just observe an ion's appearance) of m/z 91 at the harmonic frequency to the appearance potential of m/z 91 at the fundamental frequency. The coupling efficiencies calculated from simulations were the areas of the harmonic Fourier transform peaks normalized to the area of the fundamental Fourier transform peak (see Fig. 8.6).

low m/z cutoff (m/z 44) of the experiment and thus are ejected prior to detection; (2) formation of daughter ions sufficiently away from the center of the trap so that, although theoretically stable, they are inefficiently trapped; and (3) ejection of the parent ion from the trap. All of these processes are maximized at the secular frequency of the parent ion and thus all can contribute to the decrease in E_C seen in Fig. 8.9(b).

Recall that in Fig. 8.6, in addition to its fundamental secular frequency, an ion also has higher secular frequencies centered about the RF drive frequency. The molecular ion of n-butylbenzene was excited resonantly at the fundamental and each of the first two harmonic frequencies with varying resonant excitation potentials to generate energy-resolved breakdown curves.[114] The experimental frequencies yielding the maximum E_{CID} agree well with those calculated from computer simulations of ion motion (Table 8.3). From the energy-resolved breakdown curves, it was evident that the resonant excitation of ions at the harmonic frequencies was less efficient at coupling resonant excitation into CID; that is, resonant excitation at the first and second harmonic frequencies required excitation voltages of higher amplitude in order to produce qualitatively similar daughter spectra. As is seen in Table 8.3, the simulated and experimental coupling coefficients are in reasonably good agreement and indicate the inefficient resonant excitation which occurs at the harmonic frequencies. Theoretical calculations and simulations of ion motion have demonstrated that the contributions of these harmonic frequencies to the overall ion motion increase with increasing q_z of the parent ion.[37] Although higher parent ion q_z values are not utilized normally for CID studies from a practical standpoint (as the q_z value of the parent ion is increased, the range of daughter ion masses which can be

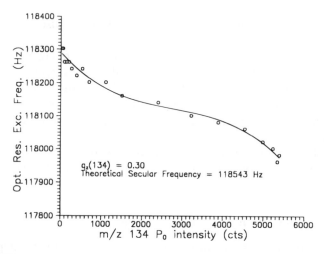

FIGURE 8.10

Effect of space charge on the secular frequency of the parent ion: shift in the resonant excitation frequency which is optimum for the maximum production of the m/z 91 daughter ion resulting from CID of the $M^{+\cdot}$ ion (m/z 134) of n-butylbenzene as a function of the P_0 (m/z 134) ion intensity. Conditions: $q_z(m/z$ 134) = 0.30; irradiation duration is 5 ms; helium admitted to the vacuum chamber to yield an indicated pressure of 9.0×10^{-5} Torr.

trapped is reduced), it would be interesting to investigate the use of harmonic frequencies for axial modulation techniques.

One of the experimental parameters which affects the optimum resonant excitation frequency is the population or, more correctly, the density of ions in the ion trap. When the density of the ions in the ion trap becomes sufficiently high that ion–ion interactions become significant, space-charge effects will become evident. Fig. 8.10 presents the results of a series of experiments in which the optimum resonant excitation frequency was determined for the m/z 134 $M^{+\cdot}$ ion of n-butylbenzene as a function of the parent ion population (varied by changing the duration of ionization). It can be seen that as the parent ion population increases, the secular frequency of the m/z 134 ion decreases (Fig. 8.10). Note that this shift to lower frequency corresponds to an apparent shift to a higher m/z ratio. Because of this shift in resonant excitation frequency with changes in ion population, a simple frequency optimization should be performed for experiments in which a constant amount of sample is being introduced, for example, in batch inlet experiments. However, in experiments where the sample concentration varies over a short time interval, for example, that of a chromatographic peak, it is not possible to construct a frequency optimization curve at each stage of the changing ion population due to the time need for frequency optimization. As such, both broadband excitation methods discussed earlier[16,17] and a method employing frequency-assignment pre-scans have been developed to overcome the difficulty of shifting secular frequencies[16] for frequency assignments during MS/MS analyses, as discussed in Chapter 4.

It should also be noted that the secular frequencies of Fig. 8.10 are lower than those observed in Figs. 8.8 and 8.9. The theoretical secular frequency (118,543 Hz) for the experiment of Fig. 8.10 is lower due to the application to the ring electrode of an RF voltage of slightly lower amplitude during the resonant excitation period (see discussion of parent ion q_z effects below). In addition to the effect of the ion population, the further decrease from the theoretical secular frequency is due to a decrease in secular frequency in the presence of a buffer gas (see discussion of buffer gas effects below).

C.   Effects of the Resonant Excitation Voltage

The shape of the frequency optimization curve is highly dependent upon the voltage and duration of the resonant excitation potential. In Fig. 8.11(a) are presented the results of a series of experiments in which the frequency optimization curves were obtained for the m/z 134 $M^{+\cdot}$ ion of n-butylbenzene using resonant excitation voltage amplitudes of 150, 500, and 1000 mV. A resonant excitation duration of 5 ms was employed at each resonant excitation amplitude. The most obvious effect of increasing the resonant excitation amplitude is an increase in the effective excitation frequency bandwidth. The frequency bandwidth (measured at $E_F = 20\%$) was obtained at a variety of resonant excitation amplitudes and then transformed into an effective mass resolution for resonant excitation. This transformation was effected by correlating the experimentally-determined bandwidth with the theoretical secular frequencies of the ions $\pm 1\,u$ on either side of the $M^{+\cdot}$ ion (1 $u \approx 922$ Hz at $q_z(m/z\,134) = 0.3$). The resulting effective resolutions for resonant excitation (expressed as a frequency bandwidth in atomic mass units) as a function of excitation amplitude are presented in Fig. 8.11(b). It can be seen that as the resonant excitation amplitude is increased, the excitation resolution increases linearly until ions in a 6 to $7u$ wide window are in resonance with the applied excitation potential. Thus, for any given CID experiment, when both the parent ion is isolated with poor efficiency and a sufficiently large resonant excitation amplitude is employed, ions adjacent to the parent ion will also be in resonance and may undergo CID. Moreover, if the parent ion fragments to form daughter ions by loss of $H^{\cdot}$ or H_2 (for example, from molecular ions of polyaromatic hydrocarbons), then these daughter ions will also be excited resonantly and may undergo dissociation subsequently. The influence of the resonant excitation amplitude on the effective frequency bandwidth should also be considered when performing mass range extension,[48,49] high resolution,[115-118] and parent and neutral loss scans[119] with the QITMS, since these techniques rely on resonant excitation for ion ejection and detection.

Fig. 8.11(a) also illustrates that as the amplitude of the resonant excitation voltage is increased, the resonant frequency which gives rise to the maximum E_{CID} shifts to values which are either greater (≈ 120.5 kHz) or less than (≈ 117.5 kHz) the theoretical secular frequency of the m/z 134 ion (118.8 kHz). With large resonant excitation potentials, E_{CID} is decreased in the center of the effective frequency band due to the decrease in collection and trapping efficiencies of parent and daughter ions. One practical aspect of radiating an ionic species at a frequency significantly removed from the theoretical value, due perhaps, to an increase in the trapped ion population, is that maximum CID efficiency can still be obtained by increasing the amplitude (or, as will be seen below, the duration) of the resonant excitation potential so as to widen the resonant frequency bandwidth. However, the widened frequency bandwidth reduces the effective mass resolution for resonant excitation. It should be noted also that, in order to determine more accurately the optimum resonant excitation frequency, normally low resonant excitation potentials are applied to yield curves similar to that obtained at 150 mV and shown in Fig. 8.11(a).

The amplitude of the resonant excitation potential also affects the observed secular frequency of the parent ion. The secular frequency of the parent ion was taken to be the resonant excitation frequency which yielded the highest m/z 91 daughter ion (that is, the frequency yielding the highest energy deposition). This resonant excitation frequency for maximum m/z 91 yield is seen to increase with increasing resonant excitation amplitude (Fig. 8.11(c)). A similar shift has been observed by other researchers.[120] This frequency shift with resonant excitation amplitude may be due to the perturbation of the trapping field by the resonant excitation potential; further work is needed to elucidate the mechanism responsible for this phenomenon. In comparison to the resonant excitation frequency corresponding to the middle of the effective resonant excitation frequency band (Fig. 8.11(c)), it is seen that the resonant excitation frequency yielding the maximum yield of the m/z 91 daughter ion is increasingly shifted towards the higher frequency side of the resonant excitation band. Thus, the curve representing the yield of m/z 91 daughter ions or the $91^{+}/92^{+}$ ratio becomes more skewed at higher resonant excitation amplitudes than the more Gaussian-shaped curves at lower resonant excitation amplitudes.

The resonant excitation amplitude-resolved breakdown graph of the m/z 134 ion M$^{+\cdot}$ ion of n-butylbenzene is presented in Fig. 8.12(a). An excitation duration of 5 ms was employed. As expected, the m/z 92 daughter ion arising from the lower energy rearrangement reaction, appears at a lower resonant excitation amplitude than that of the m/z 91 daughter ion arising from the higher energy process. Moreover, as the excitation amplitude is increased, the intensity of the m/z 92 ion decreases with a corresponding increase in the intensity of the m/z 91 daughter ion, indicating higher internal energy deposition in the m/z 134 ion at

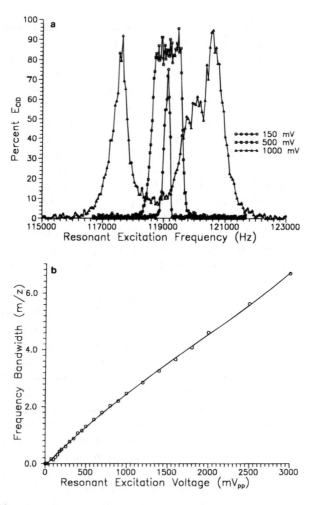

FIGURE 8.11

higher amplitudes of the excitation potential. The decrease in both the m/z 91 and 92 ion intensities above ≈800 mV is due to an increase in resonant ejection of the m/z 134 ion at these higher excitation amplitudes.

The percent E_{CID} as well as the $91^+/92^+$ ratio obtained from the data in Fig. 8.12(a) is presented in Fig. 8.12(b). Following the onset of CID, the E_{CID} rapidly reaches its maximum (≈90%) and then begins to decline, due to the anticipated decrease in the E_C of the parent and daughter ions discussed above. However, the amount of energy which can be imparted to the parent ion continues to increase until it plateaus (around 1000 mV in this case). It should be noted that the increase in "noise" of the $91^+/92^+$ ratio at higher voltages is due to the relatively low ion intensities.

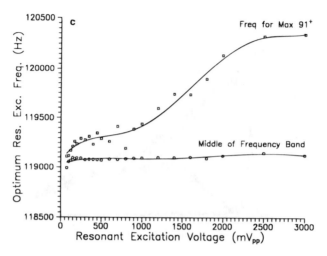

FIGURE 8.11

(a) E_{CID} of the $M^{+\cdot}$ ion of n-butylbenzene as a function of resonant excitation frequency at three resonant excitation voltages; (b) effective mass resolution obtained upon resonant excitation (expressed as a frequency bandwidth in *amu* units) as a function of the excitation voltage; (c) the resonant excitation frequency optimum for yielding the maximum m/z 91 daughter ion and the resonant excitation frequency in the middle of the resonant excitation frequency bandwidth as a function of the resonant excitation voltage. Conditions: $q_z(m/z$ 134) = 0.30; irradiation duration of 5 ms; helium admitted to the ion trap to yield an indicated vacuum chamber pressure of 1.0×10^{-5} Torr.

Resonant excitation frequency optimization curves were obtained for the $M^{+\cdot}$ ion of n-butylbenzene over a range of resonant excitation amplitudes (0 to 3000 mV); three examples of E_{CID} curves were shown in Fig. 8.11(a). The maximum E_{CID} and maximum $91^+/92^+$ ratio were obtained from each of the frequency optimization curves and are plotted *versus* the corresponding resonant excitation amplitude in Fig. 8.13. As noted above, the maximum E_{CID} was increasingly removed from the theoretical secular frequency as the amplitude of the resonant excitation was increased. However, Fig. 8.13 illustrates that the maximum E_{CID} occurred at a relatively low amplitude ($\approx$150 mV) and remained relatively constant over the range of resonant excitation voltages employed. Moreover, the maximum E_{CID} was between 85 and 95%. In contrast to the E_{CID}, the maximum $91^+/92^+$ ratio always occurred near the secular frequency of the parent ion. The amount of internal energy transferred, based on the maximum $91^+/92^+$ ratio in Fig. 8.13, was found to increase sharply and then level off around 1500 mV. A maximum $91^+/92^+$ ratio of 1.4 was obtained, which, according to the PEPICO breakdown graph in Fig. 8.4, corresponds to an internal energy of the parent ion of n-butylbenzene of $\approx$4 eV. This value is in

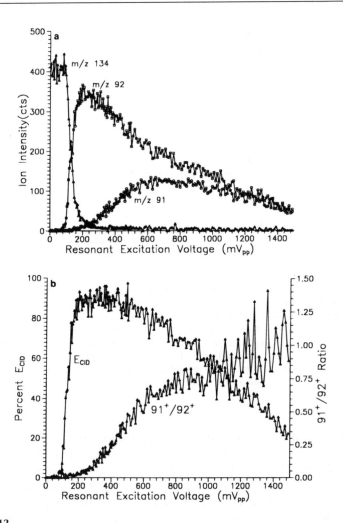

FIGURE 8.12
Resonant excitation voltage-resolved breakdown graph of the $M^{+\bullet}$ ion of n-butylbenzene (m/z 134) in terms of (a) the intensities of the m/z 91, m/z 92, and m/z 134 ions and (b) the corresponding E_{CID} and $91^+/92^+$ ratios. Conditions: $q_z(m/z\ 134) = 0.30$; duration of excitation of 5 ms; helium admitted to the ion trap to yield an indicated vacuum chamber pressure of 1.0×10^{-5} Torr.

keeping with the average internal energies for the $M^{+\bullet}$ ion of n-butylben-zene reported previously for resonant excitation CID on a QITMS.[5]

D.   Effects of the Duration of Resonant Excitation

The resonant excitation time-resolved breakdown graph of the m/z 134 $M^{+\bullet}$ ion of n-butylbenzene obtained at a resonant excitation amplitude

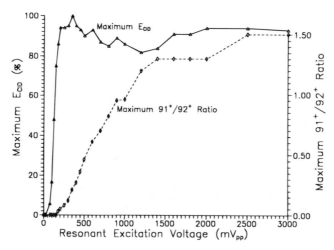

FIGURE 8.13
Maximum E_{CID} and $91^+/92^+$ ratio for the CID of the $M^{+\cdot}$ ion of n-butylbenzene (m/z 134) obtained from the frequency optimization curves constructed at a variety of excitation voltages. Conditions: $q_z(m/z$ 134$) = 0.30$; duration of excitation of 5 ms; helium admitted to the ion trap to yield an indicated vacuum chamber pressure of 1.0×10^{-5} Torr.

of 1000 mV is presented in Fig. 8.14(a). The resonant excitation time-resolved breakdown graph is qualitatively similar to the resonant excitation amplitude-resolved graph presented in Fig. 8.12(a), that is, initially (0 to 500 μs) the parent ion does not acquire sufficient kinetic energy to undergo CID. However, after a given period of irradiation, the threshold for CID is reached and the parent ion dissociates to produce daughter ions. As expected, the lower energy m/z 92 ion appears at a shorter duration of excitation (≈ 0.500 ms) than the higher energy m/z 91 ion (≈ 1.50 ms).

Fig. 8.14(b) presents the CID efficiencies (E_{CID}, E_C, E_F) and $91^+/92^+$ ratio calculated from the breakdown graph in Fig. 8.14(a). The E_{CID} obtained by increasing the resonant excitation duration is similar to that seen upon increasing the resonant excitation amplitude (Fig. 8.12(b)) and is on the order of 95 to 100%. Along with the overall increase in E_{CID} as the duration of resonant excitation is increased, the $91^+/92^+$ ratio increases, also thus indicating that more energetic CID can be brought about with prolonged excitation. It is interesting to note that it appears as if the E_C exceeds 100% in the time-resolved resonant excitation experiment resulting in E_{CID} values which are somewhat greater than 100% when excitation is maintained for a period of 2 ms. This increase in E_C can be attributed to two phenomena: (1) an increase in the trapping efficiency of the m/z 91 and m/z 92 daughter ions (ΣF_i in Eq. (8.15)) over that measured for the initial parent ion population (P_0 in Eq. (8.15)); or (2) an increase in the

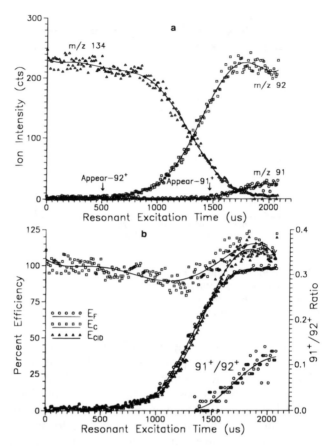

FIGURE 8.14

Resonant excitation time-resolved breakdown graph of the M$^{+\bullet}$ ion (m/z 134) of n-butylbenzene in terms of (a) the intensities of the m/z 91, m/z 92, and m/z 134 ions and (b) the corresponding E_{CID} and 91$^+$/92$^+$ ratios. Conditions: $q_z(m/z$ 134) = 0.30; resonant excitation amplitude of 1000 mV_{p-p}; helium admitted to the ion trap to yield an indicated vacuum chamber pressure of 1.0×10^{-5} Torr.

trapping efficiency of the resonantly-excited parent ions (P in Eq. (8.15)) over that measured for the initial parent ion population, P_0. Both of these phenomena have been observed experimentally and the latter has been supported by recent HYPERION simulations which show that resonant excitation can actually enhance the damping, and thus trapping efficiency, of ions.[37]

It is actually the combination of the duration of resonant excitation and amplitude which controls the CID energy for a given excitation frequency and parent ion a_z and q_z values. HYPERION simulations of resonant excitation in the absence of buffer gas indicate that the product of the resonant excitation amplitude and duration, or fluence (Vms),[38] required to achieve a particular z-excursion (and thus, kinetic energy) should remain constant. A constant fluence relationship was also predicted in the simulation studies of March et al.,[38] who found that the fluence required for the resonant ejection of an ion of m/z 100 remained constant at a q_z of 0.1955.

An experimental determination of the relationship between the duration of resonant excitation and the excitation amplitude is presented in Figs. 8.15(a)–(c). Because the z-excursion of the ion could not be monitored experimentally, the ion energy was measured by monitoring the appearance time of the m/z 91 and m/z 92 daughter ions with a variety of excitation voltages (Fig. 8.15(a)). The resulting fluences for the m/z 92 ions are plotted as a function of the corresponding excitation amplitudes in Figs. 8.15(b,c). From Fig. 8.15(a), the duration and amplitude of the resonant excitation potential required to achieve the appearance of daughter ions are related inversely. However, from Fig. 8.15(b), it can be seen that with increasing resonant excitation voltage, the fluence is not constant as predicted by the HYPERION simulations; rather, it increases steadily (line 1 of Fig. 8.15(b)). It was thought initially that this increase might be due to the differences in the durations and amplitudes requested *versus* the actual resonant excitation. Although the measured actual durations and amplitudes differed from those requested in the software, the product of the actual durations and amplitudes still shows an increase with increasing resonant excitation amplitude (line 2 of Fig. 8.15(b)), although with a smaller slope than that obtained using the requested durations and amplitudes. Thus, the experimental results indicate that in order to achieve a given ion energy, an increase in the amplitude-duration product over that predicted theoretically is required.

The error towards over-estimating the amplitude-duration product increases as the irradiation periods get shorter and the amplitudes get larger. When resonant excitation amplitudes of less than 550 mV are considered only, then lines 3 and 4 of Fig. 8.15(c) are obtained. Although there is a lot of scatter in the data from trying to measure the first appearance of an ion, a more constant fluence is found at these lower excitation voltages in agreement with the relationship predicted by the HYPERION simulations. With these lower amplitudes, longer irradiation periods are required and the resonant excitation process may be more efficient. It may be that, with resonant excitation voltages of high amplitude applied for short durations, the increase in kinetic energy of the resonated ion may lag behind that predicted theoretically. Such a lag would be more pronounced in the presence of buffer gas. Further research is required in this area to delineate fully the relationship between the resonant excitation duration and amplitude. However, it can be concluded that it is the product of the duration and magnitude of the resonant excitation potential

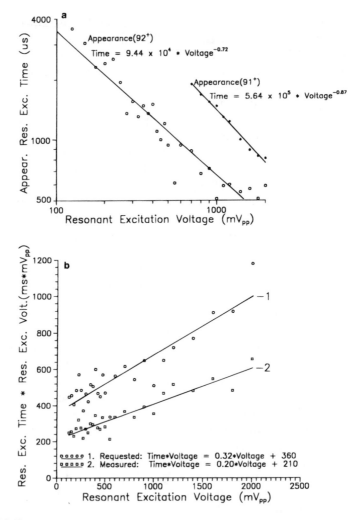

FIGURE 8.15

(that is, fluence) that determines the energetics of the resonant excitation process at specific parent ion a_z and q_z values and buffer gas pressure.

During the CID period, a number of different ion/molecule reactions can occur between the parent ion and neutral molecules: collisions resulting in loss of parent ion kinetic energy; collisions resulting in CID of the parent ion; ion/molecule reactions (for example, charge exchange) resulting in new parent ions of low kinetic energy which must be excited resonantly; and ion/molecule reactions (for example, ion/neutral adduct formation) resulting in loss of parent ion and daughter ions. The degree

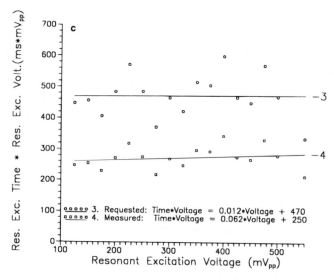

FIGURE 8.15

(a) Appearance time of the m/z 91 and m/z 92 daughter ions of the $M^{+\cdot}$ ion (m/z 134) of n-butylbenzene as a function of resonant excitation voltage; (b) the product of the resonant excitation duration and amplitude (fluence) as a function of the corresponding amplitude for the appearance of the m/z 92 ion; (c) a subset of the figure in (b). The circles represent the amplitudes and durations requested while the squares represent the measured amplitudes and durations. The equations for lines 1 and 2 were determined from linear fits to all of the circles and squares, respectively. The equations for lines 3 and 4 were determined from linear fits to only the circles and squares, respectively, shown in c. Conditions: $q_z(m/z$ 134) = 0.30; helium admitted to the ion trap to yield an indicated vacuum chamber pressure of 1.0×10^{-5} Torr.

to which each of these reactions occurs is determined by the amplitude and duration of the resonant excitation potential and the concentration and nature of ions and neutrals present in the trap. The fraction of ion/molecule reactions leading to CID of the parent ion is increased by increasing the voltage while shortening the duration of the resonant excitation potential. The fraction of ion/molecule reactions which do not result in CID is increased by increasing the duration of the resonant excitation potential. To reduce the significance of these unwanted ion/molecule reactions, CID is generally performed with resonant excitation times of 5 to 10 ms.

E.   Effects of the Parent Ion q_z Value

Two other instrumental parameters which affect the CID process are the Mathieu a_z and q_z parameters of the parent ion. The a_z and q_z values are defined by the amplitudes of the DC and RF voltages, respectively, applied to the ring electrode during excitation which, in turn, determine the secular frequency of the parent ion and the mass range of daughter ions, which may form stable trajectories in the trap following CID. The a_z and q_z values also determine the maximum amount of kinetic energy which can be imparted successfully to the ion upon excitation. Although the a_z of the parent ion can influence CID, all of the studies which are presented here were performed on the q_z-axis ($a_z = 0$) of the stability diagram (Fig. 8.1); that is, only an RF voltage trapping field was utilized during the resonant excitation CID experiments.

In Figs. 8.16(a,b) are presented the results of a series of experiments which were used to investigate the role of the q_z value of the $M^{+\cdot}$ ion of *n*-butylbenzene on the efficiencies and energetics of the CID process. Briefly, in these experiments a frequency optimization curve was obtained at a relatively low excitation amplitude for each q_z value employed. The maximum E_{CID} and the $91^+/92^+$ ratio were determined from a complete resonant excitation voltage breakdown study taken at this optimum resonant excitation frequency.

Note that in these experiments loss of the $M^{+\cdot}$ ion can occur either through ejection of the $M^{+\cdot}$ ion from the ion trap or through CID. At low q_z values, the RF trapping field is relatively weak or, in terms of the pseudo-potential well model, the well depth is relatively shallow. Thus, at low q_z values, the $M^{+\cdot}$ ion is lost principally *via* ejection under the conditions of relatively short duration (5 ms) resonant excitation. With increased q_z, the trapping field strength increases and the potential well depth deepens. Thus, with increasing q_z values, the parent ion is contained effectively and is lost principally due to CID. Note also that above a parent ion q_z value of ≈ 0.62, the trajectories of the m/z 91 and m/z 92 ions are unstable and thus these ions are not stored.

Fig. 8.16(a) indicates that the maximum E_{CID} is obtained over the range of $0.2 \le q_z \le 0.4$. The low E_{CID} at $q_z < 0.2$ can be attributed to resonant ejection of the $M^{+\cdot}$ at these relatively weak RF trapping fields (that is, the shallowness of the pseudo-potential well results in ions being ejected before they can accumulate sufficient internal energy to fragment). The maximum E_{CID} (95%) which can be obtained is reached at $q_z = 0.25$ and remains relatively constant until the major daughter ions m/z 65, and both m/z 91 and m/z 92 reached their limits of trajectory stability at q_z (m/z 134) of 0.44 and approximately 0.62, respectively. The trajectories of two minor daughter ions, m/z 105 and 119 remain stable until q_z (m/z 134) values of 0.71 and 0.81, respectively, yielding the approximately 10% E_{CID} seen in Fig. 8.16(a) above the parent ion q_z value of 0.65. The $91^+/92^+$ ratios in Fig. 8.16(b) show that for each resonant excitation voltage, the internal energy deposited upon resonant excitation steadily increases as the q_z value is increased. This effect is most likely due to the fact that

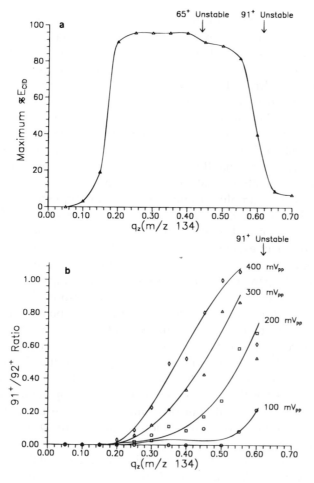

FIGURE 8.16

(a) Maximum %E_{CID} and (b) $91^+/92^+$ ratio (for four different resonant excitation voltages) obtained upon resonant excitation CID of the $M^{+\bullet}$ ion (m/z 134) of n-butylbenzene as a function of the q_z value of the $M^{+\bullet}$ ion. Conditions: duration of resonant excitation of 5 ms; helium admitted to the ion trap to yield an indicated vacuum chamber pressure of 1.0×10^{-5} Torr.

with increasing parent ion q_z, the number of cycles of resonant excitation for a set resonant excitation time increases, as will be seen below. The maximum $91^+/92^+$ ratio observed in these studies was approximately 3, obtained at a q_z (m/z 134) value of 0.5 and resonant excitation amplitude and duration of 1500 mV and 5 ms, respectively. The $91^+/92^+$ ratio of 3 corresponds to an internal energy of $\approx$5 eV, based upon the PEPICO measurements described earlier.

With the application of the QITMS to higher m/z ions,[48,49] consideration should be given to extrapolation of these parent ion q_z studies to these higher m/z ions. If efficient resonant excitation CID can be obtained only at parent ion q_z values greater than 0.25, then for the Finnigan QITMS at its maximum RF voltage (which corresponds to a low m/z cut-off of $\approx m/z$ 650), the maximum ion which can be efficiently resonated for CID would be m/z 2360. In addition, the maximum amount of energy which could be imparted to this ion is relatively small (based upon the pseudopotential well model).[39–41,63–65] Further characterization of the energetics and efficiencies of resonantly exciting more massive ions is needed.

It was shown above that the mass resolution of resonant excitation is a function of the amplitude and duration of the applied supplementary potential. The other parameter affecting the mass resolution of resonant excitation is the q_z value of the parent ion during resonant excitation. From the resonant excitation voltage breakdown curves discussed in the section above, the resonant excitation amplitude which resulted in maximum loss of the parent ion was determined. At this amplitude, a complete resonant excitation frequency optimization was performed and the width of the effective frequency band was measured at the point corresponding to a 20% loss of the m/z 134 parent ion. Note that loss of the parent ion can occur through ejection of the m/z 134 ion from the ion trap, or through CID of m/z 134 to form daughter ions. By monitoring the loss of the parent ion, examinations can be made of low q_z values (where the parent ion is ejected rather than undergoing CID) and high q_z values (where the parent ion undergoes CID but the daughter ions are unstable and are ejected, for example, above a parent ion q_z value of 0.61, m/z 91 is unstable and is not stored).

Fig. 8.17(a) presents the theoretical and experimentally-determined resonant excitation frequency bandwidths as a function of the q_z of the m/z 134 $M^{+\bullet}$ parent ion. With an increase in the q_z value, the effective frequency bandwidth of the resonant excitation voltage decreases from approximately 4400 Hz at a q_z of 0.05 to approximately 500 Hz at a q_z of 0.35. The frequency bandwidth then remains fairly constant until it begins increasing at q_z values above 0.75 with a rapid rise seen above a q_z of 0.85. HYPERION simulations suggest[37] that at these higher q_z values, the ion's trajectory increases in the axial direction, making it relatively easy to eject upon application of a resonant excitation potential. This ease of ejection at higher q_z values forms the basis of axial modulation applied during data acquisition in the Finnigan ITMS.[119] The difference in the theoretical secular frequency of two adjacent ions (1 u apart) can be considered to be the minimum frequency bandwidth for resonant excitation of unit mass/charge range. Note that in Fig. 8.17(a), the theoretical frequency bandwidth steadily increases with increasing parent ion q_z, becoming

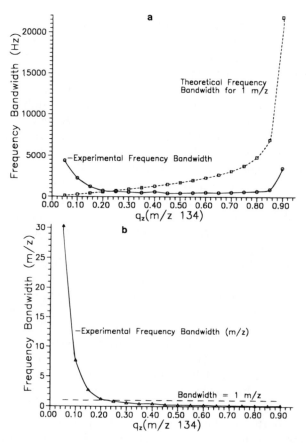

FIGURE 8.17

The resonant excitation frequency bandwidth expressed in units of (a) Hz and (b) atomic mass units as a function of the q_z value of the $M^{+\cdot}$ ion of n-butylbenzene (m/z 134). See the text for an explanation of details.

greater than the experimental frequency bandwidth above a parent ion q_z of 0.2.

The theoretical frequency bandwidth for a mass range of 1 u can be used to convert the experimental frequency bandwidth into a frequency bandwidth in terms of atomic mass units to yield the effective mass resolution of resonant excitation (Fig. 8.17(b)). Now, it is easily seen that for parent ion q_z values below 0.2, very poor mass resolution is achieved upon resonant excitation, for example, at a q_z of 0.05, ions 15 u greater than or less than the parent ion of interest are also in resonance. This result has major implications for the use of resonant excitation/ejection for mass range extension.[48,49] As mass range extension techniques use resonant excitation at low q_z values, the mass resolution which can be achieved will be limited by these ion physics considerations. However, at $q_z > 0.2$, the frequency bandwidth is less than one atomic mass unit wide and continues to decrease with increasing parent ion q_z, reaching a minimum bandwidth of 0.13 u at a parent ion q_z of 0.85. This result explains partially why the best resolution for the high resolution technique is achieved with resonant excitation at a q_z value of 0.85.[115–118] As the ion approaches its stability limit at a q_z of 0.9, even though the absolute frequency bandwidth increases (Fig. 8.17(a)), the frequency separation of two adjacent masses increases more quickly and an overall increase in mass resolution is achieved (Fig. 8.17(b)). This observation is consistent with one of the major advantages of axial modulation, that is, mass resolution of the acquired mass spectra increases dramatically.[121] Thus, for a particular combination of amplitude and duration of the resonant excitation potential, increasing the parent ion q_z results in more energetic CID with greater mass resolution for resonant excitation, but with a smaller daughter mass range.

In conclusion, the E_{CID}, the internal energy deposited upon excitation, the daughter mass range, and the mass resolution of CID all depend upon the q_z value of parent ion during resonant excitation. Generally, for a particular amplitude and duration of the resonant excitation potential, increasing the parent ion q_z results in more energetic CID, but a smaller stable daughter mass range. The characterization studies indicate that the best overall CID can be obtained at ring RF voltages corresponding to a parent ion q_z of 0.25 to 0.30. With this parent ion q_z, energetic CID can be obtained to yield the maximum E_{CID} with unit mass resolution and a reasonable daughter ion mass range. These conclusions are comparable with those reported by Paradisi $et\ al.$[14] who presented a study of the effects of the resonant excitation amplitude and q_z value of the $M^{+\cdot}$ ion of n-butylbenzene on the E_{CID} and ion internal energies deposited during CID.

F. Effects of the Buffer Gas Pressure and Nature

The effects of the buffer gas pressure and identity on the CID process in ion traps have been studied in detail.[60,99,111,114,122–124] Collisions of the ions with the neutral buffer gas reduce the kinetic energy of the ions and effectively collapse the ion cloud toward the center of the trap. This collisional damping results in more efficient ion detection and enhanced resolution.[99]

The pressure of the buffer gas determines the number of collisions the ion will experience while resident in the trap, whereas the nature of the buffer gas will determine the average energy transferred upon collision under a given set of resonant excitation conditions. Thus, higher mass buffer gases should impart more internal energy upon collision and thus effect greater fragmentation efficiency; however, as the mass of the gas is increased, scattering losses also increase, with the result that ion detection efficiency decreases. Because of this, the overall E_{CID} involves trade-offs between mass and pressure of the buffer gas species. To this end, studies using mixtures of buffer gases have been reported.[10,122–124] Morand et al.[124] found that CID of C_{60} and C_{70} buckminsterfullerenes was observed only when a massive gaseous target (Br_2, Cl_2, Kr or Xe) was mixed with the He buffer gas. CID efficiencies in excess of 65% were found with Xe added to the He buffer. Earlier studies by Glish et al.[122] also found that the addition of argon during the resonant excitation promoted the dissociation onset of the parent ion of N,N-dimethylaniline. Moreover, the product ion spectrum showed a marked increase in the relative abundance of daughter ions arising from simple cleavage over those produced by rearrangement.

Buffer gas can be introduced into the QITMS via two methods: (1) directly into the ion trap volume, and (2) into the QITMS vacuum chamber. These two introduction methods were investigated with respect to their effect upon CID. A constant amount of n-butylbenzene ($\approx 5.7 \times 10^{-7}$ Torr) was introduced into the QITMS vacuum chamber and was maintained for several days during the period of these experiments. Resonant excitation frequency optimization curves were obtained for the CID of the $M^{+\cdot}$ ion (m/z 134) of n-butylbenzene at a q_z (m/z 134) of 0.3 with a resonant excitation duration and amplitude of 10 ms and 500 mV_{p-p} for each helium pressure examined. The maximum E_{CID} and the maximum $91^+/92^+$ ratio obtained from each frequency optimization were used to construct the figures below. In addition, prior to and after the frequency optimization (but during the same acquisition file), mass spectra were acquired without the application of the resonant excitation voltage; thus, the intensity of the parent ion (P_0, m/z 134 $M^{+\cdot}$ ion of n-butylbenzene) following DC/RF isolation could be monitored.

The maximum E_{CID}, the maximum $91^+/92^+$ ratio, and the P_0 intensity as a function of the He buffer gas pressure are shown in Figs. 8.18(a) and (b) for helium introduced directly into the ion trap volume and into the QITMS vacuum chamber, respectively. The intensity of the m/z 134 P_0 was found to increase with increasing buffer gas pressure due, presumably, to the damping of the ion motion upon multiple collisions with helium, bringing the ions closer to the center of the trap. The maximum m/z 134 ion intensity is observed at indicated chamber pressures of 1×10^{-5} and 1×10^{-4} Torr for the introduction of helium into the ion trap and into the vacuum chamber, respectively. Above these pressures, the trapping efficiency of the parent ion decreases, most likely due to increased scattering losses.

The E_{CID} also increases with increasing buffer gas pressure. However, whereas the optimum buffer gas pressure had a relatively narrow maximum for the measured m/z 134 ion intensity prior to CID, the optimum buffer gas pressure exhibits a relatively broad plateau for the E_{CID}, varying between 85 and 100%. The maximum E_{CID} was obtained at an indicated pressure of 2×10^{-6} and 2×10^{-5} Torr for helium introduced directly into the ion trap and into the vacuum chamber, respectively. The behavior of the E_{CID} at pressures above these levels is not currently understood. It is interesting, however, that even though the two sets of data were obtained one day apart, the curves follow each other very well.

Whereas the effect upon the maximum m/z 134 ion intensity and E_{CID} mirror each other with increasing helium pressure, the energetics of the CID process appear to differ for the two different introduction methods. With helium introduced directly into the ion trap (Fig. 8.18(a)), a broad maximum is observed at a $91^+/92^+$ ratio of 0.5 from 8×10^{-7} to 8×10^{-6} Torr. This plateau indicates that once sufficient gas is present for CID to occur efficiently, there is a maximum amount of energy (dependent upon the amplitude and duration of the resonant excitation potential and the parent ion a_z and q_z) which can be imparted to the parent ion. With helium pressure above 8×10^{-6} Torr, a decrease in the energetics of CID or $91^+/92^+$ ratio is observed; this decrease is most likely due to increase buffering or damping action occurring at these higher pressures. That is, collisions of the parent ion with neutral helium become so frequent that sufficient kinetic energy can not be attained between collisions to impart energetic collisions, until at 4×10^{-5} Torr, no 91^+ ions are produced.

The behavior of the CID energetics ($91^+/92^+$ ratio) is more erratic with the introduction of helium into the vacuum chamber (Fig. 8.18(b)). However, above helium pressures of 1×10^{-5} Torr, the behavior essentially follows that with helium introduced into the trap. Note again that the curves are offset by a factor of ten in pressure from each other, for example, the reduction in the $91^+/92^+$ ratio occurs at 8×10^{-6} and 8×10^{-5} Torr for helium introduced into the ion trap and introduced into the vacuum chamber, respectively. As mentioned earlier, subsequent calibration of the ion gauge readings with a capacitance Baratron gauge for both introduction methods showed that when helium is introduced directly into the ion trap volume (with the electrode spacers in place), there is a ten-fold difference in pressure inside and outside of the ion trap.[13]

The buffer gas pressure was also found to affect the secular frequency and effective resonant excitation frequency bandwidth. The parent ion secular frequency was monitored by the resonant excitation frequency which yielded the highest energy CID, while the effective resonant excitation frequency bandwidth was taken as the width at 20% E_F for the E_F versus resonant excitation frequency curve. With increasing buffer gas pressure, the secular frequency of the parent ion decreased from ≈ 119.5 kHz at 6×10^{-7} Torr to 119.3 kHz at 6×10^{-5} Torr (Fig. 8.18(c)). This decrease in secular frequency is presumably due to the damping action of the buffer gas. Similarly, with increasing buffer gas pressure, the effective resonant excitation frequency bandwidth decreases from $\approx$1500 Hz at 6.0 $\times$ 10^{-7} Torr to a minimum of $\approx$1200 Hz at 5 $\times$ 10^{-6} Torr (Fig. 8.18(c)). This initial decrease in the frequency bandwidth is due, again presumably, to damping effects of the buffer gas. The magnitude of these effects may also be dependent on the "cooling" period between the RF/DC isolation and resonant excitation of the parent ion. However, with helium pressures greater than 5 $\times$ 10^{-6} Torr, the effective bandwidth broadens, due probably to scattering effects of the buffer gas.

Consideration of all three parameters in Figs. 8.18(a,b), that is, of E_{CID}, $91^+/92^+$, and m/z 134 P_0 intensity, indicates that the optimum conditions for performing CID in the Finnigan QITMS would be at indicated pressures of 8 $\times$ 10^{-6} and 8 $\times$ 10^{-5} Torr for helium introduced into the ion trap and into the vacuum chamber, respectively. These indicated pressures correspond to an absolute pressure of approximately 3 $\times$ 10^{-4} Torr helium within the ion trap.

As discussed above, buffer/CID gases more massive than helium may also be used in the QITMS in order to increase the CID energetics. In addition, in the case of direct atmospheric monitoring experiments, air may be the buffer/CID gas by default.[125] Thus, in a separate set of experiments, the effects of helium, nitrogen, and argon on CID energetics and efficiencies were investigated. An EI-MS/MS daughter scan with RF/DC isolation of the m/z 134 $M^{+\bullet}$ ion of n-butylbenzene was used with a set resonant excitation duration of 5 ms and q_z (m/z 134) of 0.30. For each pressure and gas combination, the optimum resonant excitation frequency was determined at a low resonant excitation voltage; the amplitude of the resonant excitation potential was then varied at the optimum frequency to determine the maximum E_{CID} and $91^+/92^+$ ratio. In addition, the parent ion intensity (P_0) was measured in the absence of resonant excitation. The pressures of the buffer gases were measured with a Bayard–Alpert ionization gauge and then corrected for the ionization probability of each gas. Figs. 8.19(a)–(e) present the results of these studies obtained using argon, nitrogen, and helium buffer gases.

For each of the three buffer gases, as the pressure is increased, an increase in the maximum E_{CID} is observed (Fig. 8.19(a)). Note in Fig. 8.19(a) the trend which will occur in subsequent figures: the more massive gases exert their effects at lower pressures. As in the resonant excitation voltage studies (Fig. 8.13), the maximum E_{CID} values in Fig. 8.19(a) exceed 100% for each of the gases; this behavior is attributed to an increase in the E_C due to greater trapping efficiency of the daughter and resonantly excited m/z 134 ions (see discussion of Fig. 8.13 above).

With increasing helium buffer gas pressure, the efficiency of apex isolation and trapping of the m/z 134 ion prior to resonant excitation (P_0) increases to yield a relatively broad maximum around 2 $\times$ 10^{-4} Torr (Fig.

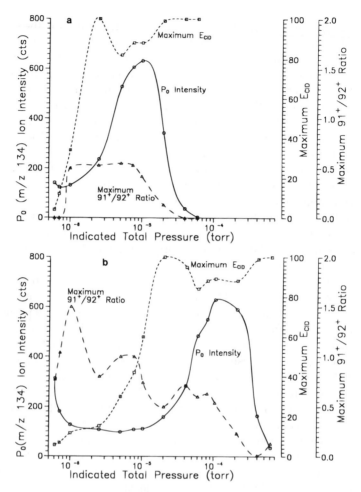

FIGURE 8.18

8.19(b)). However, with argon and nitrogen, a decrease is observed in the P_0 intensity with increasing pressure. With the heavier buffer gases, apex isolation and trapping are much less efficient. Although some of this loss in efficiency may be due to greater scattering and possibly CID during the apex isolation, some may also be attributable to space-charge effects. During the ionization period of the QITMS scan function, the ring RF voltage level is operated such that ions above m/z 10 are stored. Under such conditions, $He^{+\cdot}$ ions are rapidly ejected during ionization, whereas $N_2^{+\cdot}$ and $Ar^{+\cdot}$ ions will not be ejected and will contribute to space charge until the apex isolation of m/z 134 occurs. With increasing pressures of N_2 and Ar, space charging

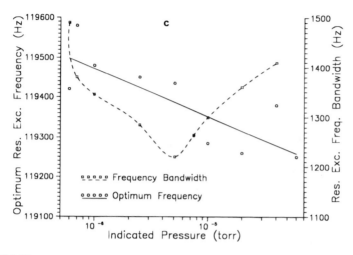

FIGURE 8.18

E_{CID}, $91^+/92^+$ ratio, and the maximum m/z 134 ion intensity as a function of the indicated He buffer gas pressure for the CID of the $M^{+\bullet}$ ion of n-butylbenzene: (a) helium introduced directly into the ion trap volume and (b) helium introduced into the QITMS vacuum chamber. (c) Effect of helium pressure on the optimum resonant excitation frequency and bandwidth for helium introduced into the ion trap. Conditions: $q_z(m/z\ 134) = 0.30$; resonant excitation amplitude of 600 mV_{p-p}; duration of excitation of 10 ms.

may lead to poorer trapping and isolation efficiency of the m/z 134 ions. The space-charge effects of buffer gas ions may be reduced by elevating the RF voltage during the ionization period such that these ions are not stored. Achievement of high sensitivity for MS/MS demands a high overall E_{CID} and also a high parent ion intensity (P_0) prior to MS/MS. Thus, with buffer gases more massive than helium, a trade-off must be made between obtaining high E_{CID} at higher pressures and high P_0 intensities at lower pressures. Optimum conditions can be gauged by monitoring the yield of daughter ions, ΣD_i^+. Thus, with decreasing P_0 and increasing E_{CID} with increasing pressure, the yield of daughter ions is maximized at intermediate pressures of 9×10^{-6}, 3×10^{-5}, and 2×10^{-4} Torr for argon, nitrogen, and helium, respectively (Fig. 8.19(c)).

As the helium buffer gas pressure is increased, the $91^+/92^+$ ratio decreases from approximately 1.5 at 1×10^{-6} Torr to approximately 0.2 at approximately 1 mTorr (Fig. 8.19(d)) due, presumably, to the increased buffering action of the gas not allowing sufficient excitation of the m/z 134 ion even though E_{CID} is 100% at these higher pressures. This explanation is supported by the fact that higher resonant excitation amplitudes are required to achieve the maximum E_{CID} as the helium pressure is increased, as seen in Fig. 8.19(e). In contrast, with increasing buffer gas pressure, the

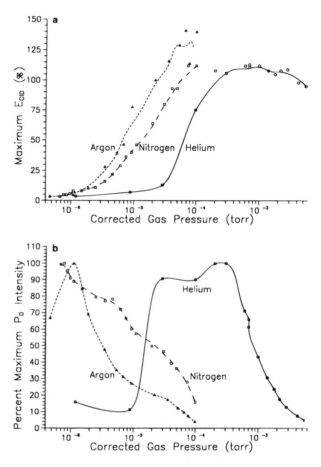

FIGURE 8.19
(a) Maximum E_{CID}, (b) m/z 134 ion intensity prior to resonant excitation (P_0), (c) yield of daughter ions (ΣD_i^+), (d) $91^+/92^+$ ratio, and (e) resonant excitation voltage (REV) for maximum E_{CID} as a function of Ar, N_2 and He buffer gas pressures for the CID of the $M^{+\cdot}$ ion of n-butylbenzene. Conditions: $q_z(m/z\ 134) = 0.30$; duration of resonant excitation of 5 ms.

$91^+/92^+$ ratio increases for both argon and nitrogen, reaching maxima of 2.8 and 3.0, respectively, in the 1 to 2×10^{-5} Torr range (Fig. 8.19(d)). More energetic CID is expected as one goes to the more massive buffer/CID gases. With a further increase in the buffer gas pressure, the $91^+/92^+$ ratio also decreases for nitrogen and argon, again, presumably due to the increased buffering action of the gases. However, unlike helium, the resonant excitation amplitude required to achieve the maximum E_{CID} remains relatively constant with increasing buffer gas pressure (Fig. 8.19(e)).

These studies indicate that the use of nitrogen or argon as the buffer gas does not lead to very good overall performance of a QITMS. Not only

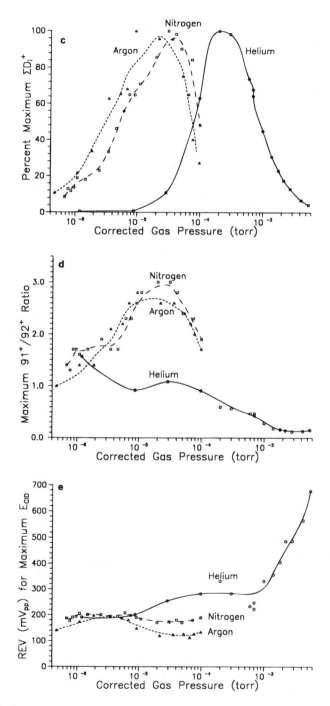

FIGURE 8.19

is the absolute sensitivity lower, but the use of nitrogen or argon leads to peak broadening during the mass-selective instability scan. However, if greater energy needs to be imparted during CID to enhance fragmentation, the use of more massive buffer gases may be necessary. The mixing of more massive gases with helium may offer a good compromise between optimum sensitivity and optimum CID energetics.[10,122–124]

VII. CONCLUSIONS

In conclusion, the ion energetics and efficiency of ionization, trapping, isolation, resonant excitation CID, and detection in the ion trap are governed by a number of experimental parameters including the frequency, amplitude and duration of the resonant excitation potential, the a_z and q_z of the parent ion, and the pressure and nature of the buffer gas present in the trap. These parameters are all interrelated and changing one affects the others. However, each can be varied to achieve the desired result, whether to change the ion energetics, enhance the efficiency, or to make a compromise between the two. Based upon the discussions above, efficient, energetic, and highly sensitive resonant excitation CID can be accomplished typically in the QITMS with parent ion trapping parameters $q_z = 0.3$ and $a_z = 0$, fluence of 2000 mV_{p-p}*ms, and helium pressure in the ion trap of 3×10^{-4} Torr (corrected).

REFERENCES

1. Mathieu, É. *J. Math. Pures Appl. (J. Liouville)*. 1868, *13*, 137.
2. Zwillinger, D. *Handbook of Differential Equations*. Academic Press, Boston, 1989, 575–579.
3. Wuerker, R. F.; Shelton, H.; Langmuir, R. V. *J. Appl. Phys.* 1959, *30*, 342–349.
4. March, R. E.; Hughes, R. J. *Quadrupole Storage Mass Spectrometry*. John Wiley & Sons, NY, 1989.
5. Louris, J. N.; Cooks, R. G.; Syka, J. E. P.; Kelley, P. E.; Stafford, G. C.; Todd, J. F. J. *Anal. Chem.* 1987, *59*, 1677–1685.
6. Louris, J. N.; Brodbelt-Lustig, J. S.; Cooks, R. G.; Glish, G. L; Van Berkel, G. J.; McLuckey, S. A. *Int. J. Mass Spectrom. Ion Processes*. 1990, *96*, 117–137.
7. Johnson, J. V.; Yost, R. A.; Kelley, P. E.; Bradford, D. C. *Anal. Chem.* 1990, *62*, 2162–2172.
8. McLuckey, S. A.; Glish, G. L.; Van Berkel, G. J. *Int. J. Mass Spectrom. Ion Processes*. 1991, *106*, 213–235.
9. Brodbelt-Lustig, J. S.; Kenttämaa, H. I.; Cooks, R. G. *Org. Mass Spectrom*. 1988, *23*, 6–9.
10. Nourse, B. D.; Cox, K. A.; Morand, K. L.; Cooks, R. G. *J. Am. Chem. Soc.* 1992, *114*, 2010–2016.
11. Dawson, P. H., Ed. *Quadrupole Mass Spectrometry and Its Applications*. Elsevier Scientific, Amsterdam, 1976.

12. Evans, C.; Traldi, P.; Mele, A.; Sottani, S. *Org. Mass Spectrom.* 1991, *26*, 347–349.
13. Basic, C. "Probing Trapped Ion Energies in a Quadrupole Ion Trap Mass Spectrometer," Ph.D. Dissertation, University of Florida, 1992.
14. Paradisi, C.; Todd, J. F. J.; Vettori, U. *Org. Mass Spectrom.* 1992, *27*, 1210–1215.
15. Vedel, F.; Vedel, M.; March, R. E. *Int. J. Mass Spectrom. Ion Processes.* 1991, *108*, R11–R20.
16. Yates, N. A.; Yost, R. A.; Bradshaw, S. C.; Tucker, D. B. *Proc. 39th ASMS Conf. Mass Spectrometry and Allied Topics.* Nashville, 1991, Bovum, P. R. 132–133.
17. Yates, N. A.; Griffin, T. P.; Yost, R. A. *Proc. 41st ASMS Conf. Mass Spectrometry and Allied Topics,* San Francisco, 1993, 444a–b.
18. *Practical Aspects of Ion Trap Mass Spectrometry,* Eds. R. E. March and J. F. J. Todd, Vols. 1 and 2, "Fundamentals and Instrumentation", Modern Mass Spectrometry series, CRC Press, Boca Raton, FL, 1994.
19. Curcuruto, O.; Fontana, S.; Traldi, P.; Celon, E. *Rapid Comm. Mass Spectrom.* 1992, *6*, 322–323.
20. Paradisi, C.; Todd, J. F. J.; Traldi, P.; Vettori, U. *Org. Mass Spectrom.* 1992, *27*, 251–254.
21. Creaser, C. S.; O'Neill, K. E. *Org. Mass Spectrom.* 1993, *28*, 564–569.
22. McLuckey, S. A.; Goeringer, D. E.; Glish, G. L. *Anal. Chem.* 1992, *64*, 1455–1460.
23. Lammert, S. A.; Cooks, R. G. *J. Am. Soc. Mass Spectrom.* 1991, *2*, 487–491.
24. Lammert, S. A.; Cooks, R. G. *Rapid Commun. Mass Spectrom.* 1992, *6*, 528–530.
25. Hughes, R. J.; March, R. E.; Young, A. B. *Can. J. Chem.* 1983, *61*, 824–833.
26. Hughes, R. J.; March, R. E.; Young, A. B. *Can. J. Chem.* 1983, *61*, 834–845.
27. Hughes, R. J.; March, R. E.; Young, A. B. *Int. J. Mass Spectrom. Ion Phys.* 1982, *42*, 255–263.
28. Hughes, R. J.; March, R. E.; Young, A. B. *Int. J. Mass Spectrom. Ion Phys.* 1983, *47*, 85–88.
29. Young, A. B.; March, R. E.; Hughes, R. J. *Can. J. Chem.* 1985, *63*, 2324–2331.
30. Louris, J. N.; Brodbelt, J. S.; Cooks, R. G. *Int. J. Mass Spectrom. Ion Processes.* 1987, *75*, 345–352.
31. Creaser, C. S.; McCoustra, M. R. S.; O'Neill, K. E. *Org. Mass Spectrom.* 1991, *26*, 335–338.
32. Stephenson, Jr. J. L.; Booth, M. M.; Shalosky, J. R.; Eyler, J. A.; Yost, R. A. *Proc. 41st ASMS Conf. Mass Spectrometry and Allied Topics.* San Francisco, May 30–June 5, 1993, 445a–b.
33. Franzen, J. *Int. J. Mass Spectrom. Ion Processes.* 1991, *106*, 63–78.
34. Wang, Y.; Franzen, J.; Wanczek, K. P. *Int. J. Mass Spectrom. Ion Processes.* 1993, *124*, 125–144.
35. Dawson, P. H.; Whetten, N. R. *J. Vac. Sci. Technol.* 1968, *5*, 1–10.
36. Pedder, R. E.; Yost, R. A. *Proc. 36th ASMS Conf. Mass Spectrometry and Allied Topics.* San Francisco, 1988, pp. 632–633.
37. Pedder, R. E. "Fundamental Studies in Quadrupole Ion Trap Mass Spectrometry", Ph.D. Dissertation, University of Florida, 1992.
38. March, R. E.; McMahon, A. W.; Londry, F. A.; Alfred, R. L.; Todd, J. F. J; Vedel, F. *Int. J. Mass Spectrom. Ion Processes.* 1989, *95*, 119–156.
39. Lawson, G.; Todd, J. F. J; Bonner, R. F. in *Dynamic Mass Spectrometry.* D. Price and J. F. J. Todd (Eds.), Heyden and Sons, Ltd., London, 1976, Vol. 4, 39–81.
40. Todd, J. F. J.; Waldren, R. M.; Bonner, R. F. *Int. J. Mass Spectrom. Ion Phys.* 1980, *34*, 17–36.
41. Todd, J. F. J.; Lawson, G.; Bonner R. G. in *Quadrupole Mass Spectrometry and Its Applications.* P. H. Dawson, Ed., Elsevier Scientific, New York, 1976, pp. 181–224.
42. Bonner, R. F.; March, R. E.; Durup, J. *Int. J. Mass Spectrom. Ion Phys.* 1976, *22*, 17–34.
43. Bonner, R. F.; March, R. E. *Int. J. Mass Spectrom. Ion Phys.* 1977, *25*, 411–431.
44. Doran, M. C.; Fulford, J. E.; Hughes, R. J.; Morita, Y.; March, R. E.; Bonner, R. F. *Int. J. Mass Spectrom. Ion Phys.* 1980, *33*, 139–159.
45. Reiser, H. P.; Julian, R. K.; Cooks, R. G. *Int. J. Mass Spectrom. Ion Processes.* 1992, *121*, 49–63.

46. Reiser, H. P.; Kaiser, R. E.; Savickas, P. J.; Cooks, R. G. *Int. J. Mass Spectrom. Ion Processes.* 1991, *106*, 237–247.

47. Stafford, G. C.; Kelley, P. E.; Syka, J. E. P.; Reynolds, W. E.; Todd, J. F. J. *Int. J. Mass Spectrom. Ion Phys.* 1984, *60*, 85–98.

48. Kaiser, R. E.; Louris, J. N.; Amy, J. A.; Cooks, R. G. *Rapid Commun. Mass Spectrom.* 1989, *3*, 225–229.

49. Kaiser, R. E.; Cooks, R. G.; Stafford, G. C.; Syka, J. E. P.; Hemberger, P. H. *Int. J. Mass Spectrom. Ion Processes.* 1991, *106*, 79–115.

50. Williams, J. D.; Reiser, H. P.; Kaiser, R. E., Cooks, R., *Int. J. Mass Spectrom. Ion Processes.* 1991, *108*, 199–219.

51. Julian, R. K.; Reiser, H. P.; Cooks, R. G. *Int. J. Mass Spectrom. Ion Processes.* 1993, *123*, 85–96.

52. March, R. E.; McMahon, A. W.; Allinson, E. T.; Londry, F. A.; Alfred, R. L.; Todd, J. F. J.; Vedel, F. *Int. J. Mass Spectrom. Ion Processes.* 1990, *99*, 109–124.

53. March, R. E.; Londry, F. A.; Alfred, R. L.; Todd, J. F. J.; Penman, A. D.; Vedel, F.; Vedel, M. *Int. J. Mass Spectrom. Ion Processes.* 1991, *110*, 159–178.

54. March, R. E.; Londry, F. A.; Alfred, R. L.; Franklin, A. M.; Todd, J. F. J. *Int. J. Mass Spectrom. Ion Processes.* 1992, *112*, 247–271.

55. March, R. E.; Tkaczyk, M.; Londry, F. A.; Alfred, R. L. *Int. J. Mass Spectrom. Ion Processes.* 1993, *125*, 9–32.

56. March, R. E.; Weir, M. R.; Tkaczyk, M.; Londry, F. A.; Alfred, R. L.; Franklin, A. M.; Todd, J. F. J. *Org. Mass Spectrom.* 1993, *28*, 499–509.

57. Fulford, J. E.; March, R. E. *Int. J. Mass Spectrom. Ion Phys.* 1978, *26*, 155–162.

58. Weber-Grabau, M.; Kelley, P. E.; Syka, J. E. P.; Bradshaw, S. C.; Brodbelt, J. S. *Proc. 35th ASMS Conf. Mass Spectrometry and Allied Topics,* Denver, 1987, pp. 1114–1115.

59. Strife, R. J.; Kelley, P. E.; Weber-Grabau, M. *Rapid Commun. Mass Spectrom.* 1988, *2*, 105.

60. Gronowska, J.; Paradisi, C.; Traldi, P.; Vettori, U. *Rapid Commun. Mass Spectrom.* 1990, *4*, 306–313.

61. Ardanaz, C. E.; Traldi, P.; Vettori, U.; Kavka, J.; Guidugli, F. *Rapid Commun. Mass Spectrom.* 1991, *5*, 5–10.

62. Yates, N. A.; Yost, R. A. *Proc. 39th ASMS Conf. Mass Spectrometry and Allied Topics,* Nashville, 1991, pp. 1489–1490.

63. Dehmelt, H. G. *Adv. At. Mol. Phys.* 1967, *3*, 53–72.

64. Major, F. G.; Dehmelt, H. G. *Phys. Rev.* 1968, *170*, 91–107.

65. Todd, J. F. J.; Freer, D. A.; Waldren, R. M. *Int. J. Mass Spectrom. Ion Phys.* 1980, *36*, 185–203.

66. Todd, J. F. J.; Freer, D. A.; Waldren, R. M. *Int. J. Mass Spectrom. Ion Phys.* 1980, *36*, 371–386.

67. Todd, J. F. J.; Waldren, R. M.; Freer, D. A.; Turner, R. B. *Int. J. Mass Spectrom. Ion Phys.* 1980, *35*, 107–150.

68. Todd, J. F. J. in *Dynamic Mass Spectrometry.* D. Price, Ed., Heyden and Sons, Ltd., London, 1981, Vol. 6, 44–70.

69. Baril, M. in *Dynamic Mass Spectrometry,* D. Price (Ed.), Heyden and Sons, London, 1981, Vol. 6, 33–43.

70. André, J. *J. de Phys.* 1976, *37*, 719–730.

71. André, J.; Vedel, F. *J. de Phys.* 1977, *38*, 1381–1398.

72. Vedel, F.; André, J.; Vedel, M.; Brincourt, G. *Phys. Rev. A,* 1983, *27*, 2321–2330.

73. Vedel, F.; André, J.; Vedel, M. *J. de Phys.* 1981, *42*, 391–398.

74. Vedel, F.; André, J.; Vedel, M. *J. de Phys.* 1981, *42*, 1611–1622.

75. Vedel, F.; André, J. *Phys. Rev. A,* 1984, *29*, 2098–2101.

76. Vedel, F.; André, J. *Int. J. Mass Spectrom. Ion Processes.* 1985, *65*, 1–22.

77. Vedel, F. *Int. J. Mass Spectrom. Ion Processes.* 1991, *106*, 3–61.
78. Church, D. A.; Dehmelt, H. G. *J. Appl. Phys.* 1969, *40*, 3421–3424.
79. Church, D. A. *Phys. Rev.* A, 1988, *37*, 277–279.
80. Iffländer, R.; Werth, G. *Metrologia.* 1977, *13*, 167–170.
81. Knight, R. D.; Prior, M. H. *J. Appl. Phys.* 1979, *50*, 3044–3049.
82. Schaaf, H.; Schmeling, U.; Werth, G. *Appl. Phys.* 1981, *25*, 249–251.
83. Cutler, L. S.; Giffard, R. P.; McGuire, M. D. *Appl. Phys.* B, 1985, *36*, 137–142.
84. Cutler, L. S.; Flory, C. A.; Giffard, R. P.; McGuire, M. D. *Appl. Phys.* B, 1986, *39*, 251–259.
85. Münch, A.; Berkler, M.; Gerz, Ch.; Wilsdorf, D.; Werth, G. *Phys. Rev.* A, 1987, *35*, 4147–4150.
86. Siemers, I,; Blatt, R.; Sauter, Th.; Neuhauser, W. *Phys. Rev.* A, 1988, *38*, 5121–5128.
87. Lunney, M. D. N.; Buchinger, F.; Moore, R. B. *J. Mod. Optics.* 1992, *39*, 349–360.
88. Bartmess, J. E. in *Structure/Reactivity, and Thermochemistry of Ions.* P. Ausloos and S. G. Lias (Eds.), NATO ASI Series C. Mathematical and Physical Sciences, D. Reidel, Dordrecht, 1987, *193*, 367–380.
89. Lawson, G.; Bonner, R. F.; Mathers, R. E.; Todd, J. F. J; March, R. E. *J. Chem. Soc., Faraday Trans.* I, 1976, *72*, 545–557.
90. Brodbelt-Lustig, J. S.; Cooks, R. G. *Talanta.* 1989, *36*, 255–260.
91. McLuckey, S. A.; Cameron, D.; Cooks, R. G. *J. Am. Chem. Soc.* 1981, *103*, 1313–1317.
92. Basic, C.; Eyler, J. R.; Yost, R. A. *J. Am. Soc. Mass Spectrom.* 1992, *3*, 716–726.
93. Lindinger, W.; Howorka, F.; Lukac, P.; Kuhn, S.; Villinger, H.; Alge, E.; Ramler, H. *Phys. Rev.* A, 1981, *23*, 2319–2326.
94. Smith, D.; Adams, N. G. *Phys. Rev.* A, 1981, *23*, 2327–2330.
95. Dotan, I.; Lindinger, W. *J. Chem. Phys.* 1982, *76*, 4972–4977.
96. Nourse, B. D.; Kenttämaa, H. I. *J. Phys. Chem.* 1990, *94*, 5809–5812.
97. Kenttämaa, H. I.; Cooks, R. G. *J. Am. Chem. Soc.* 1985, *107*, 1881–1886.
98. Rosenstock, H. M.; Draxl, K.; Steine, B. W.; Herman, J. T. *J. Phys. Chem. Ref. Data,* 1977, *6*, Suppl. 1.
99. Wu, H.-F.; Brodbelt, J. S. *Int. J. Mass Spectrom. Ion Processes.* 1992, *115*, 67–81.
100. McLuckey, S. A.; Glish, G. L.; Kelley, P. E. *Anal. Chem.* 1987, *59*, 1670–1674.
101. Brown, P. *Org. Mass Spectrom.* 1970, *3*, 1175–1186.
102. Chen, J. H.; Hayes, J. D.; Dunbar, R. C. *J. Phys. Chem.* 1984, *88*, 4759–4764.
103. Baer, T.; Dutuit, O.; Mestdagh, H.; Rolando, C. *J. Phys. Chem.* 1988, *92*, 5674–5679.
104. Mukhtar, E. S.; Griffiths, I. W.; Harris, F. M.; Beynon, J. H. *Int. J. Mass Spectrom. Ion Phys.* 1981, *37*, 159–166.
105. Griffiths, I. W.; Harris, F. M.; Mukhtar, E. S.; Beynon, J. H. *Int. J. Mass Spectrom. Ion Phys.* 1981, *41*, 83–88.
106. Harrison, A. G.; Lin, M. S. *Int. J. Mass Spectrom. Ion Phys.* 1983, *51*, 353–356.
107. Boyd, R. K.; Harris, F. M.; Beynon, J. H. *Int. J. Mass Spectrom. Ion Processes.* 1985, *66*, 185–194.
108. Griffiths, I. W.; Mukhtar, E. S.; March, R. E.; Harris, F. M.; Beynon, J. H. *Int. J. Mass Spectrom. Ion Phys.* 1981, *39*, 125–132.
109. Dawson, P. H.; Sun, W.-F. *Int. J. Mass Spectrom. Ion Phys.* 1982, *44*, 51–59.
110. McLuckey, S. A.; Sallans, L.; Cody, R. B.; Burnier, R. C.; Verma, S.; Freiser, B. S.; Cooks, R. G. *Int. J. Mass Spectrom. Ion Processes.* 1982, *44*, 215–229.
111. Johnson, J. V.; Pedder, R. E.; Kleintop B.; Yost, R. A. *Proc. 38th ASMS Conf. Mass Spectrometry and Allied Topics.* Tucson, 1990, pp. 1130–1131.
112. Berberich, D. W.; Hail, M. E.; Johnson, J. V.; Yost, R. A. *Int. J. Mass Spectrom. Ion Processes.* 1989, *94*, 115–147.
113. Yost, R. A.; Enke, C. G. *J. Am. Chem. Soc.* 1978, *100*, 2274–2275.
114. Pedder, R. E.; Yost, R. A. "Enhancements to MS/MS in a Quadrupole Ion Trap Mass

Spectrometer," *Proceedings of the 1987 U.S. Army Chemical Research, Development and Engineering Center Scientific Conference on Chemical Defense Research.* Aberdeen Proving Ground, MD. November 17–20, 1987, 1061–1067.

115. Williams, J. D.; Cox, K. A.; Cooks, R. G.; Kaiser, R. E.; Schwartz, J. C. *Rapid Commun. Mass Spectrom.* 1991, *5*, 327–329.

116. Schwartz, J. C.; Syka, J. E. P.; Jardine, I. *J. Am. Soc. Mass Spectrom.* 1991, *2*, 198–204.

117. Londry, F. A.; Wells, G. J.; March, R. E. *Rapid Commun. Mass Spectrom.* 1993, *7*, 43–45.

118. Stephenson, J. L.; Booth, M. M.; Johnson, J. V.; Yost, R. A. "Fundamentals and Applications in High Resolution Quadrupole Ion Trap Mass Spectrometry," *Florida Sections of the American Chemical Society Annual Meeting.* April 30–May 2, 1992.

119. Johnson, J. V.; Pedder, R. E.; Yost, R. A. *Int. J. Mass Spectrom. Ion Processes.* 1991, *106*, 197–212.

120. Hart, K. J.; Habibi-Goudarzi, S.; McLuckey, S. A. *Proc. 41st ASMS Conf. Mass Spectrometry and Allied Topics.* San Francisco, June 1993, 685a–b.

121. Weber-Grabau, M.; Kelley, P. E.; Syka, J. E. P.; Bradshaw, S. C.; Brodbelt, J. S. *Proc. 35th Conf. Mass Spectrometry and Allied Topics.* 1987, pp. 1114–1115.

122. Weber-Grabau, M.; Kelley, P. E.; Syka, J. E. P.; Bradshaw, S. C.; Brodbelt, J. S. *Proc. 35th Conf. Mass Spectrometry and Allied Topics.* 1987, pp. 1114–1115.

123. Morand, K. L.; Cox, A. A.; Cooks, R. G. *Rapid Commun. Mass Spectrom.* 1992, *6*, 520–523.

124. Morand, K. L.; Hoke, S. H.; Ederlin, M. N.; Payne, G.; Cooks, R. G. *Org. Mass Spectrom.* 1992, *27*, 284–288.

125. McLuckey, S. A.; Glish, G. L.; Asano, K. G. *Anal. Chim. Acta.* 1989, *225*, 25–35.

Chapter 9

ION STRUCTURE STUDIES IN THE QUADRUPOLE ION TRAP

Silvia Catinella and Pietro Traldi

CONTENTS

I. INTRODUCTION

Thomson's first mass spectrograph[1] was constructed in such a way that, when the beam of positive ions passed through the field-free regions before and between the two magnets, a novel opportunity was presented

0-8493-8251-3/95/$0.00+$.50
© 1995 by CRC Press, Inc.

313

for the ions to undergo collisions with background gases. Unfortunately, Thomson's early papers are not readily available, but an account of his contributions to this field has been presented by Beynon and Morgan.[2] Since the time of Thomson's experiments, mass spectroscopists have lived (sometimes consciously, sometimes not) with ion/molecule interactions occurring outside of the ion source.

Following the initial observation of metastable ions in conventional electron impact mass spectra by Hipple and Condon,[3] it was some years before the power of such data was recognized and applied to either fundamental studies on ion structure or to the analytical field. Once the utility of these data was appreciated, many methodologies were proposed for the detection of such ions in sector mass spectrometers so as to obtain metastable ion spectra unencumbered by any interference from conventional ions.[4]

The experimental observation that such spectra increase both the number of ions and the abundance of naturally occurring decomposition products with the presence of a collision gas in the suitable region(s) of the spectrometer,[5] led to the development of a new technique, called collision spectroscopy (or tandem mass spectrometry, or MS/MS): the extensive application of this technique constitutes ample proof of its power.

While some books[6] are exhaustive in describing both the fundamental and practical applications of all the different aspects of tandem mass spectrometry, a different and interesting approach to the application of collision spectroscopy has been presented by Yost,[7] who has quantified the utility of different mass spectrometric techniques in terms of the informing power, defined as:

$$P_{info} = \sum_{i=1}^{n} \log_2 S_i \tag{9.1}$$

where P_{info} is the informing power, n is the number of different quantities to be determined, and S_i is the number of measurable steps for a given quantity.

By such formulae, it can be calculated easily that passing from a simple low-resolution mass spectrometer (for example, a quadrupole mass filter with a mass range of 1000 Da, unit resolution, and intensity range $S = 2^{12}$) to a MS/MS system (for example, a triple quadrupole instrument) an increase of informing power from 1.2×10^4 to 1.2×10^7 is observed. This quite considerable increase of informing power illustrates clearly why collision spectroscopy currently represents one of the most powerful methods available for the fundamental study of ion structure, as well as for the detection and quantification of molecules of interest in complex matrices. The use of different instrumental approaches, such as multisector machines, triple-stage quadrupole instruments, and hybrid geometry instruments, permits the study of collision-induced decomposition pathways in different kinetic energy regimes, ranging from a few eV up to 10^4 eV.

The energy deposition mechanism operating in collisional experiments has been the object of many studies which indicate that, aside from a relatively small quantity of ions which experience electronic excitation, most of the collisionally-induced decompositions originate from vibrationally-excited precursors.[6] Interestingly, different internal energy profiles are accessed by different instrumental approaches, and more interesting results are achieved usually by low-energy collisions (1 to 100 eV) than with high-energy (keV) collisions.

However, it must be stressed that quite a high internal energy distribution is observed in all cases; such a distribution allows the activation of different decomposition channels with different activation energies. The results obtained are highly informative from the structural point of view: first of all, collision experiments are performed on ions with low internal energy content; secondly, collisional-induced activation conditions are relatively mild. Thus, isomeric compounds, for which identical electron impact mass spectra are frequently obtained, often yield clearly different collisional spectra of their $M^{+\cdot}$ ions, proving that the energy barriers to isomerization which are overcome by electron ionization (EI) within the ion source are not reached by collision.

Among the collisional approaches described above, those which originate by interaction of gaseous ions with metallic surfaces, usually called surface-induced dissociation (SID),[8] requires a separate discussion. In this latter approach, highly effective energy deposition is achieved with a narrow internal energy distribution, the mean value of which can be varied linearly by changing the kinetic energy of the mass-selected species.

II. DAUGHTER ION SPECTROSCOPY
BY ION TRAP MASS SPECTROMETRY (ITMS)

The practice of ITMS has proved to be highly suitable for collisional experiments.[9] Different approaches are available for ion excitation, mainly those based on (1) axial modulation of the ion cloud by a supplementary radio frequency (RF) voltage;[10] (2) ion acceleration by imposing a_z and q_z values which lie close to the border of the stability diagram;[11] and (3) surface-induced dissociation on the ring electrode brought about by the application of a fast direct current (DC) pulse.[12]

Furthermore, it must be considered that, due to the high efficiency of collisions in ITMS, multistep MS/MS experiments can be carried out easily. For example, MS^n experiments with $n > 8$ have been described by the groups of Cooks,[13] Glish,[14] and Todd.[15] The first step in a MS/MS experiment by ITMS consists, necessarily, in the isolation of the ion species of interest. This process was first accomplished by manipulation of the RF and DC voltages applied to the ion trap so that the working point, (a_z, q_z), of the selected ion species was moved to just below the upper apex of the stability diagram.[16] In principle, by such an approach, the trajectories of all the other ion species present in the ion trap should become unstable, and the ions are easily ejected from the trap. In practice, this situation pertains only for ions exhibiting a mass difference of >2 to 3 Da with respect to the ionic species of interest; for example, following high-efficiency retention of a selected ion species, M^+, a cluster of neighboring ions such as $[M–H]^+$ and $[M + 1]^+$ is always present, thus leading on occasion to misleading data.

This undesirable behavior can be explained by considering the real shape of the stability diagram for the commercial ITMS instrument. As determined experimentally by Yost et al.,[17] the upper apex of the Finnigan extended version of the ion trap is round-shaped rather than acute-angled, thus explaining the observed undesired behavior.

In order to avoid the retention of ionic species other than that of the ionic species of interest, a more effective method has been proposed, consisting of a two-step ion ejection process.[18,19] By inspection of the stability diagram shown in Fig. 9.1, it can be seen that, once the working point of the ion species of interest has been moved to point A in the stability diagram, the application of a negative DC voltage leads to the elimination of all the ions of lower mass-to-charge ratio. The elimination of ions of mass-to-charge ratio higher than that of the ion of interest is accomplished by the application of a positive DC voltage, once the working point of the ion species of interest is lowered to a q_z value of 0.5 (point B). The working point of the isolated ion species is returned to B, by removal of the positive DC voltage, whereupon the ionic species can be irradiated subsequently at point C. By this method, an effective cleaning of all of the unwanted ions from the ion trap is easily achieved. However, it is worth noting that the mass spectrum acquired at this point in the scan function, in the absence of any ion irradiation, still shows the presence of ions of low mass-to-charge ratio.

The observation of low mass ions at this stage is not due to the incomplete ejection of these ions at the $\beta_z = 1$ boundary, but rather they have been formed by the dissociation of the mass-selected ions once the working point has been moved vertically downwards from the q_z-axis, with $q_z = 0.5$, to a point near the $\beta_z = 0$ boundary. The mass-selected ions undergo disorderly motion under these conditions, leading to an enhancement in their kinetic energy; subsequent collisions with background helium buffer gas can cause extensive fragmentation.

This problem of interfering low mass ions can be eliminated by changing the order of the two phases of the ion isolation process or by a further cleaning phase consisting of a scan of the drive RF voltage such that the

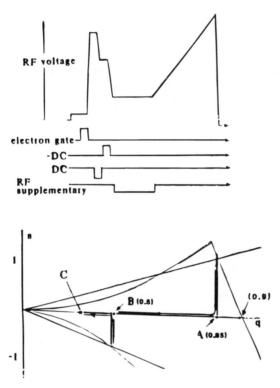

FIGURE 9.1

Scan function and stability diagram operation for the two-step isolation of ionic species for tandem mass spectrometry (MS/MS) experiments. (From Ardanaz, C. E.; Traldi, P.; Vettori, U.; Kavka, J., Guidugli, F. *Rapid Commun. Mass Spectrom.* 1991, 5, 5. Reproduced by courtesy of John Wiley & Sons.)

working point of the selected ion is moved from point C to the point at $q_z = 0.9$ followed by its return to point C.

Ion excitation by means of resonance irradiation has been described from both a theoretical[20] and a practical[21] point of view by many researchers. For this kind of experiment, a supplementary RF potential is generally employed and applied to the two end-cap electrodes, so that they are out-of-phase with each other. When such supplementary RF fields exhibit the same frequency (or frequencies) as that of the motion of a specific ion species, the ions will gain energy from the field, and the orbits of the ions will become increasingly larger due to a strong interaction with the main RF field. This interaction will lead to ion acceleration and subsequent activating collisions with the buffer gas. The axial modulation is characterized by two quantities, that is, the amplitude of the supplementary alternating current (AC) voltage and the irradiation time. The product of these quantities can be considered as a fluence.

It must be stressed that there is a substantial difference between a collisional experiment performed in a multisector instrument (MS/MS "in space") and that performed by ITMS (MS/MS "in time"). In the former, the preselected ions undergo a variety of interactions with the target gas leading (as described in the Introduction) to a family of excited species with quite a large energy distribution. Such a wide distribution necessarily enhances the feasibility of accessing decomposition channels having large differences in activation energies.

The same phenomenon is completely different in collisional experiments performed by ITMS using a supplementary RF voltage. In fact, under such conditions, the energy deposition mechanism can be considered a step-by-step one, in which low activation energy processes are usually favored.

Such behavior has been discussed by Gronowska et al.[19] who demonstrated that the step-by-step deposition of discrete amounts of internal energy necessarily leads to decomposition through the lower activation energy processes. When two or more decomposition processes exhibit comparable activation energies, they all can become operative under the influence of the single-step energy deposition mechanism. In Gronowska's study,[19] substituted analogs of aryl ketones (Fig. 9.2) were used as models so as to gain information on the intricate mechanism of energy deposition by ITMS. The effects of different target gases, He and Ar, at different pressures were also examined.

Compounds **2** to **4**, when subjected to collisional activation in the ion trap, showed similar behavior due to the existence for each compound of fragmentation pathways of low critical energy (loss of CO_2 for **2** and **3**, and loss of H_2O for **4**; Fig. 9.3) which overwhelm all other competing reactions. The resulting fragments F_1 undergo further fragmentation to an extent depending either on the critical energies E_1 and E_2, or, on the internal energy content of the activated $M^{+\bullet *}$ An increase in the F_2/F_1 ratio will be the result of collisional conditions leading to a greater energy deposition in the precursor ions. Figs. 9.4(a) and 9.4(b) show the effect of He pressure on the absolute intensities of ions F_1 and F_2 for compound **3**. A decrease in the ion abundance, together with a more or less pronounced increase in the F_2/F_1 ratio, is observed by increasing the He pressure. As expected, the gradual increase of internal energy has a more pronounced effect for **2** (the F_2/F_1 ratio varies from 8 to 70) than for compound **3** (F_2/F_1 varies from 0.3 to 0.5), due to the larger difference in the rates of $X^{\bullet}$ loss from each of the two compounds.

The same experiments using Ar as a target gas demonstrate an enhancement in the occurrence of the secondary fragmentation (Fig. 9.4(c)). Thus, for both **2** and **3**, the F_1/F_2 ratio reaches larger values than those observed with He, even under the most favorable pressure condition. Moreover with Ar, the F_2/F_1 ratio is not significantly affected by the

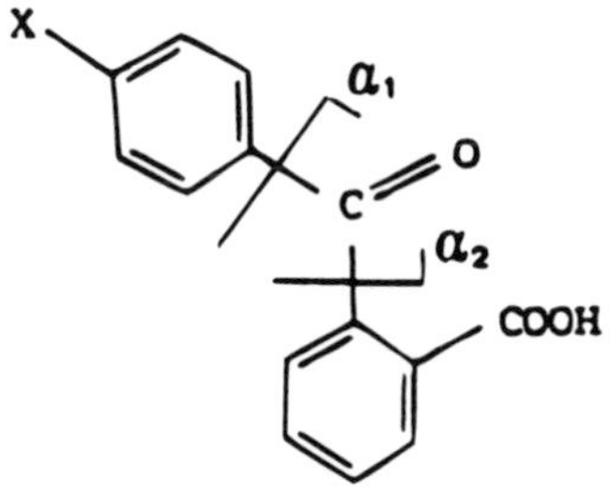

1: X = H; 2: X = Cl; 3: X = OCH₃

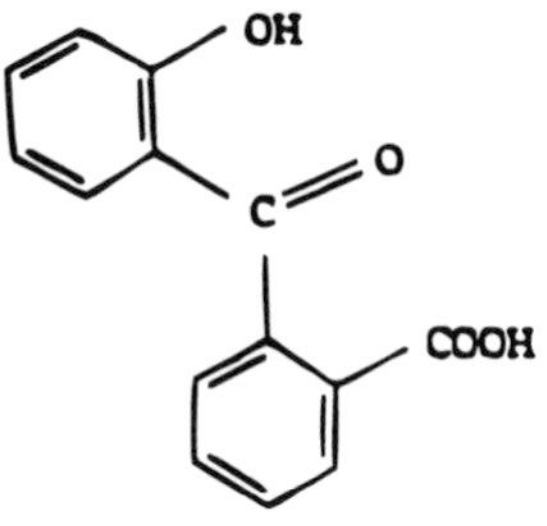

4

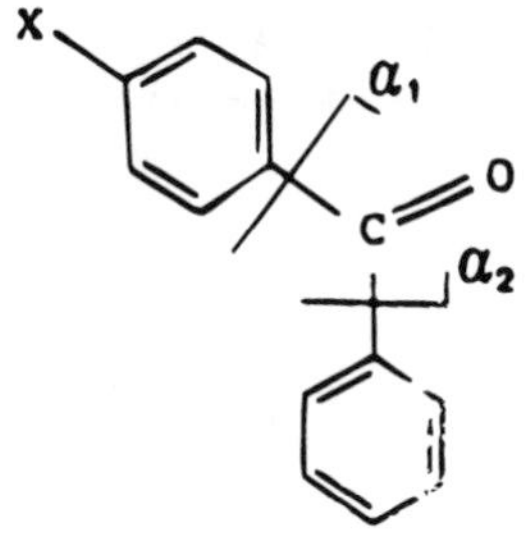

5: X = H; 6: X = Cl; 7: X = OCH₃

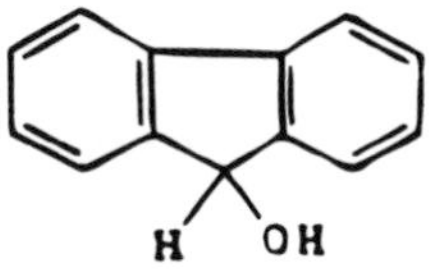

8

FIGURE 9.2
Structures for compounds **1** to **8**. (From Gronowska, J.; Paradisi, C.; Traldi, P.; Vettori, U. *Rapid Commun. Mass Spectrom.* 1990, *4*, 306. Reproduced by courtesy of John Wiley & Sons.)

pressure, while this parameter plays a critical role in determining the number of ions which can be detected. For example, let us examine the results obtained for compound **3** which are reported in Fig. 9.4(d).

These findings clearly indicate that collisional activation with Ar target gas produces parent ions with higher internal energies than those generally present with He activation. This effect can be related to the larger mass of the target, which accounts for both the higher energy

$$\underset{\mathbf{2,3}}{M^{+\cdot}} \xrightarrow{\text{collision}} M^{+\cdot *} \underset{-CO_2}{\overset{E_1}{\longrightarrow}} F_1^{+\cdot *} \underset{-X}{\overset{E_2}{\longrightarrow}} F_2^{+}$$

$$\underset{\mathbf{4}}{M^{+\cdot}} \xrightarrow{\text{collision}} M^{+\cdot *} \underset{-H_2O}{\overset{E_1}{\longrightarrow}} F_1^{+\cdot *} \underset{-CO}{\overset{E_2}{\longrightarrow}} F_2^{+}$$

FIGURE 9.3
Decomposition patterns of compounds **2** to **4** obtained by collisional activation in the ion trap. (From Gronowska, J.; Paradisi, C.; Traldi, P.; Vettori, U. *Rapid Commun. Mass Spectrom.* 1990, *4*, 306. Reproduced by courtesy of John Wiley & Sons.)

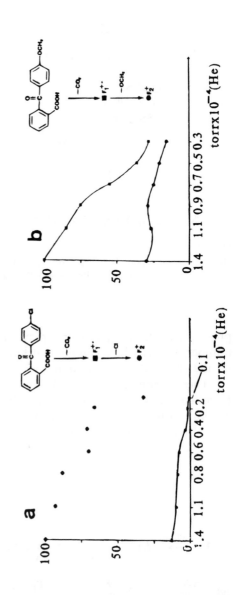

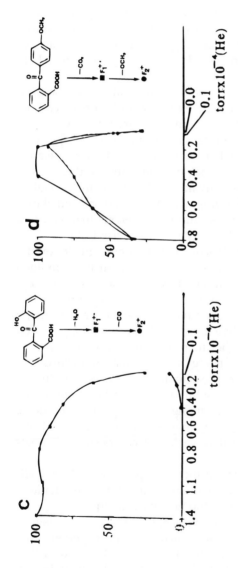

FIGURE 9.4

Effect of buffer gas pressure on ion intensities: (a), (b) and (d) effect of He pressure on the absolute intensities of ions F_1^+ and F_2^+ for compound two, three, and four; (c) effect of Ar pressure on the absolute intensities of ions F_1^+ and F_2^+ for compound three. (From Gronowska, J.; Paradisi, C.; P.; Vettori, U. *Rapid Commun. Mass Spectrom.* 1990, 4, 306. Reproduced by courtesy of John Wiley & Sons.)

deposition and the trapping efficiency observed. Furthermore, this behavior explains the progressive decrease in trapping efficiency with pressure reduction; additionally, while more energetic collisions occur, they occur less frequently, which leads to a reduction in the damping of ion motion by the target gas.

In order to obtain a rough estimate of the energetics involved in the ion trap, a comparison was made of the mass-analyzed ion kinetic energy spectrum (MIKES) of $M^{+\cdot}$ with both the high-energy collisional spectrum of $M^{+\cdot}$, and the EI mass spectrum. This comparison showed that the critical energy for α-cleavage (Fig. 9.2) is greater than that required for the loss of $X^\cdot$. Moreover, $E(\alpha_1) < E(\alpha_2)$ (Fig. 9.2) both when $X = OCH_3$ and when $X = Cl$.

It was noted that in ITMS, the main collisionally-induced decomposition channel was the loss of $X^\cdot$, with the exception of **7,** for which both the peaks due to α-cleavage and $X^\cdot$ loss were of similar intensity. By way of contrast, α-cleavage occurred to a larger extent in the MIKE spectrum of the same species, thus suggesting that activation in ITMS is soft, relative to the other systems examined. A reasonable explanation for these observations lies in the mechanism of ion activation in ITMS and in the relative energetics of the possible competing fragmentation. As discussed above, ion activation in the ion trap involves multiple collisions that lead to a gradual increase of the ion internal energy in a step-wise fashion.

One could expect that an ion will achieve increasingly greater excitation so long as the rate of energy uptake is greater than the combined rates of de-excitation and fragmentation. However, due to the step-by-step internal energy acquisition, only a single decomposition process will be observed when this reaction channel has a lower activation energy than those of all the other competitive decomposition channels, provided that its rate of fragmentation is higher than the rate of energy uptake.

Thus, the effective deposition of energy in daughter ion spectroscopy by ITMS does not necessarily lead to the activation of processes of higher activation energy; it is not only the absolute value of critical energy that matters, but also the energy spacing among the competitive reactions. The kinetics of energy acquisition are fundamental, therefore, to the determination of the relative magnitudes of the various decomposition channels of the parent ion species.

The above considerations point out the weak point of ITMS in collisional experiments and, for this reason, many efforts have been made to activate ions by techniques other than the tickling technique.

Considering that the real action of the supplementary AC voltage is to move ions to the region of space in which they can experience the high electric field, which is responsible for their acceleration and further activation by collision, it had been thought possible to obtain analogous results simply by moving the working point of the ions close to the

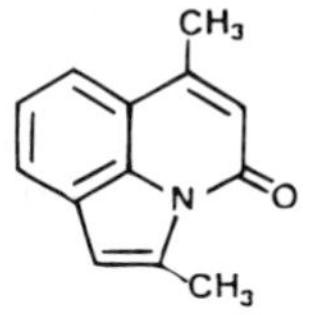

FIGURE 9.5
Structure of benzoquinolizidine derivative. (From (a) Paradisi, C.; Todd, J. F. J.; Traldi, P.; Vettori, U. *Org. Mass Spectrom.* 1992, *27*, 251. (b) Curcurcuto, O.; Fontana, S.; Traldi, P.; Celon, E. *Rapid Commun. Mass Spectrom.* 1992, *6*, 322. Reproduced by courtesy of John Wiley & Sons.)

border of the stability diagram.[11] Some data have been presented which demonstrate that such a "border effect" not only leads to daughter ion spectra with high efficiency, but that the energy deposition is higher than that achieved by tickling. Using a benzoquinolizidine derivative (Fig. 9.5) as a model compound, m/z 227, it was shown that daughter ion spectra richer in fragment ion intensity and variety were obtained by the border effect (see Table 9.1).

While only one fragment ion species, formed by primary methyl loss, is generated by tickling, new daughter ions corresponding to $CHO^{\cdot}$, $CH_3CO^{\cdot}$ and C_2H_4O losses were detected by the border effect. In order to obtain information on the kinetic energies achieved by tickling and by the border effect, the above results were compared with those obtained by triple quadrupole collision experiments (Table 9.1). It was found experimentally that the tickling data obtained with the ITMS were very close to those obtained by ions having 5 eV of kinetic energy, while the border effect led to energy deposition close to that observed from collisions of ions having 100 eV of kinetic energy.

Worthy of note here is the possibility of unequivocal discrimination between consecutive and competing collisionally-induced decompositions, obtained by the collisional activation of a parent ion and the simultaneous ejection of one of its daughter ions.[22] Such experiments were carried out by applying the border effect to the parent ion and, at the same time, applying a tickling potential suitable to eject the selected daughter ion from the ion trap. By such an approach, the possible intermediacy of

TABLE 9.1

Daughter ion mass spectra of $M^{+\cdot}$ of a benzoquinolizidine derivative

m/z	Tickling	Application of DC Voltage	Collision With He With Kinetic Energy of	
			5 eV	100 eV
212	100	94	100	97
198	—	3.4	—	1
184		0.8		2
183	—	0.4	—	—

fragment ions in the production of species of lower mass/charge ratio can be easily established.

Finally, the ion trap has been applied to studies of a novel technique for surface-induced dissociation.[12] A short (<5 μs), fast-rising (<20 ns rise time) DC pulse of high voltage applied to the end-cap electrodes of the ion trap causes the trajectories of trapped ions to become unstable in the radial direction such that they collide subsequently with the ring electrode. The daughter spectra so obtained demonstrated that high-energy decomposition channels were activated; simulations of ion motion in the ion trap provided evidence that collisions take place in the 10 to 10^2 eV kinetic energy range.

III. APPLICATIONS

A. Chalcones

Chalconoids represent an important class of natural flavonoids, exhibiting interesting properties from both biosynthetic and biomedical points of view.[23-25]

Mass spectrometry has proven to be particularly useful for the structural characterization of such compounds, and the observation of their general fragmentation patterns has been the subject of many publications. However, there is still some debate concerning specific electron impact induced decomposition pathways; in particular, there is disagreement in the literature concerning the structure of cations originating from primary H[·] loss.

Ronayne *et al.*[26] postulated that the [M–H][+] ion of chalcone (compound **1** in Fig. 9.6) originates by H[·] loss from ring A with the formation

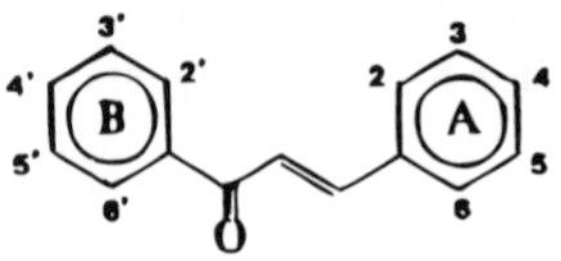

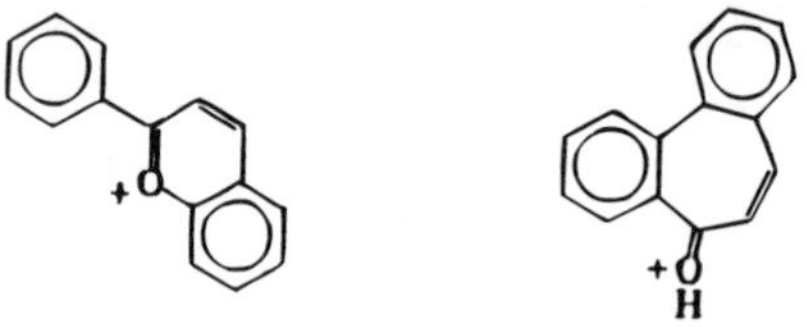

FIGURE 9.6
Structure of chalcone; (a) flavinium cation, (b) structure **b**. (From Ardanaz, C. E.; Traldi, P.; Vettori, U.; Kavka, J.; Guidugli, F. *Rapid Commun. Mass Spectrom.* 1991, 5, 5. Reproduced by courtesy of John Wiley & Sons.)

of the flavinium cation **a**; such a hypothesis was also proposed by Van de Sande et al.,[27] analogues to that proposed for the formation of [M–H]⁺ ions in the EI mass spectrum of 3-flavene.[28] An alternative mechanism which involved the formation of structure **b** was proposed by Beynon *et al.*,[29] but the related mechanism was not discussed in detail.

ITMS has proven to be highly effective in obtaining the solution to this hotly contested problem.[30] The EI generated molecular ions of chalcone (**1**) and the deuterated analogous 2′, 3′, 4′, 5′, 6′-pentadeutero-chalcone (**2**) and 2,3,4,5,6-pentadeuterochalcone (**3**) were mass-selected by the two-step isolation procedure described above, and subjected to a supplementary RF field oscillating at the appropriate frequency. The usual EI spectra indicated that hydrogen loss from ring A is involved in the formation of the [M–H]⁺. In fact, collision-induced dissociation of the molecular ion of compound **3**, deuterium-labelled on ring A, showed that the formation of [M–D]⁺ ions is highly favored. Further-more, comparison of the behavior of compounds **1** and **2** with the corresponding *cis* isomers, showed that the [M–H]⁺ ions for **1** and **2** and [M–D]⁺ ions for **3**, are 15% more abundant for *cis* than for *trans* isomers. These data are in agreement with the presence of structure **c** of Scheme 9.1.

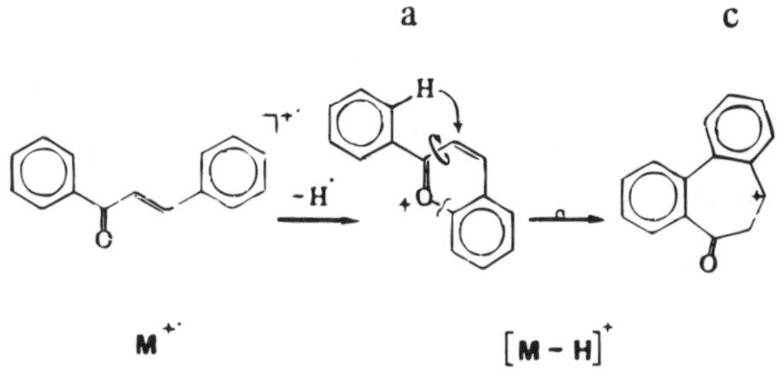

SCHEME 9.1
(From Ardanaz, C. E.; Traldi, P.; Vettori, U.; Kavka, J.; Guidugli, F. *Rapid Commun. Mass Spectrom.* 1991, 5, 5. Reproduced by courtesy of John Wiley & Sons.)

Collisionally-induced decomposition studies carried out with the ion trap were highly effective for the structural assignment of the [M–H]⁺ ion. The daughter ion spectrum of the [M–H]⁺ ion of chalcone (Fig. 9.7) contained an easily detectable ion at m/z 165, originating from the loss of ketene, which necessarily requires that the [M–H]⁺ ion have structure **c** (Scheme 9.2). Observation of the two other fragment ion species, m/z 179 and m/z 178, formed by collisionally-induced fragmentation of the [M–H]⁺ species and by loss of CO and HCO•, respectively, is also in agreement with structure **c**.

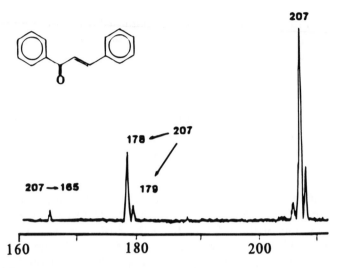

FIGURE 9.7
Daughter ion spectrum of the [M–H]⁺ ion (*m/z* 207) of chalcone. (From Ardanaz, C. E.; Traldi, P.; Vettori, U.; Kavka, J.; Guidugli, F. *Rapid Commun. Mass Spectrom.* 1991, 5, 5. Reproduced by courtesy of John Wiley & Sons.)

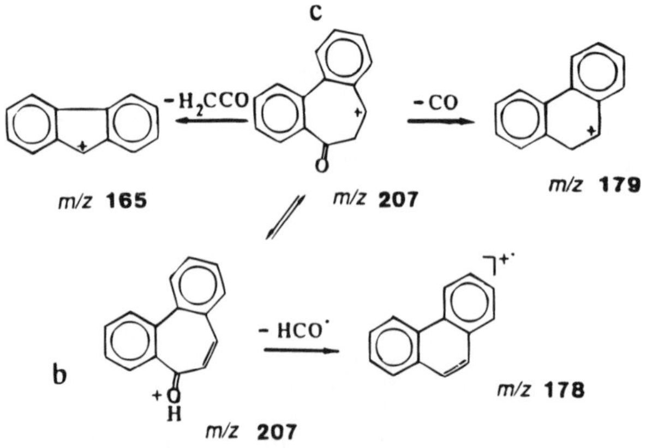

SCHEME 9.2
(From Ardanaz, C. E.; Traldi, P.; Vettori, U.; Kavka, J.; Guidugli, F. *Rapid Commun. Mass Spectrom.* 1991, 5, 5. Reproduced by courtesy of John Wiley & Sons.)

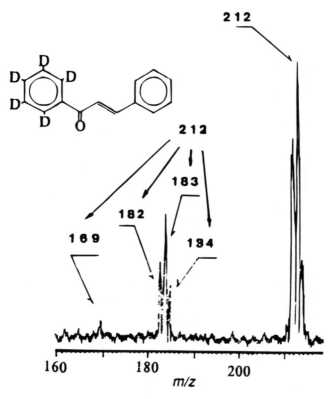

FIGURE 9.8
Daughter ion spectrum of the [M–H]⁺ ion (*m/z* 212) of 2′,3′,4′,5′,6′-pentadeuterochalcone. (From Ardanaz, C. E.; Traldi, P.; Vettori, U.; Kavka, J.; Guidugli, F. *Rapid Commun. Mass Spectrom.* 1991, 5, 5. Reproduced by courtesy of John Wiley & Sons.)

In order to obtain further experimental evidence that the H atom in the *ortho* position is involved mainly in ketene and HCO˙ elimination from [M–H]⁺ species, the [M–H]⁺ ion of **2** was subjected to collisional experiments. The related daughter ion spectrum (see Fig. 9.8 and Scheme 9.3) showed losses of HDC–CO, DCO˙, HCO˙ and CO. Furthermore, an investigation was undertaken on 2′,6′-difluoro-chalcone and showed that primary F loss is highly favored; again, collisional experiments performed on [M–F]⁺ ions showed the expected losses of ketene, HCO˙ and CO˙.

Interestingly, the [M–H]⁺ ion of 3-flavene showed perfectly identical behavior to that of the [M–H]⁺ ion of chalcone, proving that it rearranges to structure **c** prior to fragmentation (see Fig. 9.9 and Scheme 9.4).

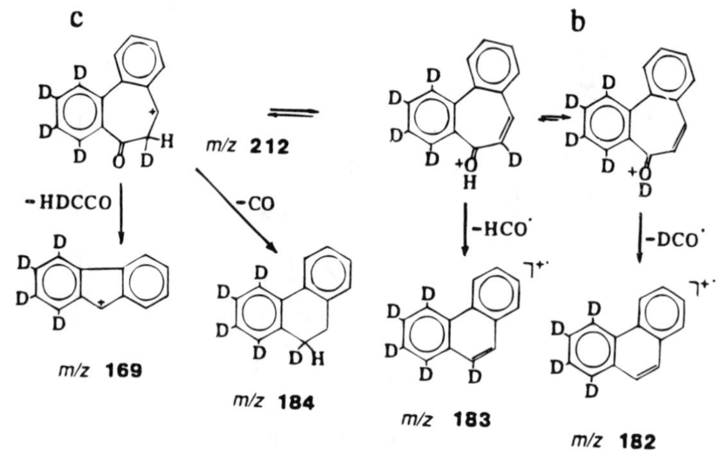

SCHEME 9.3
(From Ardanaz, C. E.; Traldi, P.; Vettori, U.; Kavka, J.; Guidugli, F. *Rapid Commun. Mass Spectrom.* 1991, 5, 5. Reproduced by courtesy of John Wiley & Sons.)

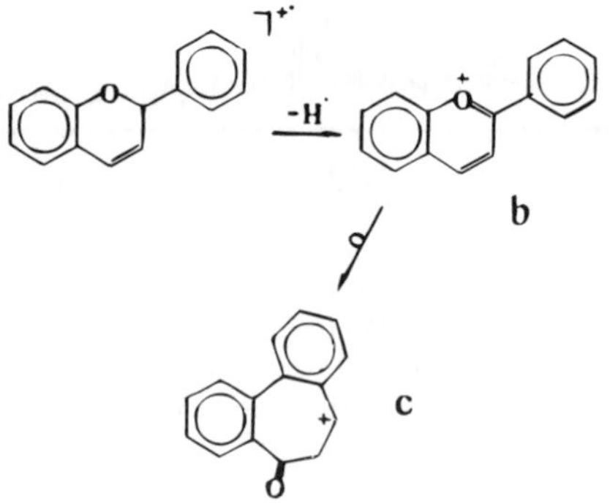

SCHEME 9.4
(From Ardanaz, C. E.; Traldi, P.; Vettori, U.; Kavka, J.; Guidugli, F. *Rapid Commun. Mass Spectrom.* 1991, 5, 5. Reproduced by courtesy of John Wiley & Sons.)

Finally, an investigation of the structure of [M–H–ketene]$^+$ ions originating from chalcone and 3-flavene was carried out by comparing the related daughter ion spectra with that of the [M–H]$^+$ ion of fluorene (Fig. 9.10). The remarkable similarity of these mass spectra proves that the ions at m/z 165, originating from fragmentation of [M–H]$^+$ ions from both chalcone and 3-flavene, have the same structure as the [M–H]$^+$ ion of fluorene.

There was still a disagreement in the literature on the formation of ions at m/z 130 of chalcone, which are considered highly diagnostic from a structural point of view. Thus, Van de Sande *et al.*[27] have suggested that structure **a,** represented in Scheme 9.5, originates from a cyclic intermedi-

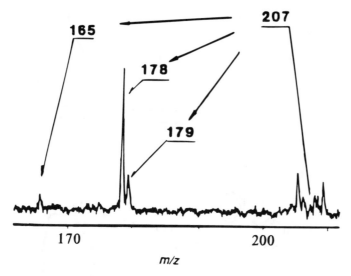

FIGURE 9.9
Daughter ion spectrum of the [M–M]$^+$ ion (m/z 207) of 3-flavene. (From Ardanaz, C. E.; Traldi, P.; Vettori, U.; Kavka, J.; Guidugli, F. *Rapid Commun. Mass Spectrom.* 1991, 5, 5. Reproduced by courtesy of John Wiley & Sons.)

ate, while Rouvier *et al.*[31] maintain that structure **b** (Scheme 9.5) is the more reasonable structure.

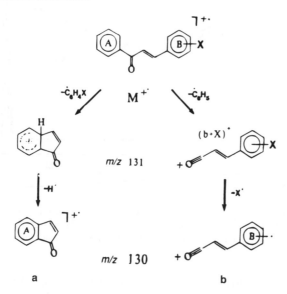

SCHEME 9.5
(From Ardanaz, C. E.; Kavka, J.; Curcuruto, O.; Traldi, P.; Guidugli, F. *Rapid Commun. Mass Spectrom.* 1991, 5, 569. Reproduced by courtesy of John Wiley & Sons.)

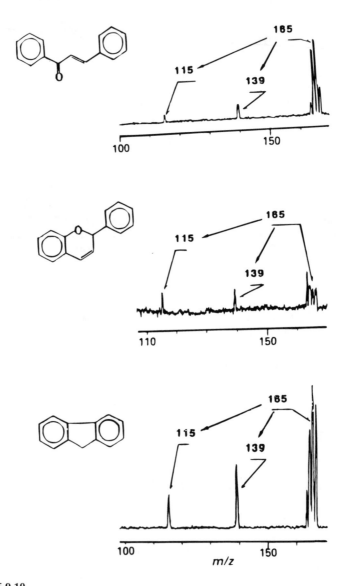

FIGURE 9.10
Daughter ion spectra of the ion of m/z 165 generated by electron ionization of (a) chalcone, (b) 3-flavene, and (c) fluorene. (From Ardanaz, C. E.; Traldi, P.; Vettori, U.; Kavka, J.; Guidugli, F. *Rapid Commun. Mass Spectrom.* 1991, 5, 5. Reproduced by courtesy of John Wiley & Sons.)

In a more recent study, a comparison of collisional data, obtained in both the high- and low-energy regimes for differently labelled (D, ^{13}C) and substituted (CH$_3$,F,Cl) compounds, led to a definitive structure assignment.[32] The fragmentation mechanism reported in Scheme 9.6 was proposed and confirmed by ^{13}C labeled compounds. Thus, the formation of the four-membered cyclic intermediate is hypothesized in the upper part of Scheme 9.6, while in the lower one an acyclic intermediate is shown. These hypotheses were based initially on the observed losses of both [12]CO and [13]CO, but more definitive data were obtained by energy-resolved mass spectrometry (ERMS) using the ITMS.[32]

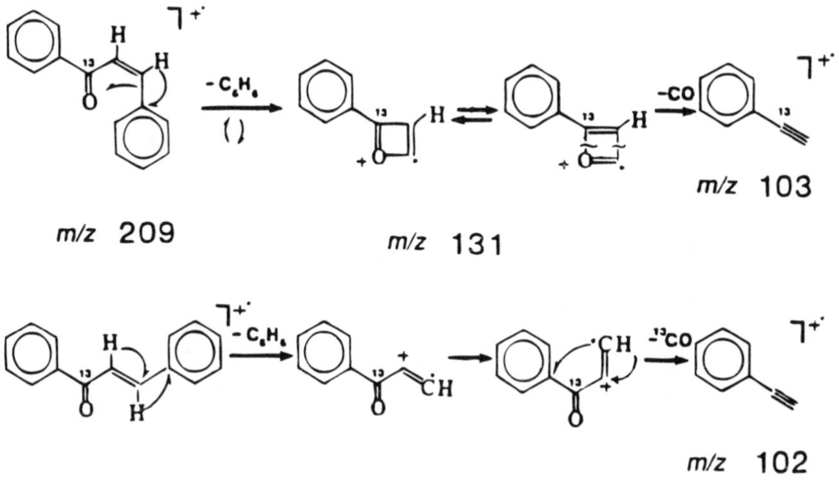

SCHEME 9.6
(From Ardanaz, C. E.; Kavka, J.; Curcuruto, O.; Traldi, P.; Guidugli, F. *Rapid Commun. Mass Spectrom.* 1991, *5*, 569. Reproduced by courtesy of John Wiley & Sons.)

The related breakdown curves, displayed in Fig. 9.11, are in agreement with the presence of two intermediates having different structures. From these data, it is evident that [13]CO loss is a process requiring high internal energy, and that the clear differences in the ERMS for the ions of m/z 102 and m/z 103 can be rationalized by invoking both of these different structures for the intermediates reported in Scheme 9.6.

B. Hydroxy- and Methoxy-Indoles

Hydroxyindoles represent important oxidative degradation products of tryptophan in man: 3-, 4-, and 5-hydroxyindoles (compounds **4** to **6**, Fig. 9.12) are present in small amounts in human urine, but considerable increases in their concentrations are observed in patients with malignant melanoma.[33] Hence, the development of analytical procedure for the characterization of isomeric 3-, 4-, and 5-hydroxyindoles is of high interest from a biomedical point of view.

The electron ionization mass spectra of compounds **4** to **6** are practically identical, with only minor and not significant differences in relative abundances of product ions (Table 9.2, Scheme 9.7). An investigation

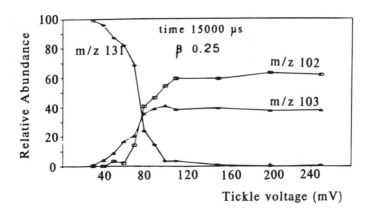

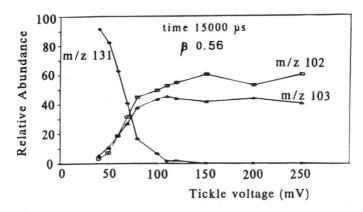

FIGURE 9.11
Breakdown curves of $[M–C_6H_6]^+$, m/z 131, obtained by variation of tickle voltage amplitude, for two values of β_z. m/z 131 was formed by EI of carbonyl-^{13}C chalcone. (From Ardanaz, C. E.; Kavka, J.; Curcuruto, O.; Traldi, P.; Guidugli, F. *Rapid Commun. Mass Spectrom.* 1991, 5, 569. Reproduced by courtesy of John Wiley & Sons.)

3-hydroxyindole 4

4-hydroxyindole 5

5-hydroxyindole 6

FIGURE 9.12
Structures of 3-, 4-, and 5-hydroxyindole, compounds **4** to **6**. (From Evans, C.; Catinella, S.; Traldi, P.; Vettori, U.; Allegri, G. *Rapid Commun. Mass Spectrom.* 1990, 4, 335. Reproduced by courtesy of John Wiley & Sons.)

TABLE 9.2

70 eV EI mass spectra of the three hydroxyindole isomers

	% Relative Abundance		
m/z	3-Hydroxyindole	4-Hydroxyindole	5-Hydroxyindole
134	9	8	9
133	100	100	100
132	15	11	12
105	14	19	18
104	32	40	39
79	—	5	7
78	2	11	19
77	15	10	14
52	—	5	8
51	6	7	9

has been carried out by the energy-resolved mass spectrometric method available with the ITMS, that is, by varying tickle time at constant tickle voltage amplitude and β_z values, or by varying tickle voltage amplitude at constant tickle time and β_z values, or by variation of β_z value at constant tickle voltage amplitude and tickle time.[34]

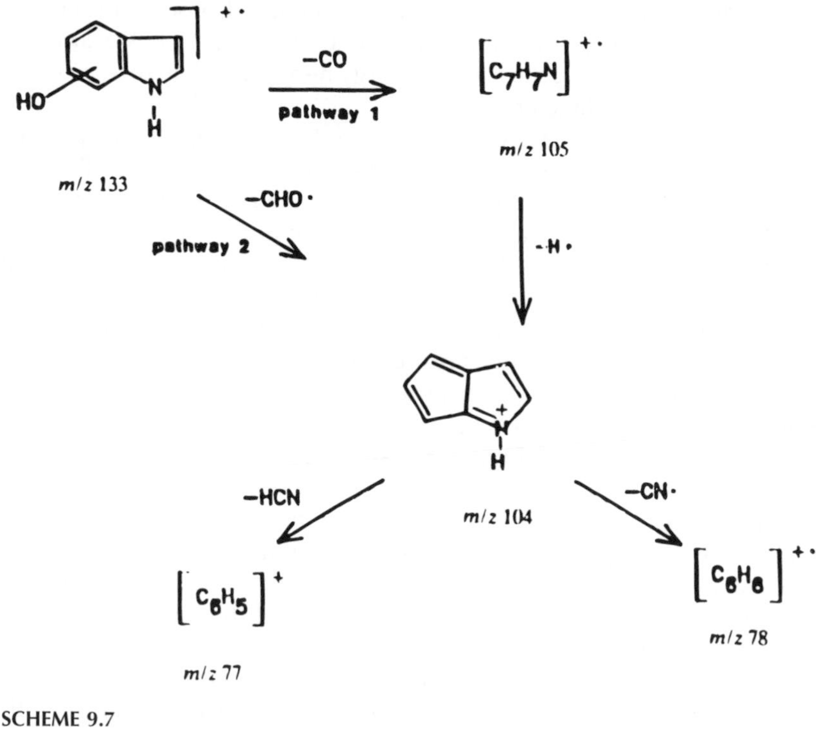

$$HO-\text{[indole]}^{+\cdot} \xrightarrow[\text{pathway 1}]{-CO} \left[C_7H_7N\right]^{+\cdot} \quad m/z\ 105$$

$m/z\ 133$

$\xrightarrow[\text{pathway 2}]{-CHO\cdot}$

$-H\cdot$

$m/z\ 104$

$\xrightarrow{-HCN} \left[C_6H_5\right]^{+} \quad m/z\ 77$

$\xrightarrow{-CN\cdot} \left[C_6H_6\right]^{+\cdot} \quad m/z\ 78$

SCHEME 9.7

(From Evans, C.; Catinella, S.; Traldi, P.; Vettori, U.; Allegri, G. *Rapid Commun. Mass Spectrom.* 1990, 4, 335. Reproduced by courtesy of John Wiley & Sons.)

As discussed in the Introduction, by increasing the supplementary RF voltage amplitude, the parent ions are excited into larger trajectories, thus experiencing greater interaction with the main quadrupolar field. Such an interaction leads to ion acceleration and more energetic collisions with the buffer gas.

The breakdown curves of molecular ions of the three isomeric compounds, 3-, 4-, and 5-hydroxyindole, obtained by varying the tickle voltage, are reported in Fig. 9.13. The relative abundances of the two primary fragment ions of m/z 105 and m/z 104 originating from CO and CHO$^{\bullet}$ loss, respectively, show completely different ratios for the three isomers examined. Thus, while ions of m/z 104 for compound 4 remain the most abundant throughout the range of tickle voltage amplitudes examined, the plots related to ions of m/z 104 and m/z 105 for both 5 and 6 show well-defined crossing points (at tickle voltage zero-to-peak amplitudes of 750 mV for 5 and 150 mV for 6), thus proving that the primary CO loss is favored at lower collision energies for such compounds.

ERMS obtained by variation of tickle time, at constant tickle voltage amplitude, was found to be less effective for isomer characterization. As can be seen in Fig. 9.14, the ratio of signal intensities of m/z 104 and m/z 105 remained constant for isomers 5 and 6, while this ratio was inverted for the same ions in the case of isomer 4. The trend observed seems to suggest that the energy distribution obtained by varying tickle times in the range 0 to 200 ms leads to a lower energy deposition than that achieved by varying tickle voltage amplitude.

Finally, breakdown curves were obtained by varying β_z. The β_z lines in the stability diagram are related to the secular frequency of oscillations of the ions in the axial direction of the ion trap. The fundamental axial oscillation frequency is related to the values of β_z and the RF drive frequency. The setting for the RF drive voltage amplitude determines the values of both q_z and, to a first approximation, β_z for each ion species stored within the ion trap. Thus, by variation of the RF drive voltage amplitude, a series of β_z values can be obtained. When a tickle voltage was applied to the device and adjusted to the fundamental axial secular frequency of the chosen ion species, it was found that larger amounts of energy were deposited at higher β_z values. The results so obtained for $M^{+\bullet}$ of the three hydroxyindole isomers are reported in Fig. 9.15.

Such breakdown curves have several features in common with those obtained by varying tickle voltage amplitude, but the energy deposition achieved by varying β_z was less than that achieved by increasing the tickle voltage amplitude. For example, the ratio of the intensities of m/z 104 and m/z 105 was found to be 0.75 for compound 5 and 1.25 for 6; these signal intensities were obtained at the highest β_z values, and correspond to the same intensity ratios obtained at a tickle voltage amplitude of $ca.$ 400 mV$_{(0-p)}$, that is, zero-to-peak amplitude.

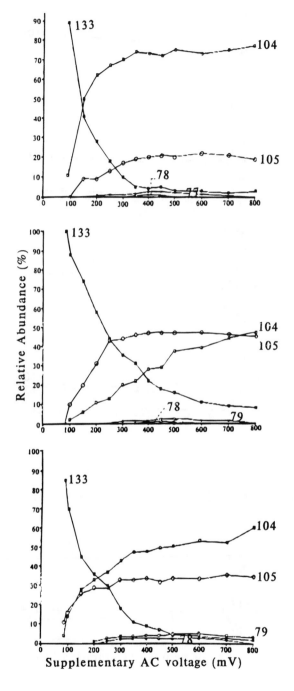

FIGURE 9.13
Energy-resolved mass spectra obtained for the molecular cation ($M^{+\cdot}$) of hydroxyindole isomers **4** to **6** at various amplitudes (zero-to-peak) of the supplementary RF voltage ("tickle voltage"). Top spectrum relates to compound **4**; center spectrum to compound **5**; bottom spectrum relates to compound **6**. (From Evans, C.; Catinella, S.; Traldi, P.; Vettori, U.; Allegri, G. *Rapid Commun. Mass Spectrom.* 1990, *4*, 335. Reproduced by courtesy of John Wiley & Sons.)

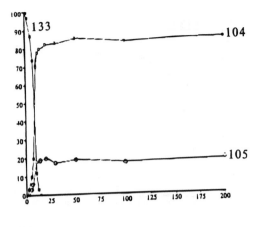

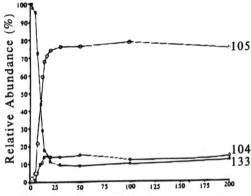

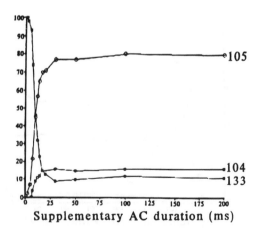

FIGURE 9.14
Energy-resolved mass spectra obtained of the molecular cation (M⁺⁺) of hydroxyindole isomers **4** to **6** by varying tickle time. Top spectrum relates to compound **4**; center spectrum to compound **5**; bottom spectrum relates to compound **6**. (From Evans, C.; Catinella, S.; Traldi, P.; Vettori, U.; Allegri, G. *Rapid Commun. Mass Spectrom.* 1990, *4*, 335. Reproduced by courtesy of John Wiley & Sons.)

Supplementary AC duration (ms)

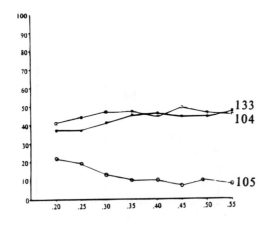

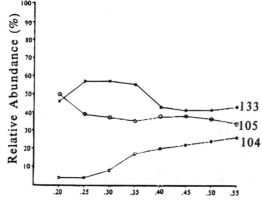

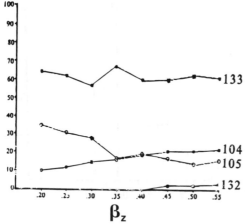

FIGURE 9.15
Energy-resolved mass spectra
obtained for the molecular
cation ($M^{+\cdot}$) of hydroxyindole
isomers 4 to 6 by varying β_z
values. (Relative abundances
are normalized to the sum of the
abundance of the ions plotted.)
Top spectrum relates to
compound **4;** center spectrum
to compound **5;** bottom
spectrum relates to compound
6. (From Evans, C.; Catinella,
S.; Traldi, P.; Vettori, U.; Allegri,
G. *Rapid Commun. Mass
Spectrom.* 1990, *4*, 335.
Reproduced by courtesy of
John Wiley & Sons.)

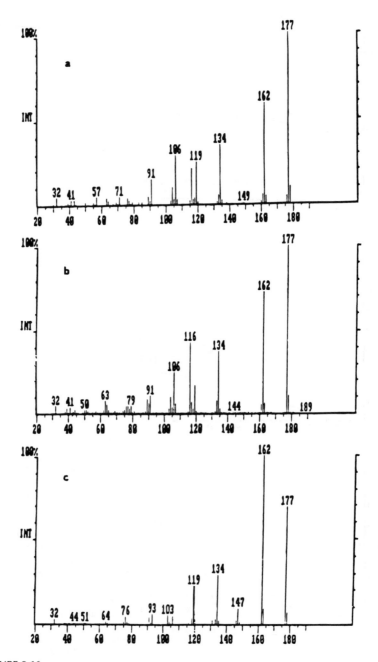

FIGURE 9.16
EI (70 eV) mass spectra of dimethoxyindoles. (a) 5,6-dimethoxyindole (**7**), (b) 6,7-dimethoxy-indole (**8**) and (c) 4,7-dimethoxyindole (**9**). (From Vettori, U.; Allegri, G.; Ferlin, M. G.; Traldi, P. *Int. J. Mass Spectrom Ion Processes.* 1990, *99*, 99. Reproduced by courtesy of Elsevier Science.)

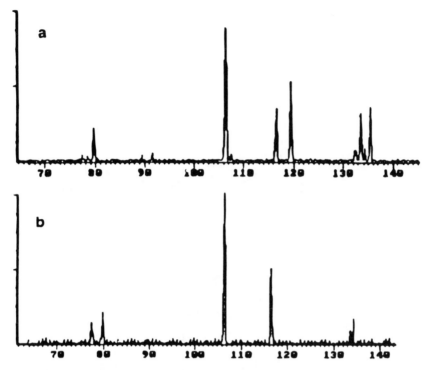

FIGURE 9.17
CAD mass spectra of m/z 134 ions from 5,6-dimethoxyindole (**7**). m/z 134 ions were generated by (a) EI; (b) collisional activation. (From Vettori, U.; Allegri, G.; Ferlin, M. G.; Traldi, P. *Int. J. Mass Spectrom Ion Processes*. 1990, *99*, 99. Reproduced by courtesy of Elsevier Science.)

An analogous investigation was carried out on some dimethoxy-indole derivatives, that is, 5,6-, 6,7-, and 4,7-dimethoxy-indoles[35] (compounds **7, 8,** and **9,** respectively).

By ITMS, a highly detailed description of the behavior of these compounds was obtained, permitting a clear characterization of the three isomers. As can be seen from Fig. 9.16, while compound **9** led to a quite different EI mass spectrum, compounds **7** and **8** led to mass spectra which were practically superimposable with the only difference being in the abundance of ions at m/z 116. The general fragmentation pattern of **7** and **8** is reported in Scheme 9.8, in which are summarized the data of both EI and collisional experiments; while in Scheme 9.9, the fragmentation pattern of **9** is shown. All three compounds showed initially similar behavior, consisting of sequential losses of $CH_3^\cdot$ and CO, responsible for the ions of m/z 162 and m/z 134, respectively. The collisional spectra of the ions of m/z 177 ($M^{+\cdot}$), and m/z 162 are not effective for isomer characterization while those of the ions of m/z 134 (see upper part of Figs. 9.17 to 9.19) are clearly different. In

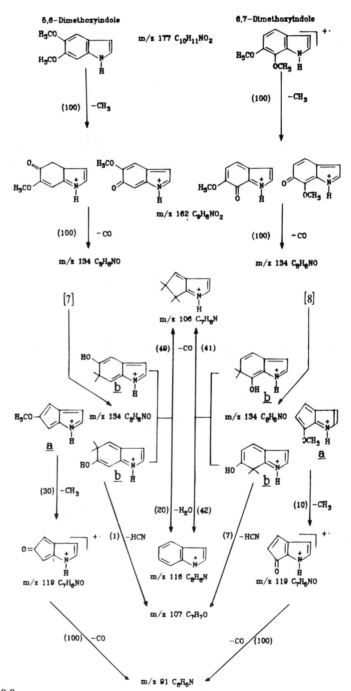

SCHEME 9.8

(From Vettori, U.; Allegri, G.; Ferlin, M. G.; Traldi, P. *Int. J. Mass Spectrom Ion Processes.* 1990, *99*, 99. Reproduced by courtesy of Elsevier Science.)

fact, such ionic species undergo for **7** and **8,** competitive losses of CH_3^*, H_2O, HCN and CO (leading to ions at m/z 119, 116, 107, and 106, respectively) with clearly different abundances. In the case of compound **9,** CH_3^* loss is highly favored. Such data can be rationalized by considering that ions of m/z 134 are obtained for **7** and **8** through different mechanisms leading to two different structures as reported in Scheme 9.8.

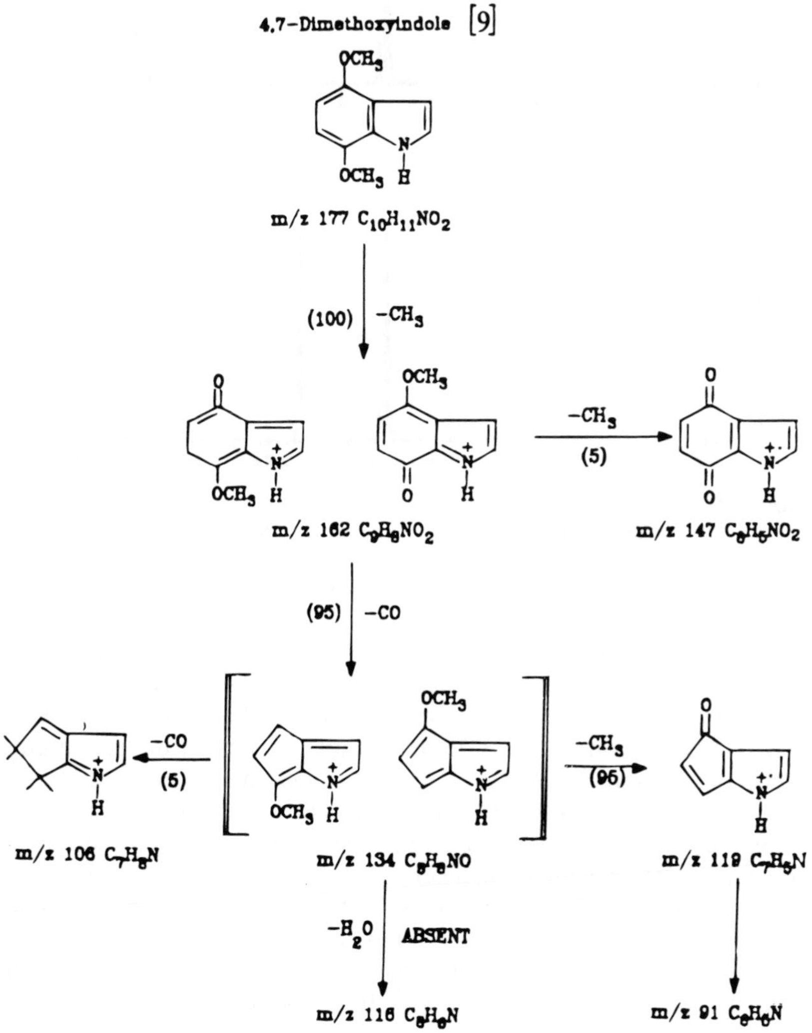

4,7-Dimethoxyindole $\boxed{9}$

SCHEME 9.9

(From Vettori, U.; Allegri, G.; Ferlin, M. G.; Traldi, P. *Int. J. Mass Spectrom Ion Processes.* 1990, *99,* 99. Reproduced by courtesy of Elsevier Science.)

Such a hypothesis has been confirmed by consecutive MS/MS experiments which can be easily and effectively performed by ITMS. The ion of

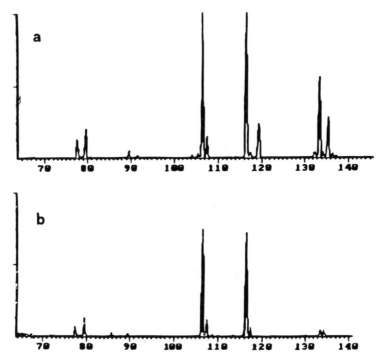

FIGURE 9.18

CAD mass spectra of m/z 134 ions from 6,7-dimethoxyindole (**8**). m/z 134 ions were generated by (a) EI; (b) collisional activation. (From Vettori, U.; Allegri, G.; Ferlin, M. G.; Traldi, P. *Int. J. Mass Spectrom Ion Processes*. 1990, *99*, 99. Reproduced by courtesy of Elsevier Science.)

m/z 134 was obtained for each of **7** and **8** through consecutive collisional-activated decomposition: $M^{+\cdot}$ (m/z 177) → m/z 162 → m/z 134 and further subjected to collisional dissociation (see lower parts of Figs. 9.17 to 9.19). For **7** and **8**, the ions of m/z 119 are completely absent in the daughter spectra of collisionally-generated m/z 134 ions. This observation implies that the mechanism leading to structure **a** of Scheme 8 involves an activation energy higher than that leading to structure **b**. Such a mechanism becomes competitive only when sufficient internal energy is deposited in the precursor m/z 162 ion; in this case, both structures **a** and **b** are formed. On the other hand, when m/z 134 ions are collisionally produced, the lower-energy deposition prevents the formation of the energetically-unfavored species **a**, and ions of only structure **b** are detected.

C.   α- and β-Hydroxycholesterol

Analytical methods devoted to the detection of steroids in biological fluids are particularly relevant for the diagnostic value of such compounds

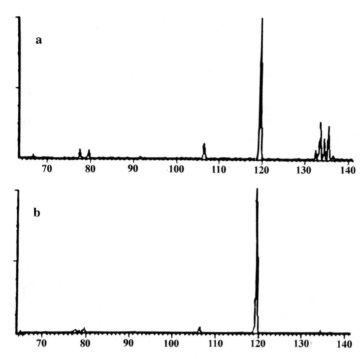

FIGURE 9.19
CAD mass spectrum of m/z 134 ion from 4,7-dimethoxyindole (**9**). m/z 134 ions were generated by (a) EI; (b) collisional activation. (From Vettori, U.; Allegri, G.; Ferlin, M. G.; Traldi, P. *Int. J. Mass Spectrom Ion Processes.* 1990, *99*, 99. Reproduced by courtesy of Elsevier Science.)

in physiological and pathological conditions.[36] In particular, the discovery of cholesterol oxides in human serum and aortic plaque of hypercholesterolemic subjects[37] led to the suspicion that cholesterol oxide may be involved in atherosclerosis.

While different approaches have been proposed for cholesterol oxide analysis, GC/MS remains the predominant technique.[38] The structural assignment of α and β isomers by GC/MS is based only on retention times, because the related EI mass spectra are wholly superimposable. However, in biological extracts, some interferences of α and β stereoisomers with other components have often been described and, consequently, their identification and quantification can be quite difficult.

An MS/MS approach which has been proposed has proven to be highly effective for the identification of 7α and 7β hydroxycholesterols[39] (compounds **10** and **11**, respectively). MS/MS characterization has been based on the comparative studies of results obtained by high- and low-energy collision as achieved by CAD-MIKE spectroscopy and by ITMS, respectively. Contrary to 70 eV EI mass spectra, the CAD-MIKE spectra of molecular ions of **10** and **11** (see upper part of Fig. 9.20) clearly show distinct differences. The primary H_2O loss represents the most favored decomposition pathway, leading to m/z 384 ions, which represents the base peak for both compounds (Scheme 9.10). From inspection of the published data on the mass spectrometry of cholesterol and its derivatives, it was proposed that the OH˙ loss originates from the OH group in position 3, so as to lead to the formation of a conjugated system which could account for the observed high stability of the product ions.

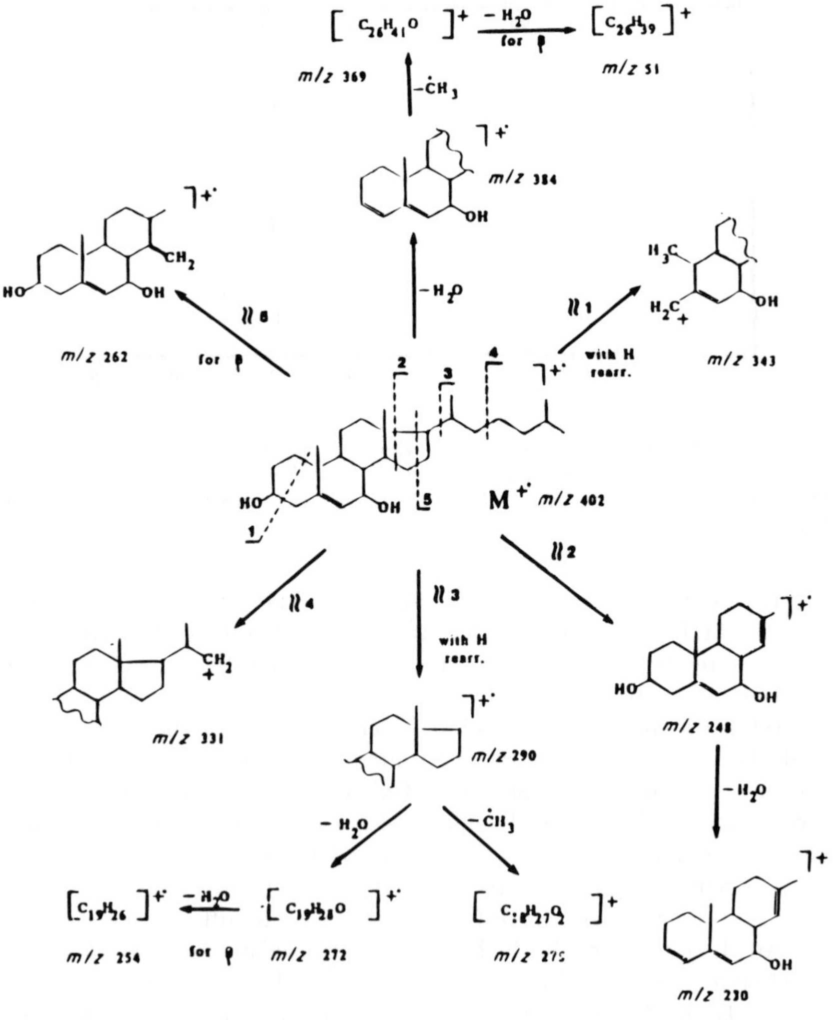

SCHEME 9.10

(From Favretto, D.; Guidugli, F.; Seraglia, R.; Traldi, P.; Ursini, F.; Sevanian, A. *Rapid Commun. Mass Spectrom.* 1991, 5, 240. Reproduced by courtesy of John Wiley & Sons.)

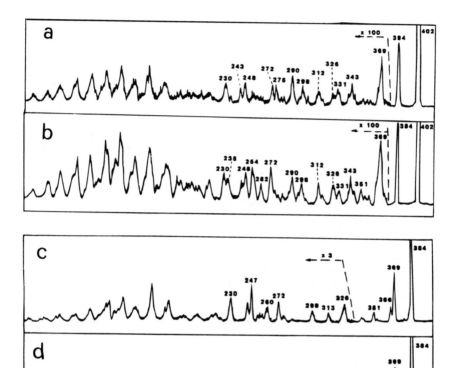

FIGURE 9.20
MIKE spectra of the molecular ion (m/z 402) and a fragment ion (m/z 384) obtained from each of α- and β-hydroxycholesterol, compounds **10** and **11**, respectively. (a) m/z 402 ($M^{+\cdot}$) generated by EI of compound **10**; (b) m/z 402 ($M^{+\cdot}$) generated by EI of compound **11**; (c) m/z 384, $[M–H_2O]^{+\cdot}$ generated by EI of compound **10**; (d) m/z 384, $[M–H_2O]^{+\cdot}$, generated by EI of compound **11**. (From Ardanaz, C. E.; Traldi, P.; Vettori, U.; Kavka, J.; Guidugli, F. *Rapid Commun. Mass Spectrom.* 1991, 5, 5. Reproduced by courtesy of John Wiley & Sons.)

Further fragmentation processes from $[M–OH]^+$ species proved to be highly diagnostic. For example, $[M–OH]^+$ ions demonstrate a further loss of $CH_3^\cdot$, giving rise to m/z 369 product ions which were found to behave differently in the two systems examined; m/z 369 ions from the β isomer (**11**) decomposed by loss of H_2O leading to m/z 351, while m/z 369 product ions from the α isomer failed to give evidence for this type of fragmentation. The complete absence of this dehydration process has been rationalized in terms of molecular geometry.

The results of high- and low-energy collision experiments on $[M–H_2O]^{+\cdot}$ ions of m/z 384 for both isomers were also informative for struc-

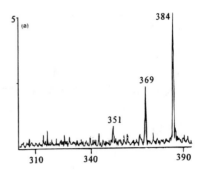

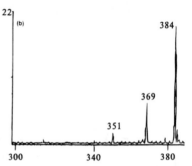

FIGURE 9.21
Daughter ion spectra obtained by ITMS of a fragment ion, $[M-H_2O]^{+\cdot}$ (m/z 384), formed from each of α- and β-hydroxycholesterol, compounds **10** and **11**, respectively. (a) m/z 384, $[M-H_2O]^{+\cdot}$ from compound **10**; (b) m/z 384, $[M-H_2O]^{+\cdot}$ for compound **11**. (From Ardanaz, C. E.; Traldi, P.; Vettori, U.; Kavka, J.; Guidugli, F. *Rapid Commun. Mass Spectrom.* 1991, *5*, 5. Reproduced by courtesy of John Wiley & Sons.)

tural investigation. The daughter ion spectra reported in Fig. 9.21 were obtained by ITMS through the application of a tickle voltage of amplitude 400 mV$_{(0-p)}$ for 10 ms. The spectra are clearly less complex than those obtained by CAD-MIKES (see lower part of Fig. 9.20), due to the lower energy deposition (either in terms of internal energy distribution or maximum energy deposition) under these conditions. The spectra for the two stereoisomers differ only in the relative abundances of fragment ions; for example, the relative abundances of m/z 369 ions are 60% and 42% for compounds **10** and **11**, respectively. The breakdown curves, obtained by varying the tickle voltage amplitude in the range of 270 to 470 mV$_{(0-p)}$ at a constant tickle time of 10 ms, are reported in Fig. 9.22. They show that the differences in the relative abundances increase by increasing the tickle voltage amplitude.

Thus, a clear distinction between α and β isomers was achieved through the use of ITMS. A comparison among the different collisional experiments was also carried out in terms of the discrepancy factor, defined as $D_{ij} = \Sigma_{ij}|m_i-m_j|$, where m_i and m_j represent the absolute abundances of isobaric ions, m, in the spectra of compounds i and j, respectively. High-energy collisional data reported in Table 9.3, show that the best characterization of the structures of **10** and **11** was obtained with the ITMS; although the mass spectra was comprised of fewer ion species, that is, for a lower value for m, the D_{ij} value obtained was of the same order of magnitude.

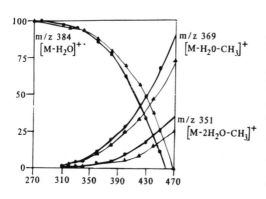

FIGURE 9.22

Plots of absolute abundances of precursor and daughter ions *versus* tickle voltage amplitude ($mV_{(0-p)}$) for each of α- and β-hydroxycholesterol, compounds **10** and **11**, respectively. Compound **10** (bold line); compound **11** (normal line). (From Ardanaz, C. E.; Traldi, P.; Vettori, U.; Kavka, J.; Guidugli, F. *Rapid Commun. Mass Spectrom.* 1991, 5, 5. Reproduced by courtesy of John Wiley & Sons.)

TABLE 9.3

Discrepancy factors

CAD MIKE M$^+$	33.8
CAD MIKE [M–H$_2$O]$^+$	24.6
ITMS [M–H$_2$O]$^+$	14

(From Favretto, D.; Guidugli, F.; Seraglia, R.; Traldi, P.; Ursini, F.; Sevanian, A. *Rapid Commun. Mass Spectrom.* 1991, 5, 240. Reproduced by courtesy of John Wiley & Sons.)

D. Coumarins, Triazolocoumarins, and Pyranocoumarins

Coumarins are a class of naturally occurring compounds,[40] some of them exhibiting pharmaceutical activity. For such reasons many efforts were, and still are, devoted to the synthesis of new derivatives and their further pharmacological testing. In particular, furano-, pyrano-, and triazolocoumarins are of wide interest because of their photobiological properties.[41] A number of new systems related to 8-methoxypsoralen, 5-methoxypsoralen and the isomeric angelicins have been synthesized in order to obtain more effective drugs free of the side effects noted in model compounds.[42] In parallel, a series of methyl-pyrano-chromones were also prepared in the expectation that they would possess repigmenting activity similar to that exhibited by the structurally related compound, Khellin.[43] ITMS has been shown to be effective in the characterization of isomeric coumarins.[44]

In a mass spectrometric investigation devoted to isomer characterization[44] of coumarins, two sets of isomers were examined (Fig. 9.23). The first set (**12** to **14**) consisted of two methylated positional isomeric coumarins and 1-methylchromone, all having the same skeleton of Seseline; more precisely, these compounds are:

2H,8H-benzo[1,2-b:3,4-b']dipyran-2-one (**12**),

5-methyl-4H,8H-benzo[1,2-b:3,4-b']dipyran-2-one (**13**), and

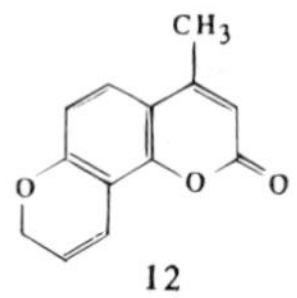

FIGURE 9.23
Structures of coumarins and chromones, compounds **12** to **18**. (From Kiremire, B.; Traldi, P.; Guiotto, A.; Pastorini, G.; Chilin, A.; Vettori, U. *Int. J. Mass Spectrom. Ion Processes.* 1991, *106*, 283. Reproduced by courtesy of Elsevier Science.)

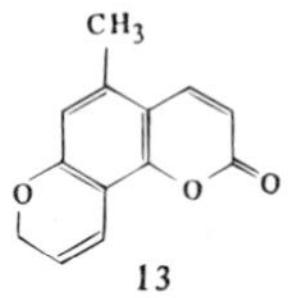

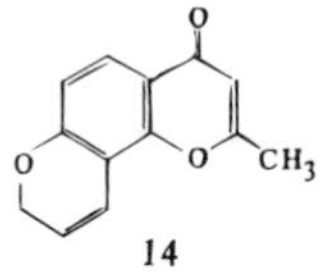

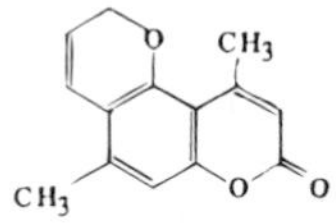

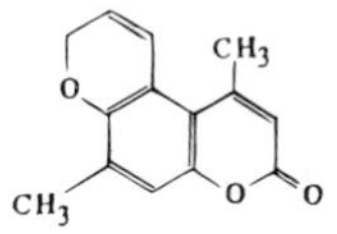

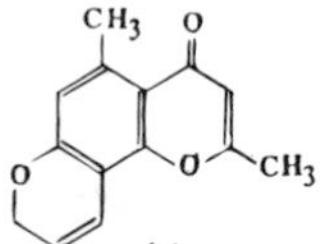

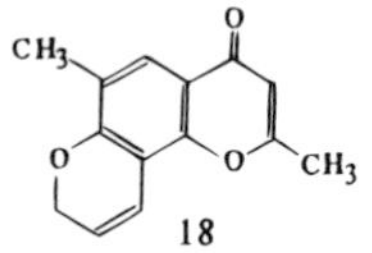

2-methyl-4H,8H-benzo[1,2-b:3,4-b']dipyran-4-one (**14**).

The second set of isomers (**15** to **18**) contained two coumarins and two chromones; the coumarins examined were:

1,6-dimethyl-3H,8H-benzo[1,2-b:3,4-b']dipyran-8-one (**15**) and

2,10-dimethyl-2H,8H-benzo[1,2-b:3,4-b']dipyran-3-one (**16**),

with a dissimilar annulation to that of seseline, but diversely methylated (Fig. 9.23); the chromones investigated were:

2,5-dimethyl-4H,8H-benzo[1,2-b:3,4-b']dipyran-4-one (**17**) and

2,6-dimethyl-4H,8H-benzo[1,2-b:3,4-b']dipyran-4-one (**18**),

having the same annulation as, and being isomeric to, seseline but methylated in various positions.

The compounds were studied by ITMS EI and collisional experiments were performed on selected molecular ions. The EI spectra of isomers **12** to **14** (Table 9.4) show abundant molecular ions together with major fragments at m/z 213, 199, 186, 185, 171, and 157, corresponding to $[M–H]^+$, $[M–CH_3]^+$, $[M–CO]^{+\cdot}$, $[M–CHO]^+$, $[M–C_2H_3O]^+$, and $[M–C_2HO_2]^+$, respectively.

TABLE 9.4

EI spectra for compounds 12 to 14 (From Kiremire, B.; Traldi, P.; Guiotto, A.; Pastorini, G.; Chilin, A.; Vettori, U. *Int. J. Mass Spectrom. Ion Processes.* 1991, *106*, 283. Reproduced by courtesy of Elsevier Science.)

m/z	Relative Abundance (%)			Ionic Species
	12	13	14	
39	—	7	11	
41	6	7	6	
43	6	6	6	
44			6	
50			6	
51	8	7		
55			3	
63	6	10	7	
69		6		
74			4	
77	6	8		
78		6		
84	2		2	
85		3	5	
89	9		11	
92	6			
101	4			
102		6		
103	5			
107	6	6		
115	14	6		
117			5	
118			6	
127	13	7		
128	12	15	5	
129	5	9	2	
130	4		4	[M–3CO]$^+$
131	4			
145			7	
146			7	
149		3		
155	5			
157	6	5	3	[185–CO]$^+$
158	11	6		[186–CO]$^+$
167	2			[185–H$_2$O]$^+$
168	3			[185–OH]$^+$
169		6		[M–COOH]$^+$
171	9	5	2	[M–C$_2$H$_3$O]$^+$
173			50	[M–C$_3$H$_5$]$^+$
174			17	[M–C$_3$H$_4$]$^+$
185	55	47	5	[M–CHO]$^+$
186	22	17	6	[M–CO]$^+$
199	4	6	1	[M–CH$_3$]$^+$
213	95	100	100	[M–H]$^+$
214	100	81	97	M$^+$

TABLE 9.5

Daughter spectra for selected parent ions in compounds 12 to 14 (From Kiremire, B.; Traldi, P.; Guiotto, A.; Pastorini, G.; Chilin, A.; Vettori, U. *Int. J. Mass Spectrom. Ion Processes.* 1991, *106*, 283. Reproduced by courtesy of Elsevier Science.)

Parent/Daughter m/z	Relative Abundance (%)			Ionic Species
	12	13	14	
214	—			M^+
213	100	100	100	$[M–H]^+$
199		8		$[M–CH_3]^+$
186	32	44	20	$[M–CO]^+$
185	6		5	$[M–CHO]^+$
174			20	$[M–C_3H_4]^+$
173			5	$[M–C_3H_5]^+$
171			3	$[M–C_2H_3O]^+$
170	15			$[M–CO_2]^+$
169		18		$[M–COOH]^+$
213				$[M–H]^+$
185	100	100		$[213–CO]^+$
173			100	$[213–C_3H_4]^+$
185				$[M–CHO]^+$
184	22	38	25	$[185–H]^+$
168	10			$[185–OH]^+$
157	50	85	72	$[185–CO]^+$
129	46	63	53	$[185–2CO]^+$
128	100	100	100	$[185–C_2HO_2]^+$
127	35	35	21	$[185–C_2H_2O_2]^+$
115	15		13	$[C_9H_7]^+$

In spite of the observation that these fragment ions are common to all three compounds, it is evident from the EI spectra that compound **14** shows additional and peculiar fragmentation pathways, leading to ions at m/z 174, 173, 146, and 145, which are completely absent from the EI spectra for compounds **12** and **13**. The first two ion species, m/z 174 and m/z 173, are due to primary losses of C_3H_4 and C_3H_5 from the pyranone ring.

On the other hand, compounds **12** and **13** could not be distinguished by their EI spectra and, consequently, some collisional experiments were performed by ITMS in an attempt to differentiate between these compounds. The collisional spectra of $M^{+\cdot}$, $[M–H]^+$ and $[M–CO]^+$ ions of compounds **12** to **14** are reported in Table 9.5. The molecular ions, $M^{+\cdot}$, of the three compounds show elimination of H˙ and CO as the only pathways in common but, again, the behavior of the molecular ion of **14** differs markedly from those observed for the molecular ions of **12** and **13**.

By collisional spectroscopy, several characteristic daughter ions all permit ready differentiation of **12** and **13**. Thus, while a loss of primary CH˙₃ is observed only for **13**, for compound **12** the collisionally-induced loss of CHO˙ is equally characteristic. Furthermore, CO_2 loss, leading to the formation of fragment ions of m/z 170, is observed exclusively in the case of **12**, while $CHO_2˙$ loss is specific for **13**. Collisional spectra of $[M–H]^+$ species showed the formation of but a single fragment ion; yet clear differences were evident among these spectra for **12**, **13**, and **14**. For **14**, the loss of C_3H_4 was observed, while for **12** and **13**, CO loss was evident. The daughter spectra of ions of m/z 185, originating from EI-induced CHO˙ loss from the molecular ions of **12**, **13**, and **14**, showed less significant differences, suggesting that these ions may have a common structure.

Compounds **15** to **18** showed common fragment ions of m/z 227, 213, 199, and 185 which can be attributed to the elimination of H˙, CH˙$_3$, CHO˙, and $C_2H_3O˙$, respectively (Table 9.6). For **15** to **17**, the base peak is due to $M^{+˙}$, the same ion showing an abundance of 72% for **18**. While the two chromones **17** and **18** produced EI spectra by which these compounds could be easily differentiated, the two coumarins **15** and **16** showed practically superimposable EI spectra. The two chromones could be easily distinguished from the coumarins on the basis of diagnostic fragments of m/z 173, 161, and 121 in compound **17** and of m/z 173, 160, and 84 in compound **18**.

The MS/MS spectra of $M^{+˙}$ could be more readily used to distinguish between the two isomeric chromones, in that they showed a variation in the relative abundance of the common fragments $[M–CO]^{+˙}$ (m/z 200), $[M–CHO]^+$ (m/z 199) and $[M–C_3H_4]^+$ (m/z 188) (Fig. 9.24). The daughter spectra of ions of m/z 199 showed the presence of peaks at m/z 184 and m/z 181 for isomer **17** only; these fragment ions suggest that the sequential elimination of the formyl radical (to form m/z 199) followed by CH˙$_3$ and H_2O losses occurs for **17** but does not occur for compound **18**.

A similar rationalization, on the basis of the daughter ion spectra of $M^{+˙}$ can be made for coumarins **15** and **16**. The major fragment channel was that of CO loss, but the relative weighting of this channel differed for **15** and **16** so that the relative abundances of the fragment ion differed for the two compounds; furthermore, minor fragment ions corresponding to CHO˙ and $C_2H_3O˙$ losses were negligible for compound **15** but significant for compound **16**.

Six triazolocoumarins, namely,

8-methylpyrano[2,3-f]benzotriazol-6-one (**19**),

7,8-dimethylpyrano[2,3-f]benzotriazol-6-one (**20**),

4,8-dimethylpyrano[2,3-f]benzotriazol-6-one (**21**),

6-methylpyrano-[2,3-e]benzotriazol-8-one (**22**),

6,7-dimethylpyrano[2,3-e]benzotriazol-8-one (**23**), and

4,6-dimethylpyrano[2,3-e]benzotriazol-8-one (**24**)

TABLE 9.6

EI mass spectra for compounds 15 to 18 (From Kiremire, B.; Traldi, P.; Guiotto, A.; Pastorini, G.; Chilin, A.; Vettori, U. *Int. J. Mass Spectrom. Ion Processes.* 1991, *106*, 283. Reproduced by courtesy of Elsevier Science.)

m/z	Relative Abundance (%)				Ionic Species
	15	16	17	18	
39	13	10	46	24	
41	15	4	15	19	
43	24	4	22	10	
51		7	9	9	
56				25	
57	9		8		
63		6	3		
65			13		
69	14	4	17	16	
77	6	6	11	20	
78	4	7	4		
83	6		7	9	
84	3	6	6	36	
89			6	7	
91	7		5		
93				6	
97	7		4		
99	6	5	4		
103	4	6	8	7	
104				4	
115	7	9	4	4	
116			6		
121			36		
122			6		
127	7	7	4		
128	17	15	8		
129	10	11	6		
131	4		9	7	
132			7		
141	5	5			
143	4	8	6		
144	4	3	6		
145	3	3	4		
149	4				
153	2		3		
157	12	8	11		
158	2	3	4		
159			6	6	
160			6	7	
161			40		
166			6		
167		2			
169	2	3			
171	9	8	5		
172	7	4	4		
173			6	4	

TABLE 9.6

EI mass spectra for compounds 15 to 18 (From Kiremire, B.; Traldi, P.; Guiotto, A.; Pastorini, G.; Chilin, A.; Vettori, U. *Int. J. Mass Spectrom. Ion Processes.* 1991, *106*, 283. Reproduced by courtesy of Elsevier Science.)

| m/z | Relative Abundance (%) | | | | Ionic Species |
	15	16	17	18	
183	5				
184		5			
185	26	27	26	6	$[M-C_2H_3O]^+$
186	5	5	6		$[M-C_2H_2O]^+$
187		5	14	30	$[M-C_3H_5]^+$
188			11	15	$[M-C_3H_4]^+$
199	41	38	44	7	$[M-CHO]^+$
200	32	27	13		$[M-CO]^+$
201	4	5	6		$[H-C_2H_3]^+$
211			4		$[M-OH]^+$
213	17	9	8	13	$[M-CH_3]^+$
227	43	27	90	100	$[M-H]^+$
228	100	100	100	72	M^+

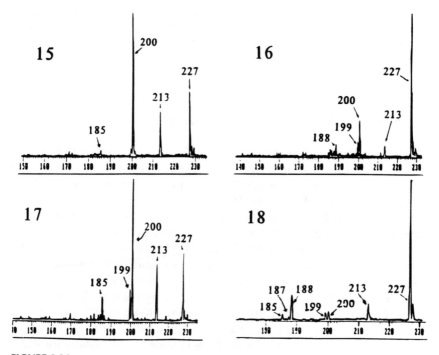

FIGURE 9.24

MS/MS daughter ion spectra of $M^{+\cdot}$ (*m/z* 228) for two coumarins and two chromones, compounds **15** to **18**. (From Kiremire, B.; Traldi, P.; Guiotto, A.; Pastorini, G.; Chilin, A.; Vettori, U. *Int. J. Mass Spectrom. Ion Processes.* 1991, *106*, 283. Reproduced by courtesy of Elsevier Science.)

(see Fig. 9.25), were studied by both high- and low-energy collision experiments, the latter obtained by ITMS.[45]

The monomethyl isomers 19 and 22 could be easily distinguished on the basis of diagnostic differences in the collisional spectra of their $M^{+\cdot}$ and selected fragment ions, as shown in Fig. 9.26. The molecular ion for each isomer showed sequential losses of moieties of 28 Da which were shown to be due to the loss of N_2 and CO, leading to ions of m/z 173 and m/z 145. While these collisional spectra of $M^{+\cdot}$ were somewhat ineffective for isomer differentiation, the daughter ion spectra of m/z 173 and m/z 145 exhibited unique features. Thus the daughter ion spectrum of $[M-N_2]^{+\cdot}$ ion of 22 showed the formation of ions of m/z 118, 117 and 90, which were completely absent for compound 19. The ion of m/z 145 led to fragment ions of m/z 105 and m/z 90, which were more abundant for 19 than for 22.

It was observed that collision of $M^{+\cdot}$ ions of the dimethyltriazolocoumarins (m/z 215) led to many cleavage processes which appeared to be sensitive to structural differences. While the base peak at m/z 187 was common to all isomers (Table 9.7), each compound displayed a different set of daughter ions: thus, 20 demonstrated the ready loss of a neutral moiety of 43 Da giving rise to an abundant ion of m/z 172; this process was observed to a minor extent only for 21. The same ion of m/z 172 was formed also by collision of the m/z 187 ion for compound 23 only.

E. Pyrrolquinolinones and Benzoquinolizinones

The structure and reactivity of molecular ions of isomeric pyrroloquinolinones and benzoquinolizinones (Fig. 9.27) were investigated by the use of both unimolecular decomposition studies by MIKES and by low-energy collisions performed by ITMS.[46]

Compounds 25 to 27 originate from the condensation of the pyrrole ring on the quinoline skeleton in the appropriate position. Such a method is highly versatile, allowing the direct synthesis of the aromatic tricyclic system and the introduction of alkyl groups in different positions of the tricyclic skeleton. During the final cyclization, the concurrent formation of three isomeric compounds 25 to 27, that is, two benzo[ij]quinolizin-5-ones, (25 and 26) and one pyrrol[3,2,1-ij]quinolin-4-one (27) takes place.

The 70 eV EI mass spectra of 25 to 27 are reported in Table 9.8, while the decomposition patterns of 25 and 27, as obtained by metastable ion studies, are reported in Schemes 9.11 and 9.12, respectively. The fragmentation pattern of 26 exhibits the same decomposition channels as does 25 and, consequently, is not reported in detail here.

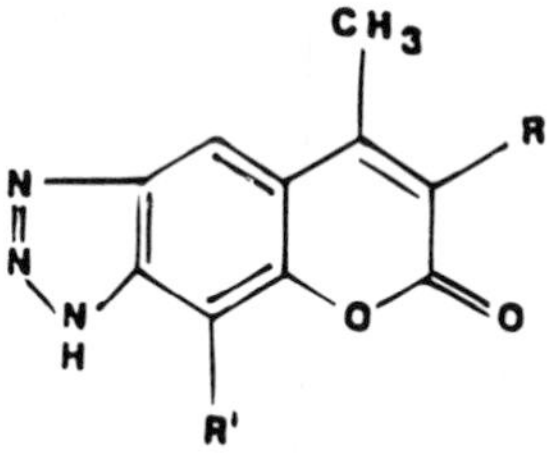

Compound

19. R = R' = H

20. R = CH₃ R' = H

21. R = H R' = CH₃

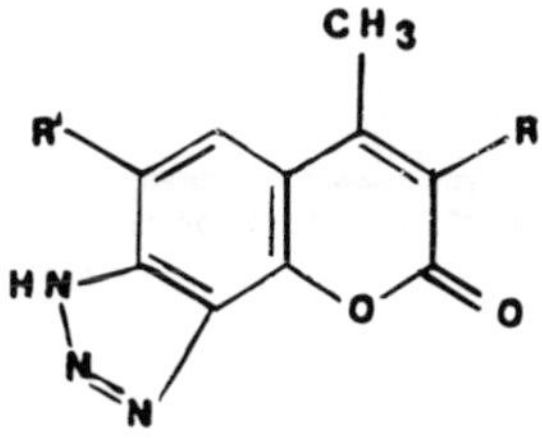

Compound

22. R = R' = H

23. R = CH₃ R' = H

24. R = H R' = CH₃

FIGURE 9.25
Structures of six triazolocoumarins, compounds **19** to **24**. (From Evans, C.; Traldi, P.; Chilin, A.; Pastorini, G.; Rodighiero, P. *Org. Mass Spectrom.* 1991, 26, 688. Reproduced by courtesy of John Wiley & Sons.)

From Table 9.8, it can be seen that compounds **25** and **26** behave quite differently: whereas the $[M–H]^+$ ions of m/z 226 for **26** show a relative abundance of but 9%, for **25**, this ion forms the base peak. Compound **27** behaves differently from both **25** and **26**. The primary CH_3 loss, which for **25** was highly favored, for **27** is still present, but with lower yield. The base peak for **27** is the m/z 211 ion, which is due to sequential losses of H˙ and CH_3; this observation contrasts sharply with those for **25** and **26** for which the m/z 211 ion was barely detected.

The clear differences existing in the EI spectra are the results of fragmentation processes having their origin in ions which possess a wide range of internal energies, while MIKE spectra originate from ions with lower and narrower internal energy distributions. The MIKES data of EI-

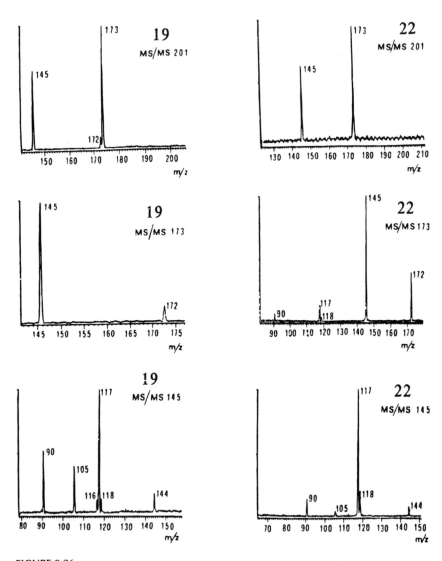

FIGURE 9.26
Low-energy CAD daughter ion spectra of $M^{+\cdot}$ (m/z 201) and fragment ions of m/z 173 and m/z 145 obtained from each of 8-methylpyrano[2,3-f]benzotriazol-6-one (**19**) and 6-methylpyrano[2,3-e]benzotriazol-8-one (**22**). (From Evans, C.; Traldi, P.; Chilin, A.; Pastorini, G.; Rodighiero, P. *Org. Mass Spectrom.* 1991, 26, 688. Reproduced by courtesy of John Wiley & Sons.)

TABLE 9.7

Low-energy CAD daughter ion spectra of M$^{+\cdot}$ of compounds 20, 21, 23, and 24 upon EI (From Evans, C.; Traldi, P.; Chilin, A.; Pastorini, G.; Rodighiero, P. *Org. Mass Spectrom.* 1991, *26*, 688. Reproduced by courtesy of John Wiley & Sons.)

m/z		Fragment Relative Abundance (%)			
Parent	M$^+$	20	21	23	24
215	187	100	100	100	100
	186	—	9	—	—
	172	46	8	—	—
	159	96	60	55	6
	144	—	1	—	—
	132	—	18	8	—
	131	—	9	14	—
	130	—	—	3	—
187	186	—	—	24	10
	172	—	—	25	—
	160	—	—	7	—
	159	100	100	100	100
	158	—	—	2	—
	144	11	3	—	—
	132	—	40	20	—
	131	3	31	24	9
	130	—	3	30	12
159	158	4	29	40	100
	157	—	—	—	5
	144	100	83	27	20
	142	—	—	—	9
	132	35	100	80	31
	131	6	52	89	40
	116	2	—	—	16
	104	2	—	—	16

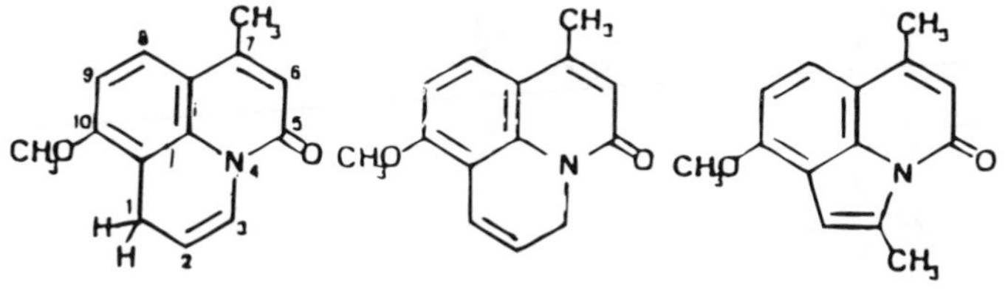

25   **26**   **27**

FIGURE 9.27

Structures of pyrrolquinolinones and benzoquinolizinones, compounds **25** to **27**. (From Fontana, S.; Curcuruto, O.; Traldi, P.; Castellin, A.; Chilin, A.; Rodighiero, P.; Guiotto, A. *Org. Mass Spectrom.* 1992, *27*, 1255. Reproduced by courtesy of John Wiley & Sons.)

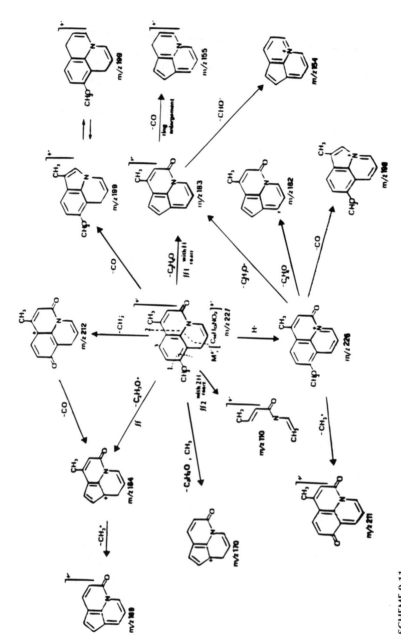

SCHEME 9.11

(From Fontana, S.; Curcuruto, O.; Traldi, P.; Castellin, A.; Chilin, A.; Rodighiero, P.; Guiotto, A. *Org. Mass Spectrom.* 1992, 27, 1255. Reproduced by courtesy of John Wiley and Sons.)

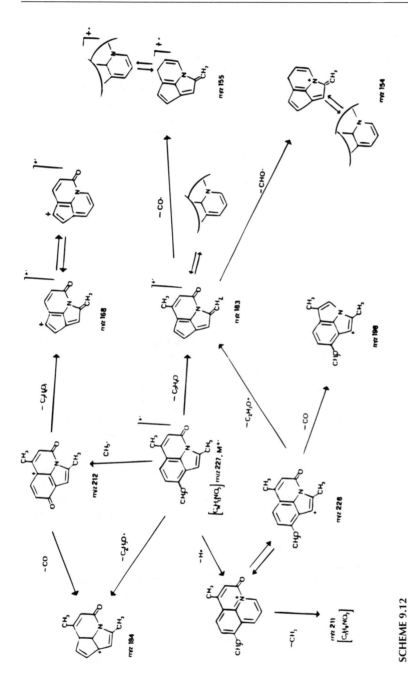

SCHEME 9.12

(From Fontana, S.; Curcuruto, O.; Traldi, P.; Castellin, A.; Chilin, A.; Rodighiero, P.; Guiotto, A. *Org. Mass Spectrom.* 1992, 27, 1255. Reproduced by courtesy of John Wiley and Sons.)

TABLE 9.8

EI mass spectra of compounds 25 to 27 (From Fontana, S.; Curcuruto, O.; Traldi, P.; Castellin, A.; Chilin, A.; Rodighiero, P.; Guiotto, A. *Org. Mass Spectrom.* 1992, 27, 1255. Reproduced by courtesy of John Wiley & Sons.)

Ionic Species	m/z	Relative Abundance (%)		
		25	26	27
$M^{+\cdot}$	227	65	70	19
$[M-H]^+$	226	100	9	98
$[M-CH_3]^+$	212	92	100	41
$[(M-H)-CH_3]^{+\cdot}$	211	14	6	100
$[M-CO]^{+\cdot}$	199	2	7	2
$[(M-H)-CO]^+$	198	1	18	5
$[M-C_2H_3O]^+$	184	8	19	12
$[M-C_2H_4O]^{+\cdot}$	183	33	20	11
$[(M-H)-C_2H_4O]^+$	182	22	14	16
$[M-C_2H_2O-CH_3]^+$	170	2	1	3
$[(M-C_2H_3O)-CH_3]^{+\cdot}$	169	8	8	1
$[M-C_2H_4O-CO]^{+\cdot}$	155	4	9	12
$[M-C_2H_2O-CHO]^+$	154	27	16	16

generated $M^{+\cdot*}$ ions, where * indicates an ion which is kinetically unstable on the time-scale of the mass spectrometer, are reported in Table 9.9. These data seem to indicate that compounds 25 and 27 partially isomerize in the time-interval required for reaching the field-free region of the apparatus.

In contrast to the observations of $M^{+\cdot*}$ species, $[M-H]^{+*}$ ions seem to retain their original structural identities. As it can be seen in Table 9.9, the MIKE spectra of $[M-H]^{+*}$ ions differ either in the relative abundances of common decomposition products or in the presence of specific fragmentation routes. The same argument can be invoked also for $[M-CH_3]^{+*}$ ions.

The results obtained by EI and MIKE techniques can be summarized as follows. First, under EI conditions, compounds 25 to 27 lead to different spectra, with clearly different relative abundances of $M^{+\cdot}$, $[M-H]^+$, $[M-CH_3]^+$ and $[M-H-CH_3]^{+\cdot}$ ions. Second, the MIKE spectra of $M^{+\cdot*}$ for 25 to 27 are, somewhat surprisingly, more similar to each other than are the corresponding EI spectra. This similarity of the MIKE spectra can be explained by the existence and magnitude of the isomerization barrier which is to be surmounted: in terms of kinetics, it means that the partial isomerization of $M^{+\cdot*}$ of 25 to 27 to a common structure takes place in the flight time from the source to the second field-free region. Third, the MIKE spectra of $[M-H]^{+*}$ and $[M-CH_3]^{+*}$ ions show peculiar fragmentation routes, indicating that the time required for isomerization of these ions must be greater than that of the corresponding $M^{+\cdot*}$.

The confirmation of this hypothesis was carried out by ITMS. First of all, the mass spectra of 25 to 27, as obtained under 70 eV EI conditions (see left side of Fig. 9.28), were, surprisingly, practically superimposable,

TABLE 9.9

MIKE spectra of compounds 25 to 27 (From Constantin, E.; Schnell, A.; Guidugli, F.; Traldi, P. *Org. Mass Spectrom.* 1992, 27, 174. Produced by courtesy of John Wiley & Sons.)

Parent Ion	Daughter Ions m/z	Absolute Abundance 25	26	27
$[M]^{+\cdot}$ (m/z 227)	226	41	25	10
	212	46	46	64
	199	4	14	—
	198	—	—	20
	184	2	8	2
	183	4	—	2
	170	2	3	—
	169	—	3	—
	168	—	—	1
	155	—	—	1
	154	1	1	—
$[M-H]^{+}$ (m/z 226)	225	21	31	23
	211	57	49	46
	199	—	7	—
	198	13	10	17
	183	5	3	14
	182	4	—	—
$[M-CH_3]^{+}$ (m/z 212)	211	37	18	17
	184	63	23	68
	183	—	29	—
	168	—	—	6
	156	—	15	—
	154	—	15	9

showing that isomerization of **25** to **27** to a common structure takes place virtually instantaneously. The ionization conditions were identical with those pertaining to the experiments described above, that is, a 70 eV electron beam interacting with the neutral molecules in the gas phase. Therefore, the isomerization must be related to the only two parameters which are unique to the ITMS, that is, the presence of helium buffer gas (at a pressure of 10^{-4} Torr) and/or the residence time in the ion trap.

In the absence of buffer gas, compounds **25** and **26** produced EI spectra (Fig. 9.28(d); (e)) comparable to those obtained with the ZAB (Fig. 9.28(a),(b)), yet the behavior of **27** (Fig. 9.28 (f)) was changed; there was no longer an ion of m/z 211, which had been the base peak as shown in Table 9.8, and the mass spectrum strongly resembled that of **26** (Fig. 9.28(e)). These results suggest that not only is the buffer gas responsible for the isomerization processes, but the residence time also must have some role to play.

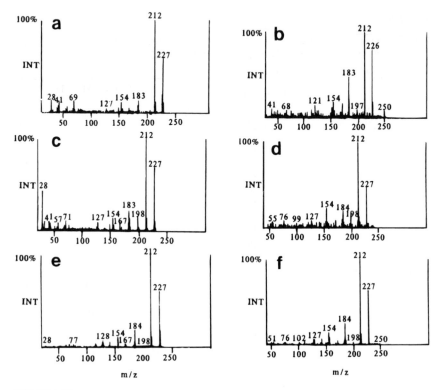

FIGURE 9.28
70 eV EI mass spectra obtained with the ion trap. (a–c) compounds **25** to **27**; (d–f) compounds **25** to **27**, spectra were obtained in the absence of helium.

The results obtained by ITMS indicate that, in the ion trap, ionization conditions are milder than those of EI. These conditions lead to a lower-energy deposition in the molecular species.

Further experiments on compounds **25** to **27** were carried out by low-energy collision of $M^{+\cdot}$ (m/z 227), $[M-H]^+$ and $[M-CH_3]^+$ ions (Table 9.10), in which supplementary RF voltages of identical amplitude and time were applied. While the molecular ion, $M^{+\cdot}$, for compounds **25** to **27** led to $[M-H]^+$ and $[M-CH_3]^+$ ions only, the preselected $[M-H]^+$ and $[M-CH_3]^+$ species showed many decomposition channels, highly diagnostic from a structural point of view, proving the effectiveness of low-energy collision spectroscopy by ITMS in structural studies.

F.   Phenothiazine

Phenothiazine (Fig. 9.29), due to its pharmacological relevance, has been the object of many mass spectrometric studies.[47–49]

TABLE 9.10

Low-energy daughter ion spectra of $[M]^{+\cdot}$, $[M-H]^+$ and $[M-CH_3]^+$ ions of compounds 25 to 27 (From Fontana, S.; Curcuruto, O.; Traldi, P.; Castellin, A.; Chilin, A.; Rodighiero, P.; Guiotto, A. *Org. Mass Spectrom.* 1992, 27, 1255. Reproduced by courtesy of John Wiley & Sons.)

Parent Ions	Daughter Ions m/z	Absolute Abundance		
		25	26	27
$[M]^{+\cdot}$	226	1	—	3
	212	99	100	97
$[M-H]^+$	212	—	8	—
	211	79	58	53
	198	7	12	36
	196	—	9	2
	183	14	11	7
	180	—	1	1
	170	—	1	1
$[M-CH_3]^+$	211	5	22	1
	184	42	38	46
	183	12	12	11
	169	7	5	3
	167	2	2	2
	166	1	1	3
	156	21	16	19
	154	5	3	5
	129	3	1	7
	128	2	—	3

FIGURE 9.29
Structure of phenothiazine. (From Constantin, E.; Schnell, A.; Guidugli, F.; Traldi, P. *Org. Mass Spectrom.* 1992, 27, 174. Reproduced by courtesy of John Wiley & Sons.)

An extensive study of its fundamental decomposition processes has been carried out by Halberg et al.,[49] they identified four main primary fragmentations, due to losses of S, HS·, H·, and CHS·. ITMS has been applied successfully to this compound to determine the collision-induced decomposition pathways of $M^{+\cdot}$ of phenothiazine, as well as of the main fragment ions.[50] From this particular application, important information on the versatility of ITMS in structural investigations was obtained, in particular, with respect to energy deposition brought about under the different operating conditions employed in MS^n experiments.

For such multiple MS experiments, that is, MS^n, the scan function shown in Fig. 9.30 was employed. The parent ions, generated by EI, were selected in the first step by application of negative and positive DC voltages to the ring electrode. By applying a supplementary RF voltage between the end-cap electrodes with an amplitude of 250 $mV_{(0-p)}$ for 15 ms, collisional spectra were generated. In the second step, a collisionally-generated ion was selected and its daughter ion spectrum was obtained by excitation under the tickle conditions described above. Finally, the ion generated in the second step was selected to produce a collisional-dissociation spectrum by the application of suitable tickling conditions.

The collisional spectrum of $M^{+\cdot}$ of phenothiazine, reported in Fig. 9.31, shows the formation of primary ions of m/z 198, 167, 166, 155, and 154, as shown in Scheme 9.13. Such results are in complete agreement with the findings of Halberg et al.[49] except for the presence in our experiments of m/z 155 primary ions formed due to the loss of CS.

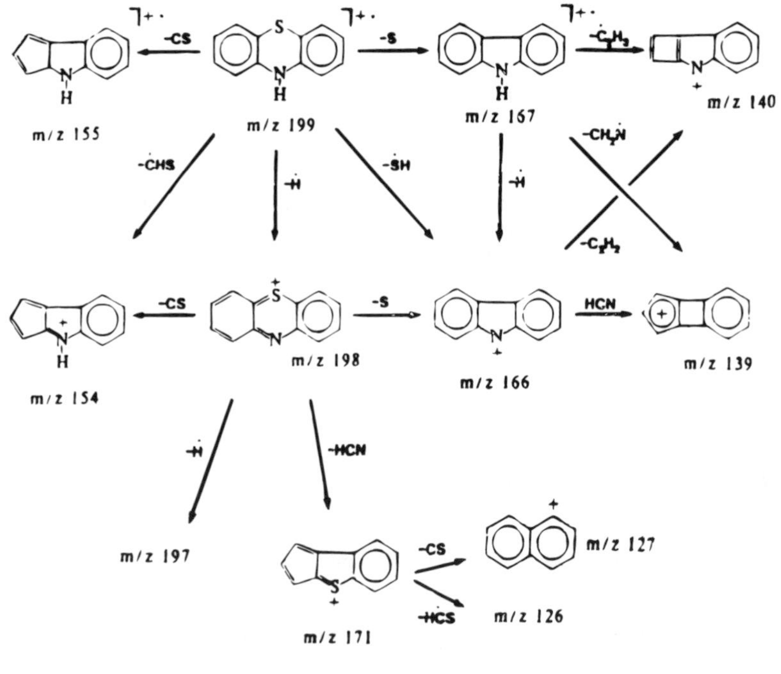

SCHEME 9.13
(From Constantin, E.; Schnell, A.; Guidugli, F.; Traldi, P. *Org. Mass Spectrom.* 1992, 27, 174. Produced by courtesy of John Wiley & Sons.)

The breakdown curves of $M^{+\cdot}$ obtained by varying tickle voltage amplitude, tickle time, and β_z values are reported in Fig. 9.32. While the first two instrumental approaches led to identical results, proving that S loss is the most favored decomposition channel in each energy regime, the plots of absolute ion abundance versus β_z exhibit an interesting behavior. With increasing β_z, an increase in energy deposition occurs: for values of β_z up to 0.4, m/z 167 ions remain the most abundant species but, at β_z = 0.41, the situation changes and a crossing point appears in the plots related to the intensities of m/z 167 and m/z 166.

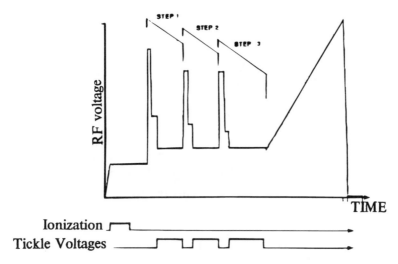

FIGURE 9.30
Scan function for multiple, (MS)^n, experiments. (From Constantin, E.; Schnell, A.; Guidugli, F.; Traldi, P. *Org. Mass Spectrom.* 1992, *27*, 174. Produced by courtesy of John Wiley & Sons.)

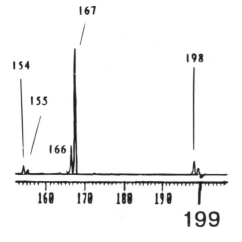

FIGURE 9.31
Daughter ion spectrum of M$^{+\cdot}$ (*m/z* 199) of phenothiazine. Tickle voltage, 220 mV$_{(0-p)}$; $q_z = 0.23$; tickle time = 10000 μs. (From Constantin, E.; Schnell, A.; Guidugli, F.; Traldi, P. *Org. Mass Spectrom.* 1992, *27*, 174. Produced by courtesy of John Wiley & Sons.)

These results can be explained by considering that the HS$^\cdot$ loss channel has a higher activation energy than does that of the S loss channel and/or that, in high-energy collision regimes, isomerization of M$^{+\cdot}$ possibly takes place, leading to a structure which more easily leads to HS$^\cdot$ loss. Alternatively, a contribution to the increase of the abundance of ions of *m/z* 166 could originate from H$^\cdot$ loss from ions of *m/z* 167.

The daughter ion spectrum of the *m/z* 171 ion species is reported in Fig. 9.33. This ion species, which has its origin in sequential losses of H$^\cdot$ and HCN, shows a ready loss of CS. The *m/z* 167 ion undergoes collision-

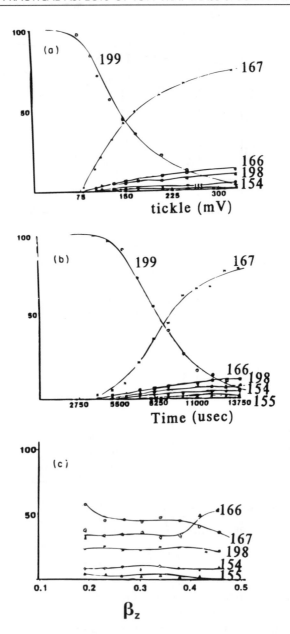

FIGURE 9.32

Breakdown curves of $M^{+\cdot}$ (m/z 199) of phenothiazine. Variation of (a) tickle voltage amplitude, with $q_z = 0.23$ and tickle time = 10000 μs; (b) tickle time, with $q_z = 0.23$ and tickle
voltage amplitude = 280 mV$_{(0-p)}$; (c) β_z, with tickle voltage amplitude = 280 mV$_{(0-p)}$ and
tickle time = 15000 μs. (From Constantin, E.; Schnell, A.; Guidugli, F.; Traldi, P. *Org. Mass
Spectrom.* 1992, *27*, 174. Produced by courtesy of John Wiley & Sons.)

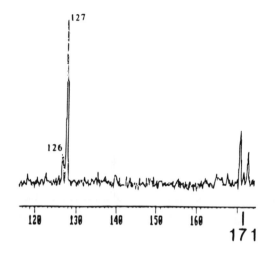

FIGURE 9.33
Daughter ion spectrum of the fragment ion [M–H–HCN]⁺ (m/z 171) of phenothiazine. Tickle voltage amplitude = 240 $mV_{(0-p)}$; tickle time = 10000 μs; q_z = 0.27). (From Constantin, E.; Schnell, A.; Guidugli, F.; Traldi, P. *Org. Mass Spectrom.* 1992, 27, 174. Produced by courtesy of John Wiley & Sons.)

induced losses of H⋅, C_2H_3, and CNH_2 leading to ions of m/z 166, 140, and 139, respectively (Fig. 9.34). Finally, m/z 166 ions decomposed through equally favored HCN and C_2H_2 loss channels giving rise to fragment ions of m/z 140 and m/z 139 (Fig. 9.35). The ERMS plots of the daughter ions of these EI-generated ions are reported in Fig. 9.36(a). The absolute abundances of m/z 140 ions always remain higher than those of m/z 139 ions over the entire range of tickle voltage amplitude, indicating that the former is the more energetically favored decomposition channel.

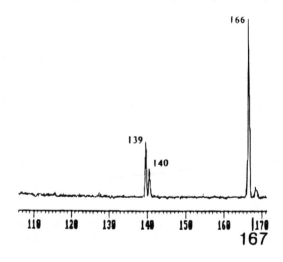

FIGURE 9.34
Daughter ion spectrum of the fragment ion [M–S]⁺⋅ (m/z 167) of phenothiazine. Tickle voltage amplitude 240 $mV_{(0-p)}$; tickle time = 10000 μs; q_z = 0.27. (From Constantin, E.; Schnell, A.; Guidugli, F.; Traldi, P. *Org. Mass Spectrom.* 1992, 27, 174. Produced by courtesy of John Wiley & Sons.)

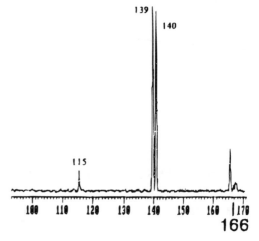

FIGURE 9.35
Daughter ion spectrum of the
fragment ion $[M–HS]^+$ (m/z 166)
of phenothiazine. Tickle voltage
amplitude = 240 mV$_{(0–p)}$; tickle
time = 10000 μs; q_z = 0.39. (From
Constantin, E.; Schnell, A.;
Guidugli, F.; Traldi, P. *Org. Mass
Spectrom.* 1992, 27, 174.
Produced by courtesy of John
Wiley & Sons.)

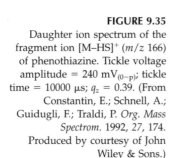

When the m/z 166 ionic species is generated by multiple MS/MS,
that is, by means of the sequential collision route:

$$M^{+\cdot} \xrightarrow{-S} [M - S]^{+\cdot} \xrightarrow{-H^{\cdot}} [M - S - H]^{+} \tag{9.2}$$

the ERMS plots shown in Fig. 9.36(b) are obtained; these plots show clearly
an inversion of the absolute abundances of ions of m/z 139 and m/z
140.

This latter result may be explained by the precursor ions having
different degrees of internal energy depending on their origin; in principle,
those originating from EI should exhibit an internal energy higher than
those generated by multiple sequential collisions. In order to investigate
further this aspect, we performed some experiments on collisionally-generated m/z 166 ions at a series of β_z values. Such an increase in β_z should
reflect an enhancement of energy deposition in the preselected species.
The ERMS plots obtained with β_z = 0.27 and β_z = 0.36 are shown in Fig.
9.36(c) and 9.36(d), respectively. In the first case, a crossing point for the
absolute abundances of m/z 139 ions and m/z 140 ions occurs at a tickle
voltage amplitude of 240 mV$_{(0–p)}$, while for β_z = 0.36, the crossing point
occurs at about 160 mV$_{(0–p)}$.

These data demonstrate that by using ITMS, highly effective experiments can be performed such that the energy deposition in the preselected
species can be finely regulated.

The ERMS data shown in Fig. 9.36 demonstrate further that,
depending on the internal energy of m/z 166 ions, different structures
can be elucidated as being more favorable than others, (Scheme 9.14),

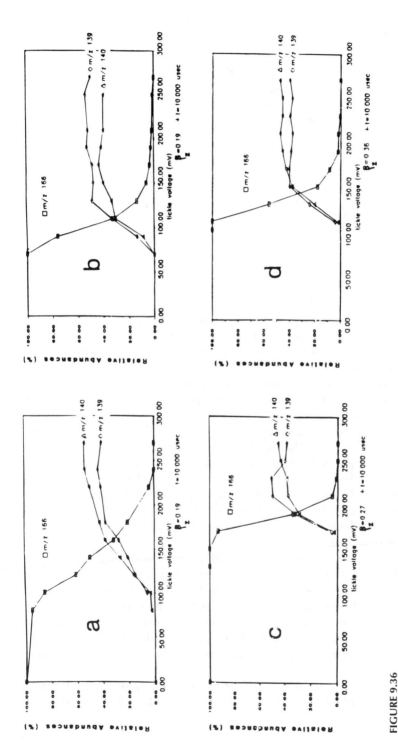

FIGURE 9.36

Breakdown curves of the fragment ion [M–HS]⁺ (m/z 166) obtained with the ITMS. (a) m/z 166 obtained by EI and irradiated at $\beta_z = 0.19$; (b) m/z 166 obtained by MS² and irradiated at $\beta_z = 0.19$; (c) m/z 166 obtained by MS³ and irradiated at $\beta_z = 0.27$; (d) m/z 166 obtained by MS⁴ and irradiated at $\beta_z = 0.36$. (From Constantin, E.; Schnell, A.; Guidugli, F.; Traldi, P. *Org. Mass Spectrom.* 1992, 27, 174. Reproduced by courtesy of John Wiley & Sons.)

such that in this instance, the formation of structure **a** is most favored under conditions of high energy deposition.

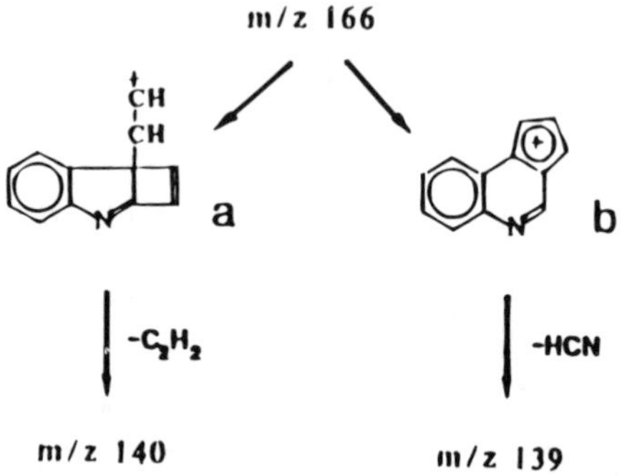

SCHEME 9.14
(From Constantin, E.; Schnell, A.; Guidugli, F.; Traldi, P. *Org. Mass Spectrom.* 1992, 27, 174. Reproduced by courtesy of John Wiley & Sons.)

G.   Morphine and Cocaine

Hair represents an interesting substrate for the investigation of drug abuse.[51] In fact, several different analytical approaches have been described in the literature for the determination of opiates, cocaine and its metabolites in this substrate.

In an investigation into heroin abuse carried out in this laboratory (that is, the investigation, not the abuse), morphine was determined as the major heroin metabolite in samples that had proved positive by radioimmunoassay. That is, samples which had tested positive by radioimmunoassay were analyzed further by GC/MS or HPLC/MS as comfirmatory assays.

In a recent paper, tandem mass spectrometry was shown to be highly effective for the identification of morphine in untreated hair extract.[52] Using high-energy collision experiments, performed on a high performance sector mass spectrometer, the molecular ion of morphine yielded daughter ion spectra highly diagnostic of the morphine structure, thus showing readily the presence of such a metabolite.

While this procedure was advantageous in that it was time-saving and required only minor sample treatment without derivitization, the instrumental complexity inherent in a double focusing instrument of reverse geometry and the high capital cost of the instrument made it unsuitable for a routine procedure in forensic laboratories. For these reasons, an analogous study was undertaken using ITMS for the identification, by means of daughter ion spectroscopy, of morphine and cocaine in hair extracts of heroin addicts.[53]

All the samples (100 mg of hair samples) were prepared according to a procedure described previously in the literature, which is summarized as follows: (1) wash with 10 mL of diethyl ether and 12 mL of 0.01 M HCl on a porous glass filter; (2) treat with 2 mL of 0.5 M HCl at 45°C for 16 h; (3) neutralize with an equimolar amount of NaOH; (4) pH buffer and extract with Toxitubes® A (Analytical Systems, Laguna Hills, CA); (5) dry in a stream of air.

The dried extracts were dissolved in 1 mL of methanol, to yield an appropriate concentration, and examined by mass spectrometry. Pure samples of morphine and cocaine were purchased from Carlo Erba (Milan, Italy). Ecgonine, ecgonine methyl ester, and benzoyl ecgonine were synthesized according to a procedure published previously.[54]

The daughter ion spectra of $M^{+\cdot}$ of morphine (m/z 285) as obtained by application of a tickle voltage of 200 $mV_{(0-p)}$ for 15 ms, is reported in Fig. 9.37.

Fragment ions at m/z 162, 215, 228, 256, and 268 were easily detected. Such ionic species, present also in the high-energy collision spectrum of the same ion, are highly diagnostic for the morphine structure, as shown in Scheme 9.15.

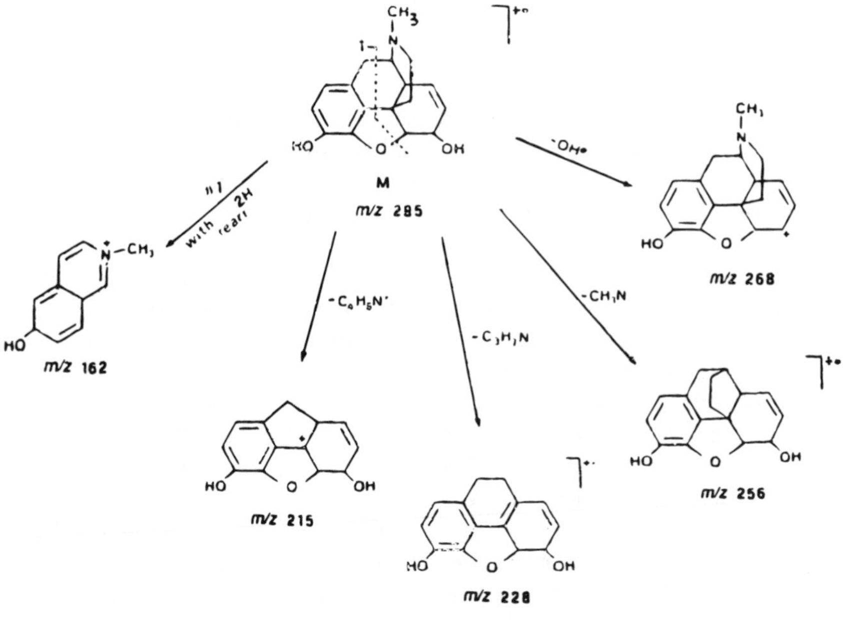

SCHEME 9.15
(From Curcuruto, O.; Guidugli, F.; Traldi, P.; Sturaro, A.; Tagliaro, F.; Marigo, M. *Rapid Commun. Mass Spectrom.* 1992, *6*, 434, Reproduced by courtesy of John Wiley & Sons.)

The normal EI mass spectrum of a blank sample of hair is highly complex in that it shows ionic species corresponding to every m/z value. The subsequent selection and collision of m/z 285 ions from this EI mass

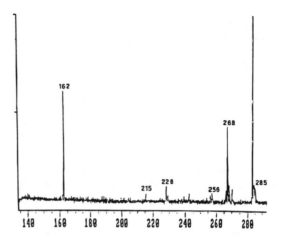

FIGURE 9.37
Daughter ion spectrum of the molecular cation, $M^{+\cdot}$, of morphine (m/z 285). Tickle voltage amplitude = 200 $mV_{(0-p)}$, tickle time = 15 ms. (From Curcuruto, O.; Guidugli, F.; Traldi, P.; Sturaro, A.; Tagliaro, F.; Marigo, M. *Rapid Commun. Mass Spectrom.* 1992, 6, 434. Reproduced by courtesy of John Wiley & Sons.)

spectrum did not lead to a spectrum of any significance, there being present only a few peaks having signal-to-noise ratios in excess of 3:1.

By spiking blank hair samples with known amounts of morphine, the detection limits for such a molecule in the complex substrate was found to lie in the range of 1 ppb. In this range, significant spectra were obtained showing m/z 268, 256, 228, and 162 ions with acceptable signal/noise ratios, proving that this analytical approach exhibits a sensitivity sufficient for this specific analytical problem. The performance of the ITMS is well-demonstrated by the daughter ion spectrum produced by m/z 285 ions obtained from a sample of hair extract from a heroin addict (Fig. 9.38). In this daughter ion spectrum, ions of m/z 268, 256, 228, 215, and 162 are present, proving that MS/MS experiments by ITMS can be successfully employed as confirmative assays for heroin abuse.

In the case of cocaine, three main metabolic pathways are present leading to ecgonine, ecgonine methyl ester, and benzoyl ecgonine, the last being the most favored one (Scheme 9.16). Preliminary attempts by high-energy collision for identification of cocaine and its metabolites in hair extracts failed, due to the presence of many species, which interfered with the diagnostic fragment ions. Consideration was given to the somewhat different ionization conditions which pertain to the ITMS, in that there might be a reduction, in the ion trap, in the abundances of the many interfering species present in high-energy collision measurements of morphine. It was, therefore, of interest to attempt to use ITMS for the identification of cocaine and its metabolites in hair extracts of cocaine addicts.

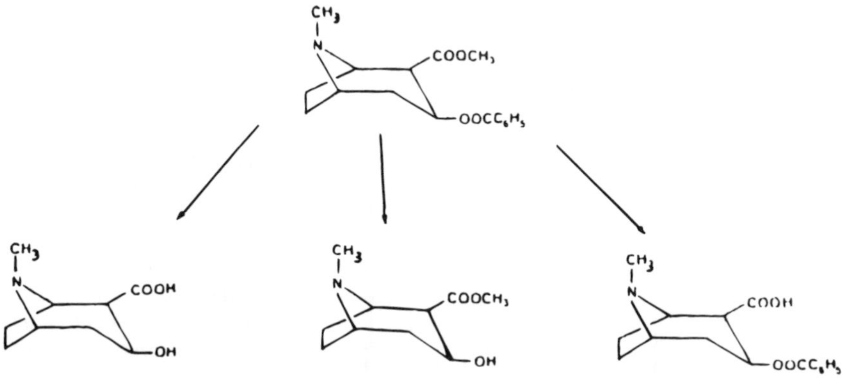

SCHEME 9.16
(From Curcuruto, O.; Guidugli, F.; Traldi, P.; Sturaro, A.; Tagliaro, F.; Marigo, M. *Rapid Commun. Mass Spectrom.* 1992, 6, 434, Reproduced by courtesy of John Wiley & Sons.)

First, the daughter ion spectra of $M^{+\cdot}$ (m/z 303) of cocaine, ecgonine, ecgonine methyl ester, and benzoyl ecgonine were studied. The data so obtained for $M^{+\cdot}$ (m/z 303) of cocaine are reported in Fig. 9.39 and Scheme 9.17; $M^{+\cdot}$ of cocaine decomposes mainly through the loss of benzoyloxy radical and benzoic acid, giving rise to m/z 182 and m/z 181 ions, respectively. Ionic species of m/z 198 are due to benzoyl radical loss, while the ions at m/z 83 and 82 are formed by a ring cleavage process (Scheme 9.17).

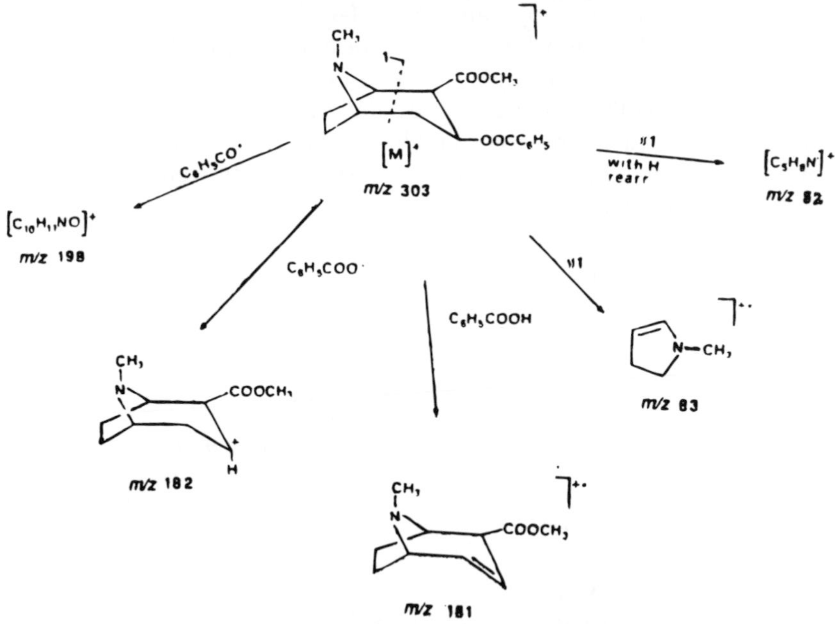

SCHEME 9.17
(From Curcuruto, O.; Guidugli, F.; Traldi, P.; Sturaro, A.; Tagliaro, F.; Marigo, M. *Rapid Commun. Mass Spectrom.* 1992, 6, 434, Reproduced by courtesy of John Wiley & Sons.)

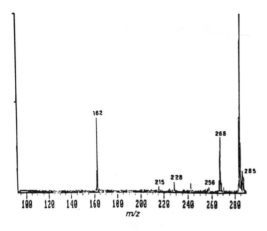

FIGURE 9.38
Daughter ion spectrum of the ion of m/z 285 present in a sample of hair extract; the hair was obtained from a heroin addict. (From Curcuruto, O.; Guidugli, F.; Traldi, P.; Sturaro, A.; Tagliaro, F.; Marigo, M. *Rapid Commun. Mass Spectrom.* 1992, 6, 434. Reproduced by courtesy of John Wiley & Sons.)

The same process is present for the $M^{+\cdot}$ for ecgonine, leading to the ions of m/z 82. For this compound, fragment ions corresponding to losses of primary methyl and H_2O were observed at m/z 170 and m/z 167, respectively. Further decomposition due to heterocyclic ring cleavage with skeletal rearrangement gave rise to ions of m/z 157 ($[M–C_2H_4]^{+\cdot}$), m/z 129 ($[M–C_3H_4O]^{+\cdot}$) and m/z 87 ($[C_4H_9N]^{+\cdot}$).

The daughter ion spectrum of ecgonine methyl ester is composed mainly of fragment ions of m/z 83 and m/z 82, as described above. Minor peaks at m/z 87 ($[C_4H_9NO]^{+\cdot}$) and 167 ($[M–CH_3OH]^{+\cdot}$) were present also.

Finally, the $M^{+\cdot}$ ions of benzoylecgonine gave rise to only one fragment ion of m/z 168 due to the highly favored loss of $C_6H_5COO^{\cdot}$.

Hence, the collisionally-generated daughter ions of $M^{+\cdot}$ of cocaine, ecgonine, ecgonine methyl ester, and benzoyl ecgonine can be considered valid fingerprints of the molecules under study, being well related to the respective structures.

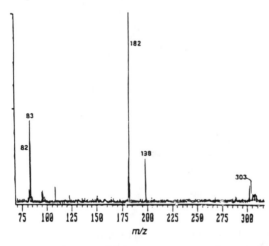

FIGURE 9.39
Daughter ion spectrum of the molecular cation, $M^{+\cdot}$ (m/z 303) of cocaine. (From Curcuruto, O.; Guidugli, F.; Traldi, P.; Sturaro, A.; Tagliaro, F.; Marigo, M. *Rapid Commun. Mass Spectrom.* 1992, 6, 434. Reproduced by courtesy of John Wiley & Sons.)

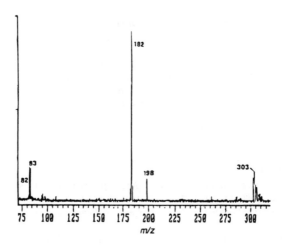

FIGURE 9.40
Daughter ion spectrum of the ion of m/z 303 present in a sample of hair extract; the hair was obtained from a cocaine addict. (From Curcuruto, O.; Guidugli, F.; Traldi, P.; Sturaro, A.; Tagliaro, F.; Marigo, M. *Rapid Commun. Mass Spectrom.* 1992, *6*, 434. Reproduced by courtesy of John Wiley & Sons.)

By the direct introduction of a hair extract from cocaine addicts and collision of m/z 303 ions, the daughter ion spectrum reported in Fig. 9.40 was obtained. It is readily seen that this spectrum is practically superimposable with that of the standard compound, as shown in Fig. 9.39, while the spectrum of the same ionic species from a blank hair does not show the formation of any diagnostic daughter ion.

However, the $M^{+\cdot}$ collisional spectra of each of ecgonine, ecgonine methyl ester, and benzoyl ecgonine are not as diagnostic as that of $M^{+\cdot}$ of cocaine. The daughter ions of $M^{+\cdot}$ of the three metabolites are present, but with abundances comparable with those of other ionic species derived by collisionally-induced decomposition of interfering ions isobaric with $M^{+\cdot}$

In conclusion, ITMS can be proposed consequently as a valid analytical tool of investigation of drug abuse.

REFERENCES

1. Thomson, J. J. *Rays of Positive Electricity and the Application to Chemical Analysis.* Longmans Green, London, 1913, p. 56.
2. Beynon, J. H.; Morgan, R. P. *Int. J. Mass Spectrom. Ion Phys.* 1978, *27*, 1–30.
3. Hipple, J. A. and Condon, E. U., *Phys. Rev.* 1945, *68*, 54.
4. (a) Craig, R. D.; Green, B. N.; Waldron, J. D. *Chimia.* 1963, *17*, 33. (b) Bruins, A. P.; Jennings, K. R.; Evans, S. *Int. J. Mass Spectrom. Ion Phys.* 1978, *26*, 395.

5. (a) Rosenstock, H. M.; Melton, C. E. *J. Chem. Phys.* 1957, *26*, 314. (b) Jennings, K. R. *Int. J. Mass Spectrom. Ion Phys.* 1968, *1*, 227. (c) Hadlon, W. F.; McLafferty, F. W. *J. Am. Chem. Soc.* 1968, *90*, 4745.

6. (a) Cooks, R. G.; Beynon, J. H.; Caprioli, R. M.; Lester, R. G. *Metastable Ions.* Elsevier, Amsterdam, 1973. (b) Lundquist, R. T.; Ruby, A. *Appl. Spectrosc.* 1966, *20*, 258. (c) Busch, K. L.; Glish, G. L.; McLuckey, S. A. *Mass Spectrometry/Mass Spectrometry. Techniques and Applications of Tandem Mass Spectrometry.* VCH, New York, 1988.

7. Yost, R. A. *Spectra.* 1983, *9*, 2.

8. Mabud, M. A.; Dekrey, M. J.; Cooks, R. G. *Int. J. Mass Spectrom. Ion Processes.* 1987, *67*, 285.

9. March, R. E.; Hughes, R. J. *Quadrupole Storage Mass Spectrometry.* John Wiley & Sons, New York 1989.

10. Vedel, F.; Vedel, M.; March, R. E. *Int. J. Mass Spectrom. Ion Processes.* 1990, *99*, 125.

11. (a) Paradisi, C.; Todd, J. F. J.; Traldi, P.; Vettori, U. *Org. Mass Spectrom.* 1992, *27*, 251. (b) Curcuruto O.; Fontana, S.; Traldi, P.; Celon, E. *Rapid Commun. Mass Spectrom.* 1992, *6*, 322.

12. Lammert, S. A.; Cooks, R. G. *J. Am. Soc. Mass Spectrom.* 1991, *2*, 487.

13. Louris, J. N.; Brodbelt-Lustig, J. S.; Cooks, R. G.; Glish, G. L.; Van Berkel, G. J.; McLuckey, S. A. *Int. J. Mass Spectrom. Ion Processes.* 1990, *96*, 117.

14. McLuckey, S. A.; Glish, G. L.; Van Berkel, G. J. *Int. J. Mass Spectrom. Ion Processes.* 1991, *106*, 213.

15. Todd, J. F. J.; March, R. E.; Franklin, A. M.; Penman, A. D. *9th Informal Meeting Mass Spectrom.* Padova, Italy, April 1991, 22–24.

16. Strife, R. J.; Kelley, P. E.; Weber-Grabau, M. *Rapid Commun. Mass Spectrom.* 1988, *2(6)*, 105.

17. Johnson, J. V.; Pedder, R. E.; Yost, R. A. *Rapid Commun. Mass Spectrom.* 1992, *6(12)*, 760.

18. Ardanaz, C. E.; Traldi, P.; Vettori, U.; Kavka, J.; Guidugli, F. *Rapid Commun. Mass Spectrom.* 1991, *5*, 5.

19. Gronowska, J.; Paradisi, C.; Traldi, P.; Vettori, U. *Rapid Commun. Mass Spectrom.* 1990, *4*, 306.

20. (a) March, R. E.; McMahon, A. W.; Londry, F. A.; Alfred, R. L.; Todd, J. F. J.; Vedel, F. *Int. J. Mass Spectrom. Ion Processes.* 1989, *95*, 119. (b) March, R. E.; McMahon, A. W.; Allinson, E. T.; Londry, F. A.; Alfred, R. L.; Todd, J. F. J.; Vedel, F. *Int. J. Mass Spectrom. Ion Processes.* 1190, *99*, 109. (c) March, R. E.; Londry, F. A.; Alfred, R. L.; Todd, J. F. J.; Penman, A. D.; Vedel, F.; Vedel, M. *Int. J. Mass Spectrom. Ion Processes.* 1991, *110*, 159.

21. Williams, J. D.; Syka, J. E. P.; Kaiser, Jr., R. E.; Cooks, R. G. *Proc. 38th ASMS Conf. Mass Spectrometry and Allied Topics.* Tucson, 1990, p. 864.

22. Paradisi, C.; Todd, J. F. J.; Traldi, P.; Vettori, U. *Rapid Commun. Mass Spectrom.* 1992, *6*, 641.

23. Bohm, B. A. *The Flavonoids; Advances in Research.* J. B. Harbone and T. J. Mabry Eds., Chapman and Hall, London, 1982, p. 313.

24. Bohm, B. A. *The Flavonoids.* J. B. Harbone, T. J. Mabry and H. Mabry Eds., Chapman and Hall, London, 1975.

25. Murakami, T.; Tamaka, N. *Progress in the Chemistry of Organic Natural Products,* Springer-Verlag, New York, 1988, p. 54.

26. Ronayne, J.; Williams, D. H.; Bowie, J. H. *J. Am. Chem. Soc.* 1966, *88*, 4980.

27. Van de Sande, S.; Serum, J. W.; Vandewalle, M. *Org. Mass Spectrom.* 1972, *6*, 1333.

28. Pelter, A.; Stainton, P. *J. Chem. Soc.* 1967, (c) 1933.

29. Beynon, J. H.; Lester, G. R.; Williams, A. E. *J. Phys. Chem.* 1959, *63*, 1861.

30. Ardanaz, C. E.; Traldi, P.; Vettori, U.; Kavka, J.; Guidugli, F. *Rapid Commun. Mass Spectrom.* 1991, *5*, 5.

31. Rouvier, E.; Medina, H.; Cambon, A. *Org. Mass Spectrom.* 1976, *11*, 800.
32. Ardanaz, C. E.; Kavka, J.; Curcuruto, O.; Traldi, P.; Guidugli, F. *Rapid Commun. Mass Spectrom.* 1991, *5*, 569.
33. Girard, M. L. *Problemes Actuels de Biochimie Appliques.* Masson, Paris, 1967, p. 121.
34. Evans, C.; Catinella, S.; Traldi, P.; Vettori, U.; Allegri, G. *Rapid Commun. Mass Spectrom.* 1990, *4*, 335.
35. Vettori, U.; Allegri, G.; Ferlin, M. G.; Traldi P. *Int. J. Mass Spectrom Ion Processes.* 1990, *99*, 99.
36. Fisher, R. T.; Trazaskos, J. M. Mass Spectrometry of Biological Material, in *Practical Spectroscopy, Vol. 8.* C.N. McEwen and B.S. Larsen (Eds.), Marcel Dekker, New York, 1990, pp. 287–296.
37. (a) Brooks, C. J. N.; McKenna, R. M.; McLachlan, W. J.; Lawrie, T. D. V. *Biochem. Soc. Trans.* 1983, *11*, 700. (b) Koopman, B. J.; Van der Molen, J. C.; Wolthers, B. G. *J. Chromatogr.* 1987, *416*, 1. (c) Brooks, C. J.; Steel, G.; Gilbert, J. D.; Harland, W. A. *Atherosclerosis.* 1971, *13*, 223. (d) Smith, L. L.; Van Lier, J. E. *Atherosclerosis.* 1970, *12*, 1.
38. Won Park, S.; Addis, P. B. *Anal. Biochem.* 1985, *149*, 275.
39. Favretto, D.; Guidugli, F.; Seraglia, R.; Traldi, P.; Ursini, F.; Sevanian, A. *Rapid Commun. Mass Spectrom.* 1991, *5*, 240.
40. Murray, R. D.; Mendez, J.; Brown, S. A. *The Natural Coumarins.* John Wiley & Sons, NY, 1982.
41. (a) Hepstein, J. *New Engl. J. Med.* 1979, *300*, 852. (b) Regan, J. D.; Parrish, J. A. *The Science of Photomedicine.* J. A. Parrish, R. S. Stern, M. A. Pathak, and T. B. Fitzpatrick (Eds.), Plenum Press, New York, 1982, p. 595. (c) Musajo, L.; Rodighiero, G. *Photophysiology, Vol. 7.* A. C. Giese (Ed.), Academic Press, New York, 1972, p. 115. (d) Dall'Acqua, F., *Research in Photobiology.* A. Castellani (Ed.), Plenum Press, New York, 1972.
42. Dall'Acqua, F.; Bordin, F.; Vedaldi, D.; Recher, M.; Rodighiero, G. *Photochem. Photobiol.* 1979, *29*, 283.
43. Pastorini, G.; Rodighiero, P.; Manzini, P.; Conconi, M. T.; Chilin, A.; Guiotto, A. *Gazz. Chim. Ital.* 1989, *119*, 481.
44. Kiremire, B.; Traldi, P.; Guiotto, A.; Pastorini, G.; Chilin, A.; Vettori, U. *Int. J. Mass Spectrom. Ion Processes.* 1991, *106*, 283.
45. Evans, C.; Traldi, P.; Chilin, A.; Pastorini, G.; Rodighiero, P. *Org. Mass Spectrom.* 1991, *26*, 688.
46. Fontana, S.; Curcuruto, O.; Traldi, P.; Castellin, A.; Chilin, A.; Rodighiero, P.; Guiotto, A. *Org. Mass Spectrom.* 1992, *27*, 1255.
47. Gilbert, J. N. T.; Millard, R. J. *Org. Mass Spectrom.* 1969, *2*, 17.
48. Audier, L.; Azzaro, M.; Cambon, A.; Guedj, R. *Bull. Soc. Chim. Fr.* 1968, 1013.
49. Halberg, A.; Al-Showaier, I.; Martin, A. R. *J. Heterocyclic Chem.* 1984, *21*, 841.
50. Constantin, E.; Schnell, A.; Guidugli, F.; Traldi, P. *Org. Mass Spectrom.* 1992, *27*, 174.
51. Yinon, J. *Mass Spectrom. Rev.* 1991, *10*, 179.
52. Pelli, B.; Traldi, P.; Tagliaro, F.; Lubli, G.; Marigo, M. *Biomed. Environ. Mass Spectrom.* 1987, *14*, 63.
53. Curcuruto, O.; Guidugli, F.; Traldi, P.; Sturaro, A.; Tagliaro, F.; Marigo, M. *Rapid Commun. Mass Spectrom.* 1992, *6*, 434.
54. Sturaro, A.; Doretti, L.; Parvoli, G.; Seraglia, R. *Laboratorio 2000.* 1989, *3*, 50.

Chapter 10

ATMOSPHERIC GLOW DISCHARGE/ION TRAP MASS SPECTROMETRY

Marian L. Langford and John F.J. Todd

CONTENTS

0-8493-8251-3/95/$0.00+$.50
© 1995 by CRC Press, Inc.

I. INTRODUCTION

Whereas inductively coupled plasma (ICP) spectrometry evolved as a method of analysis of liquid samples, much of the initial theoretical and experimental development of glow discharge mass spectrometry[1-3] was directed at the advancement of a technique for the elemental analysis of solids. Currently there is a plethora of direct current (DC) and radio frequency discharge (RF) sources, coupled to a wide variety of mass spectrometers, involved in elemental analysis. Several such instruments are available commercially.

Most glow discharge/ion trap mass spectrometry has been directed at the real-time analysis of trace vapors. However, an ion trap has been coupled to a RF glow discharge source for the purpose of the elemental analysis of metals[4] and, very recently, an ion trap has been interfaced successfully to an ICP source.[5]

This chapter covers theoretical and experimental topics in glow discharge/ion trap mass spectrometry. In Section II, a theoretical overview of the physical processes involved in the glow discharge mechanism is presented, together with a discussion of the distribution and the formation of charged particles within a discharge, particularly negative ions. In Section III, the glow discharge/ion trap instrumentation presently in use is discussed and the relevant literature is reviewed.

II. THEORETICAL FEATURES OF THE GLOW DISCHARGE
MECHANISM

A brief introduction is presented of the physics and energetics of both direct current and radio frequency glow discharges. While positive ions are formed readily in a glow discharge, the formation of negative ions is strongly dependent upon the discharge environment.

A.   The Direct Current Glow Discharge

1. DC Glow Discharge Physics

When a low voltage V is applied between two parallel plates a small current ($\sim 10^{-4}$ A) is caused to flow between the electrodes. This current consists of electrons emitted from the cathode and drawn to the anode by the applied electric field. The cathodic emission of electrons may be due to a number of processes such as thermionic emission, or the bombardment of ions, neutral particles, and/or photons. When the voltage across the electrodes is increased then, at some critical voltage, the breakdown voltage (V_{br}), the gas changes from a virtual insulator to a conductor. This change occurs when the emitted electrons are accelerated to an energy which is sufficient to cause the ionization of colliding gas particles. These collisions result in the formation of a positively charged ion (which is drawn towards the cathode) and an additional electron. Subsequent acceleration of the electrons by the field leads to further ionizing collisions (an "avalanche" effect) and a discharge current, I, of typically several mA is produced. This current can be described by

$$I = \frac{k_1 e^{k_2 d}}{(1 - k_3[e^{k_2 d} - 1])} \tag{10.1}$$

where k_1 is the initial current liberated from the cathode into the gap, k_2 is the coefficient for electron ionization, k_3 is the coefficient for positive ion ionization and d is the distance separating the electrodes.

The above equation describes the discharge current which is obtained when ionization by collision is the only method of ion creation. In fact, there are many secondary ionization processes which can affect the intensity of the discharge current. Such processes include, for instance, the creation of ions at the cathode by sputtering following bombardment by positive ions, photons which may be released when collisionally excited atoms or molecules decay, and/or neutral particles. These processes increase the intensity of the discharge current. Another secondary process, electron capture, removes electrons from the swarm, thus forming negative ions which are much heavier than electrons and so will not be accelerated to energies that can cause the ionization of gas particles. This process, therefore, effectively reduces the discharge current. The generalized equation for the discharge current is

$$I = \frac{k_1 e^{k_2 d}}{\left(1 - \dfrac{k_4}{k_2}[e^{k_2 d} - 1]\right)} \tag{10.2}$$

where k_4 is a function of the primary and secondary ionization coefficients.[6]

The formation of a discharge is dependent ultimately on two variables, the energy of the electron swarm and the number of ionizing collisions. These variables are related directly to the applied voltage and the gas

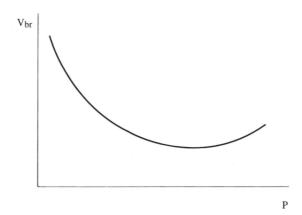

FIGURE 10.1
The Paschen curve: breakdown voltage as a function of discharge gas pressure.

pressure, respectively. When the gas pressure is too low, then the gas is essentially an insulator and collisions are too infrequent to create a discharge. When the pressure is too high, then the path of the electrons between collisions is very short and a high voltage is needed in order to produce ionizing collisions. The resulting graph of V_{br} *versus* Nd, where N is the gas number density and d is the interelectrode gap, shows the relationship between these two important variables (see Fig. 10.1) and is known as the Paschen curve. When d is constant, then the Paschen curve is generated by V_{br} *versus* the discharge gas pressure, P.

2. Regions of the Glow Discharge

The glow discharge itself can be differentiated into several regions (see Fig. 10.2). In the cathode fall region, the difference in electron and ion mobilities produces a high density layer of positive ions in front of the cathode and a scarcity of electrons. This condition creates a rapid fall in potential in the cathode region (the cathode fall). Electrons emitted from the cathode are accelerated quickly in this region and collide with gas atoms causing their ionization or excitation; the characteristic glow seen in a glow discharge is created by photons emitted when excited particles decay to lower electronic states. After collision, the electrons have insufficient energy to excite the gas and so a dark space appears. However, electrons are soon reaccelerated by the field to energies sufficient to cause gas excitation leading to the formation of another glow region. At low pressures and voltages, the positive column may consist of a number of glow and dark space regions as the electrons expend their energy and are reaccelerated. Within this region, the density of electrons and positive ions is high and approximately equal, creating a quasi-neutral plasma.

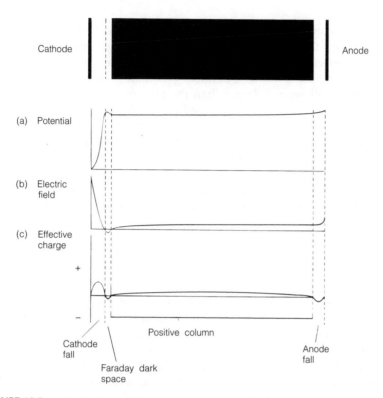

FIGURE 10.2
The DC glow discharge showing the cathode fall, Faraday dark space, positive column, and anode fall regions where (a) shows the variation of electric potential, (b) shows the electric field, and (c) shows the distribution of the effective charge throughout the discharge.

B. The Radio Frequency Glow Discharge

1. RF Glow Discharge Physics

A radio frequency glow discharge source is usually constructed by capacitively coupling one electrode to a RF drive voltage and the other electrode to ground. The action of capacitively coupling one electrode to the RF drive effectively prevents any net current passing through the system; this arrangement, together with the difference in mobility between electrons and positive ions within the plasma, causes a self-bias to develop (V_{DC}) on the excitation electrode around which the drive potential (V) operates

$$V_c(t) = V_{DC} + V\sin\omega t \tag{10.3}$$

where $V_c(t)$ is the effective voltage of the excitation electrode.

The degree of self-bias is governed by the geometry of the source itself.[2] When the area of the excitation electrode equals the area of the grounded electrode plus any area of the source walls in contact with the plasma, then no self-bias is generated. When the grounded electrode plus contacting wall areas is larger than the area of the excitation electrode, then the excitation electrode is biased positively. When, as is usually the case, the geometry of the source is such that the area of the excitation electrode is small in comparison with the area of the grounded electrode plus contacting walls, then a negative self-bias is generated. For this reason, the excitation electrode is usually termed the cathode and the grounded, or floating electrode, the anode. For a plasma in contact with electrodes only, the self-bias can be related approximately to the RF amplitude by

$$V_{DC} = \frac{V(C_c - C_a)}{(C_c + C_a)} \tag{10.4}$$

where C_c and C_a are the capacitances of the sheaths at the cathode and the anode surfaces, respectively.[2] The magnitude of these capacitances is dependent on the area of the electrode in contact with the plasma.

The general form of a RF plasma (see Fig. 10.3) is similar to that of a DC plasma in that a quasi-neutral plasma, the bulk plasma region, extends over most of the inter-electrode gap with areas of high electric field and pronounced potential drop near the electrodes. These drops in potential are the anode and cathode falls and are regions of low electron density known as the anode and cathode sheaths. In the more usual geometry, the cathode fall and sheath is larger due to a negative self-bias. The majority of electrons are found in the bulk of the plasma outside the sheaths. These electrons, which have low energies and high collision frequencies, determine the current and charge characteristics of the plasma. Due to high electron mobility, the electron density is very low in the sheaths, but those electrons found in these regions are accelerated quickly by the strong electric fields, and are responsible for the majority of ionization reactions at the sheath/plasma interface.

A radio frequency discharge is sustained by two processes involving "primary" and "secondary" electrons. Primary electrons are generated by the ionization of the parent species in the bulk plasma region. These electrons gain energy either from field fluctuations in the bulk plasma region or by "sheath heating." Sheath heating is caused by the passage of primarily low energy primary electrons into the sheaths. Initially, the electron is slowed to a halt by the sheath fields and then accelerated back into the bulk plasma. The sheath varies in thickness throughout the RF cycle so that when the sheath is expanding, the accelerating electron

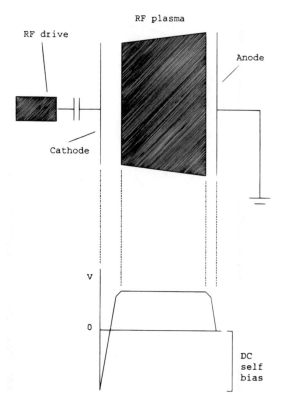

FIGURE 10.3
The distribution of potential across a radio frequency plasma.

spends more time in the field than when decelerating, and so there is a net gain in electron energy. Sheath heating depends strongly on the RF frequency and is most important at high frequencies. The energy gained by the bulk electrons is balanced by energy lost by excitation and ionization reactions and the diffusion of some electrons to the walls.

Secondary electrons are emitted from the surface of the electrodes following their bombardment by metastable species, fast neutrals, ions and, to a lesser extent, electrons and photons. The emitted electrons are low in energy, typically less than 5 eV, but they are accelerated to high energies by the sheath fields and deposited in the bulk plasma. These highly energetic electrons, which may have energies of several hundred electronvolts, cause ionization and excitation reactions at the sheath/plasma interface. This process is most important at high discharge voltages.

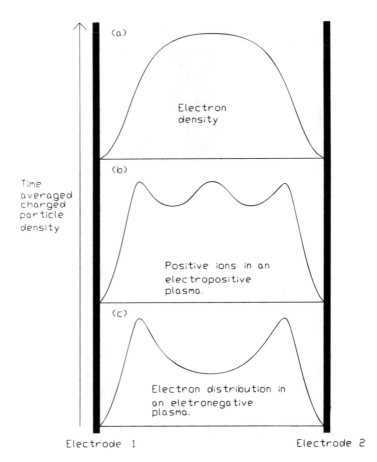

FIGURE 10.4

Charged particle distribution in a radio frequency plasma where (a) shows the distribution of electrons between the electrodes, (b) shows the distribution of positive ions in a plasma formed in an electropositive discharge gas, and (c) shows the distribution of electrons in a plasma formed in an electronegative discharge gas.

2.   The Distributions and Energetics of Charged Particles in a RF Glow Discharge

The electron density profile in a radio frequency discharge is very low in the sheaths (but modulated by the RF fields) and is high in the bulk plasma region (see Fig. 10.4(a)); the densities of positive and negative ion profiles depend on the creation and loss reactions in which the ions are involved.[7]

In a discharge containing a proportion of metastable neutrals, metastable species created at the edges of the sheaths diffuse into the bulk where they may ionize other species by Penning ionization. In an electropositive plasma, positively charged ions are created near the sheath edges and move into the plasma bulk where they are removed by neutralization processes, causing a drop in the density of positive ions. This phenomenon effectively traps remaining positive ions in the bulk center (see Fig. 10.4(b)). In an electronegative plasma, the relative proportions of electrons to negative ions change with source pressure in order to maintain the quasi-neutrality of the plasma bulk. At some source pressures, the density of the negative ions increases, causing a drop in the density of electrons. As the electrons must still carry the discharge current, they oscillate back and forth during the RF cycle, creating high electron densities by the sheath edges and thus trapping the negatively charged ions in the plasma bulk (see Fig. 10.4(c)).

The energy distribution of ions impacting on an electrode is of great importance in sputtering experiments. This distribution is governed primarily by the difference between the plasma potential (V_p) and the surface potential (V_s) where the surface may be DC biased, DC self-biased, grounded, or floating. When the time taken for the ion to pass is of the order of one RF cycle, then the ion energy distribution will range from $(V_p - V_s) + V$ to $(V_p - V_s) - V$. When the time for passage through the sheath is over many RF cycles, then the ion energy modulation due to the drive amplitude averages out. The ion energy distribution (E_i) due to RF modulation alone can be described as

$$E_i = \frac{8eV}{3\omega d} \sqrt{\frac{m}{2eV_{DC}}} \qquad (10.5)$$

where d is the thickness of the sheath and m is the ionic mass.[8]

Although many ions have an energy distribution characteristic of the potential difference between the plasma and the surface, a proportion of ions strike the surface with lower energies due to energy losses incurred through collisions with neutrals within the sheath. On entering the sheath, the ions have a Maxwellian spread of energies of approximately 0.5 eV. Once within the sheath, they receive energy from the rapidly varying field. Fast ions may be slowed by collisions with neutrals within the sheath, but they can create also a subset of fast neutrals with energies up to a maximum of $V_p - V_s$ by charge exchange reactions.[9]

C. The Formation of Negative Ions in a Glow Discharge Source

The formation of positive ions within a glow discharge source is a straightforward consequence of the discharge mechanism. Electrons are accelerated in the electric field and may collide with an atom or molecule. When the energy transferred from the fast electron to the target is sufficient to remove an electron, then a positive ion will be formed. The formation of negative ions, however, is a relatively complex process. Much of the work described in the next section involves the trapping and analysis of negative ions; thus, a discussion of the formation of negative ions in a glow discharge source is appropriate.

Negative ions are formed by the interaction of electrons with neutral molecules. The process is strongly dependent on the energies of the electrons in the ion source and on the source pressure. In a glow discharge ion source, the pressure in the region of the plasma is high and the energy of the electrons ranges from thermal (immediately following electron impact) to an electron energy defined by the source voltage, source pressure, and the ionization energy of the discharge gas. An inelastic collision between a fast electron and a neutral particle in the discharge can cause the excitation or ionization of the neutral, depending on the amount of internal energy gained by the neutral particle during the collision. Both of these processes may result in the dissociation of the neutral. Thus, the excitation and ionization processes, which are at the heart of the discharge mechanism, can produce an abundance of neutral species which can then capture an electron in one of the negative ion formation mechanisms discussed below.

There are a number of different negative ion formation mechanisms, including resonant electron capture followed by collisional stabilization:

$$AB + e \rightleftarrows [AB^-]^* \tag{10.6}$$

$$[AB^-]^* + C \rightarrow AB^- + C \tag{10.7}$$

where C is an electron, an atom, or a molecule. This process is important at high pressures where collisional stabilization reactions involving neutral molecules can occur. At thermal energies, dissociative electron capture

$$AB + e \rightarrow A + B^- \tag{10.8}$$

and, if the electron affinity of AB exceeds that of C, charge-transfer reactions

$$AB + C^- \rightarrow AB^- + C \tag{10.9}$$

may take place. Ion pair formation,

$$AB + e \rightarrow A^+ + B^- + e \tag{10.10}$$

where the incident electron is not captured but merely serves to excite the orbital electrons of the molecule, is primarily a low-pressure higher-electron energy process and occurs over a wide range of electron energies above 10 eV, whereas proton-transfer reactions,

$$AH + B^- \rightarrow A^- + BH \qquad (10.11)$$

where B^- is a strong gas-phase base which is capable of abstracting a proton from gas-phase acids, occur primarily at higher pressures. Ion/molecule reactions can also lead to the generation of negative ion species.

III. GLOW DISCHARGE/ION TRAP MASS SPECTROMETERS

At least two kinds of glow discharge sources have been coupled to an ion trap; an air sampling glow discharge ionization (ASGDI) source which is a DC glow discharge ion source, and a RF glow discharge ion source. These instruments are described in detail below, together with a review of work carried out on the resultant glow discharge/ion trap apparatus.

A. The Air Sampling Glow Discharge Ionization (ASGDI) Source/ITMS Combination

The majority of glow discharge/ion trap work has been realized using the ASGDI source which was developed originally by McLuckey *et al.*[10] It was coupled initially to a quadrupole mass filter/time-of-flight instrument where it was used primarily for the real-time analysis of trace organics in ambient air. The source was noted to be both rugged and sensitive. An identical source was coupled later to an ion trap mass spectrometer (ITMS) in order to utilize the mass storage and MS^n capabilities of the ion trap.[11]

The ASGDI source samples the atmosphere directly through a 200 μm aperture. Air and sample molecules are drawn into the source (see Fig. 10.5) by the pressure gradient between atmospheric pressure and that in the discharge chamber, which is in the order of 0.1 to 2 Torr. Within the source, air and sample are subjected to an electrical field which is applied across two steel plates, A_1 and A_2. As air is allowed to pass into the source at an increasing rate, the pressure in the source increases and, at some combination of source pressure and applied voltage, electrical breakdown will occur and a DC glow discharge is formed. The Paschen curve for an ASGDI source is shown in Fig. 10.6.[12] As air and sample molecules are drawn through the plasma, ionization, excitation, fragmentation, electron capture, and recombination events can all occur. A proportion of ions (positive or negative) formed in the discharge is drawn out

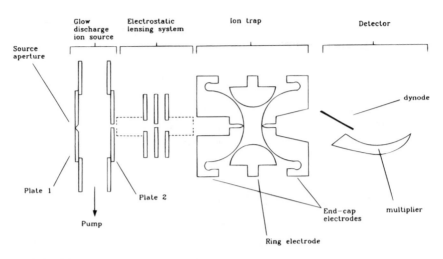

FIGURE 10.5
Diagram of the ASGDI source coupled to an ion trap mass spectrometer.

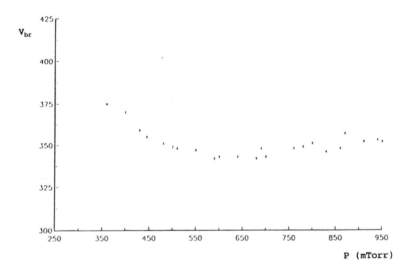

FIGURE 10.6
Initial breakdown voltages as a function of source pressure obtained using an ASGDI source.

of the source by a pressure gradient between the source and the inner chamber (10^{-4} to 10^{-3} Torr) where a lensing system directs and focuses the ions into an ITMS. Trace pollutants in the air can be sampled directly by this method or, alternatively, gaseous sample molecules can be introduced into the airstream, or air can be drawn over a solid or liquid substance placed in the sampling chamber.

The efficiency of ion injection from the ASGDI source into an ITMS was examined in some detail by McLuckey *et al.*[11] Efficiency was studied in relation to the nature of the buffer gas in the ion trap and the level of RF voltage during ion injection, and also with respect to collision-induced dissociation (CID), charge exchange and, in the case of negative ions, electron detachment reactions. Because of the nature of the instrument, there is always a background pressure of air in the ion trap, particularly at high source pressures. Topping up the air background with a number of different buffer gases showed that gases of low mass gave the best results; the most intense signals were achieved using hydrogen as the buffer gas. The efficiency of ion injection was shown to be very sensitive to the amplitude of the RF trapping field during the period when ions are gated into the trap. An examination of the optimum trapping RF voltage (V_{RF}) as a function of injected ion mass (m) showed a linear relationship. Estimating the constants from the graph published in this investigation gives[11]

$$V_{RF} = 188 + 0.935m \tag{10.12}$$

CID, charge exchange, and electron detachment are all reactions that can affect the injection efficiency. CID was found to be reduced at low RF injection voltages and high bath gas pressures.

1. Detection of Explosives

Further work[13] illustrated the ability of the ASGDI/ITMS combination to detect traces of vapor from explosive substances in ambient air. The headspace vapor over a sample of trinitrotoluene (TNT) was sampled by means of an injection period of 0.3 ms. The molecular anion, M^-, and two fragment ions, $(M-OH)^-$ and $(M-NO)^-$, were observed. Assuming a concentration of TNT in air of a few ppb, the quantity of TNT drawn into the system during this period was calculated to be just 50 fg. In addition, the $(M-NO_2)^-$ anion from RDX was shown to fragment to $(CH_2NNO_2)^-$ with 65% efficiency and the base peaks in ethylene glycol dinitrate (EGDN), nitroglycerine, and pentaerythritol tetranitrate (PETN) mass spectra were found to be the $(NO_3)^-$ anion.

The detection of explosives using the NO_3^- anion (the base peak in the mass spectra of several explosives) is problematic, as this anion is generated also by a glow discharge in air. It was shown, however, that NO_3^- has two stable configurations;[14] the D_{3h} structure which fragments to NO_2^- and the less stable peroxy structure which dissociates to form O_2^-. Experiments on the ASGDI/ITMS combination[14] demonstrated that whereas NO_3^- generated by ion/molecule reactions in the glow discharge source fragments to O_2^-, NO_3^- formed by the dissociation of PETN and

EGDN dissociates to give the NO_2^- fragment; thus, through MS/MS experiments, it is possible to differentiate between the sources of the anion.

2. Ejection of Matrix Ions

To enhance the signal and to improve the sensitivity of the instrument with regard to the detection of small quantities of organic molecules, an effort was made to reduce the effects due to the storage of matrix ions in the ion trap.[15] In an air sampling ion source, the matrix ions necessarily form the bulk of the injected ion current. The storage of an excessive number of ions in the ion trap causes space-charge effects which lead to a reduction in spectral resolution; however, the storage of the ion species of interest can be maximized without deleterious space-charge effects when the matrix ions are ejected selectively. The ejection of such ions can be achieved by resonant ejection using the axial modulation signal.

Two methods were used to realize the ejection of matrix ions. Either the axial modulation signal was held constant and the RF drive amplitude was ramped or, alternatively, the axial modulation signal was ramped and the RF drive amplitude was held constant. In the first method, the working points of ion species were ramped through a point on the Mathieu stability diagram at which the secular frequencies of the matrix ions became resonant with the axial modulation signal; consequently, the excursions of the ion trajectories exceeded the physical dimensions of the ion trap and the ions were lost. In the second method, the axial modulation frequency was ramped through a range of secular frequencies resonant with those of the unwanted ions. Ramping the RF drive amplitude through one or more points of instability also causes ejection of lower mass ions when their working points are made to pass through the $\beta_z = 1$ stability boundary at $q_z = 0.908$. In this way, the separate isolation of several discrete masses can be achieved in a single scan.

3. Negative Chemical Ionization

It is difficult to perform negative chemical ionization (NCI) experiments in an ion trap because of a dearth of thermal electrons. However, the abundance of electrons in a glow discharge, and the concomitant relatively high probability of negative ion formation, makes the ASGDI source attractive as an external ion injection source for NCI experiments carried out in an ion trap. Experiments of this nature were performed recently.[16] Samples of TNT and RDX were introduced on an unheated probe into the trap cavity where NCI experiments were carried out; a number of different reagent ions were used which had been injected into the trap from the ASGDI source. The ASGDI discharge was formed in air, and in gaseous mixtures of N_2O/CH_4 and $N_2O/H_2/He$ in order to form the reagent ions OH^-, O_2^-, NO_2^- and NO^-. Fragmentation of TNT as a function of the various reagent anions showed that NCI reactions involving O_2^- produced primarily the molecular ion while those NCI reactions involving OH^- produced the most fragmentation.

The ASGDI/ITMS combination has been used also for a number of electrospray and ion spray studies[17,18] (see also Chapters 5 and 6). In these experiments, the plates A1 and A2 were used to aid lensing, but no glow discharge was employed.

The instrument has been used also by the present authors for the real-time analysis of trace vapors from explosive compounds. The experiments described below utilize the ability of the ASGDI source to sample directly from the atmosphere, and the facility of the ion trap to store injected ions and to perform MS^n experiments on the ions of interest.

4.   Tandem Mass Spectrometry of 2,4,6-Trinitrotoluene

A few grains of 2,4,6-TNT were placed in a sampling chamber just outside the source aperture and air was drawn over the sample into the source chamber containing the DC glow discharge.[19] To maximize the signal, the source pressure and current were maintained at the relatively low values of 750 mTorr with a discharge current of 5 to 7 mA; however, the production of lower mass fragments of TNT was found to maximize at the higher source pressure of 1 Torr and a discharge current of 12 mA.

A scan function was constructed to enable MS/MS/MS experiments to be performed in order to study fragmentation reactions down to the third generation; fourth generation fragment ions were determined by MS/MS/MS experiments performed on the relevant ions formed in the primary mass spectrum.

The isolation and dissociation of the molecular anion, $M^{-\bullet}$, (see Fig. 10.7) led to the production of two fragment ions with the masses 197 u and 210 u in agreement with earlier work on the same instrument.[13] The isolation and dissociation of the daughter ions disclosed that the two ions initiate two completely separate fragmentation pathways, as is illustrated in Fig. 10.8. Branch A was initiated by the loss of a NO fragment and appeared to be the main fragmentation channel for the $M^{-\bullet}$ precursor. Figures 10.9 and 10.10 show the granddaughter ions generated by the dissociation of the $(M-OH)^-$ and the $(M-NO)^-$ fragments, respectively. The major channel for the dissociation of this ion appears to be the loss of a further NO fragment and 2 (NO) fragments. Fourth generation ions were found at m/z 109 $(M-3(NO)-CO)^-$, m/z 137 $(M-3(NO))^-$ and m/z 139 $(M-2(NO)-CO)^-$.

Branch B was initiated by the loss of an OH fragment and may be a manifestation of the o-effect where an oxygen atom from an *ortho* nitro group is lost together with a hydrogen from the methyl group. Ions were

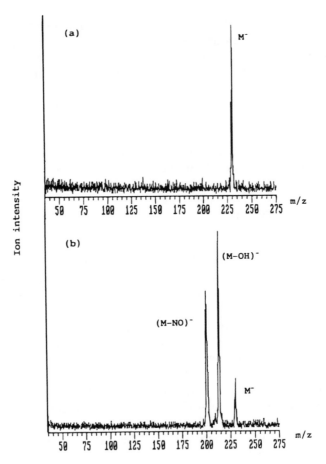

FIGURE 10.7
The MS/MS spectrum of the molecular anion where (a) shows the isolated parent anion, and (b) shows the mass spectrum generated by applying a supplementary "tickle" voltage.

observed at m/z 152 (M–OH–NO–CO)⁻, m/z 124 (M–OH–NO–2 (CO))⁻ and m/z 94 (M–OH–2 (NO)–2 (CO))⁻. These ions and assignments are listed in Table 10.1. The mass spectrum of long-lived negative ions generated in the glow discharge source is shown in Fig. 10.11.

It is apparent that some of the peaks occurring in the mass spectrum of TNT shown in Fig. 10.11 do not appear in the fragmentation tree depicted in Fig. 10.8. The formation of negative ions, by "tickling," within the ion trap follows the dissociation of selected trapped ions which, in this investigation, originate from TNT molecules. Only stable negative ions ($\tau > 10^{-3}$s) are stored in the trap. These ions are caused to fragment by translational excitation and subsequent collision-induced dissociation

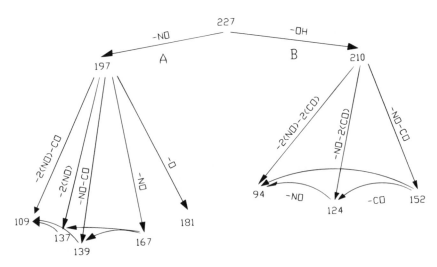

FIGURE 10.8
The negative ion fragmentation pathways of the 2,4,6-trinitrotoluene anion.

with the buffer gas, in this case, helium. The dissociation process is detected only when it is of the form

$$AB^- \rightarrow A + B^- \qquad (10.13)$$

where A is a neutral fragment(s) and where B^- is stable.

In contrast, the formation of negative ions in the ASGDI source is extremely complex. A major characteristic of a glow discharge is the range of kinetic energies of the electron swarm as a function of their axial position between the electrodes. Electrons successively gain energy from the field and lose it in collisions with the discharge gas (air, in this case). Electrons which have gained energy can impact upon neutral species, leading to the ionization and fragmentation of sample and atmospheric molecules; low-energy and thermalized electrons are available for resonant dissociative or non-dissociative electron capture.

The sample vapor concentration in the air stream entering the ASGDI source is extremely low; thus, sample anions form but a very small proportion of the total number of anions that are formed. The interaction of atmospheric ions and neutral fragments in the source leads to the appearance of ions in the mass spectrum which do not, however, appear in the fragmentation tree. The residence times of neutral fragments in the source, which are determined solely by the pumping speed, as such species are not affected by the electrostatic fields applied to the source, can be sufficiently long to permit neutral–neutral interactions. The variety of products

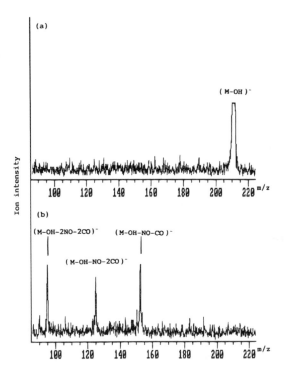

FIGURE 10.9

The MS/MS spectrum of $(M–OH)^-$ where (a) shows the isolated parent anion, and (b) shows the mass spectrum generated by applying a supplementary "tickle" voltage.

formed by such interactions may undergo electron capture by which a wide variety of negative ions can be produced.

Fig. 10.12 shows a negative ion mass spectrum which was obtained when laboratory air alone was introduced into the source. The spectrum, which is dominated by a number of low-mass peaks, shows some peaks of lower intensity present in the mid-range of the mass spectrum. The identities of the high intensity low-mass ions are given in Table 10.2. In order to assign the lower intensity middle-mass range peaks, the mass spectrum was recorded using "zero rated" air (<3ppm CO_2 and <1 ppm H_2O) which had been passed through a molecular sieve in order to remove all molecular contaminants originating in the gas inlet system; this mass spectrum is shown in Fig. 10.13(a). This mass spectrum was compared with that obtained when a small amount of water was placed in the sampling chamber and used in conjunction with the "zero air" stream (so that CO_2 only was removed from the source), and is shown in Fig. 10.13(b), and with the mass spectrum (Fig. 10.13(c)) observed when laboratory air was introduced into the source. Fig. 10.13 shows that the peaks

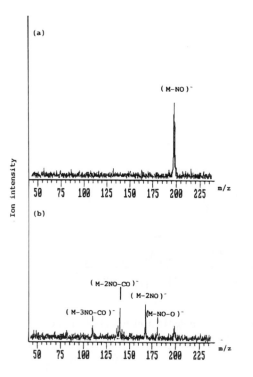

FIGURE 10.10
The MS/MS spectrum of (M–NO)⁻ where (a) shows the isolated parent anion, and (b) shows the mass spectrum generated by applying a supplementary "tickle" voltage.

TABLE 10.1

An assignment of the negative ions generated by the fragmentation of 2,4,6-trinitrotoluene

Generation	m/z	Assignment
Parent ion	227	$M^{-\bullet}$
Daughter ions	210	$(M-OH)^-$
	197	$(M-NO)^-$
	46	NO_2^-
Granddaughter ions	181	$(M-NO-O)^-$
	167	$(M-2NO)^-$
	152	$(M-OH-NO-CO)^-$
Great-granddaughter ions	139	$(M-2(NO)-CO)^-$
	137	$(M-3NO)^-$
	124	$(M-OH-NO-2(CO))^-$
	109	$(M-3NO-CO)^-$
	94	$(M-OH-2(NO)-2(CO))^-$

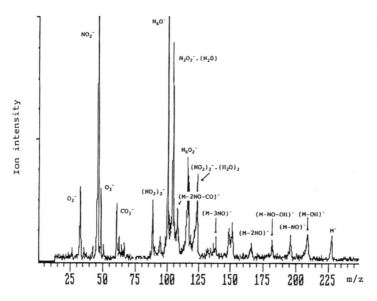

FIGURE 10.11
The mass spectrum of long-lived negative ions generated by a mixture of 2,4,6-trinitrotoluene (M) vapor and air matrix in the glow discharge ion source.

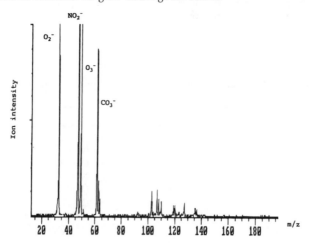

FIGURE 10.12
The negative ion mass spectrum generated by the passage of laboratory air only through the glow discharge ion source.

at m/z 60, 106, 107, and 124 disappear when CO_2 and H_2O are removed, but that the peaks at m/z 106, 107, and 124 are restored when moisture is re-introduced into the system.

To assist in a final assignment, MS/MS experiments were carried out on the middle-mass range peaks and on the ion at m/z 60. The results of

TABLE 10.2

An assignment of the negative ions generated by the air matrix.

m/z	Assignments
I. The low-mass anions which were observed in the background mass spectrum at relatively high intensities	
32	O_2^-
46	NO_2^-
48	O_3^-
60	CO_3^-
61	CHO_3^-
62	NO_3^-
II. Higher mass anions which were observed at low intensities in the background mass spectrum	
88	$(N_2O)_2^-$
100	N_6O^-
104	$N_2O_2^{-\cdot}(N_2O)$
106	$(N_2O)^{-\cdot}(H_2O)$
116	$N_6O_2^-$
124	$(N_2O)_2^{-\cdot}(H_2O)$
132	$N_6O_3^-$

these experiments are shown in Fig. 10.14 and the assignments for these peaks are given in Table 10.2. The assignments were made on the basis of the molecular mixture which was present in the source during the experiments and on the negative ion fragmentation pathways observed. Pathway (a), which is observed only when CO_2 is present in the source, shows the fragmentation of CO_3^- to O_3^-. Pathway (b), which is observed only when moisture is present, is assigned to the presence of water clusters around $(N_2O)_2^-$ anions. This structure is known to be stable in that it has an electron affinity of 0.95 eV measured by negative ion photoelectron spectroscopy.[20] Pathways (c) and (d) remained unchanged when CO_2 and H_2O were removed from the sample. Pathway (c) has been assigned as $N_2O_2^-\cdot(N_2O) \rightarrow (N_2O)_2^-$. The cluster $N_2O_2^-\cdot(N_2O)$ has been observed previously in low energy electron attachment studies.[21,22] Pathway (d) has been assigned tentatively as the dissociation of $N_6O_3^-$ leading to the formation of $N_6O_2^-$ and N_6O^-. This assignment is in partial agreement with a previous report.[23]

B. The Coupling of a Radio Frequency Glow Discharge Source to an ITMS

RF glow discharge sources are coupled frequently with quadrupoles and other mass spectrometers to facilitate analyses of solid samples. This type of ion source is particularly valuable for the study of nonconducting solids which cannot be investigated easily using a DC glow discharge.

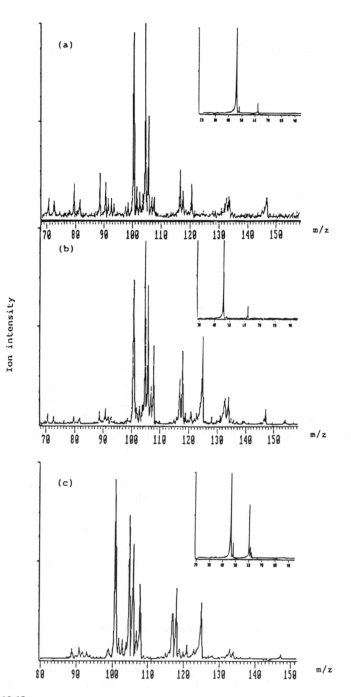

FIGURE 10.13
Negative ion mass spectra of air generated using (a) "zero" air, (b) "zero" air plus moisture, and (c) laboratory air. The insets show the low-mass regions of the respective negative ion mass spectra.

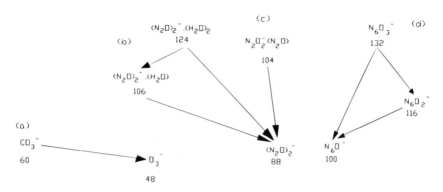

FIGURE 10.14
The negative ion fragmentation pathways of some background anions.

Recently, an RF glow discharge ion source has been coupled to an ITMS[4] in order to utilize the facilities of the ion trap for mass-selective ion isolation and tandem mass spectrometry.

The source consists of a probe-type assembly fabricated from a central steel cylinder to which the RF power (30 to 40 W) is applied. The cylinder is enclosed in insulators apart from a small circular area facing the grounded anode into which the sample is pressed; the exposed sample surface acts as the cathode. The support gas for the RF discharge was Ar at 300 to 500 mTorr. To perform the experiments, the probe was inserted into the instrument casing so that the sample lay between 1 and 2 cm from the aperture leading into the inner chamber housing, the lensing system, and the ion trap. The sample analyzed in this investigation was NIST brass.

The mass spectra of lead, tin, and nickel (all constituents of the brass sample) contained metal oxides, hydroxides, and argides where the most abundant polyatomics were metal hydroxides and, on occasion, metal oxides. Comparative experiments using the same RF ion source coupled to a quadrupole mass spectrometer indicated that the hydroxides were generated in the ion trap, whereas the metal oxides were formed in the ion source. MS/MS of these ions produced bare metal ions with high efficiencies.

The intensities of many of the plentiful ions which are derived from the discharge support gas such as ArH^+, Ar^+, and Ar^{2+} were reduced considerably when long injection times were used; the reduction in intensity of these ions was due to charge transfer reactions with N_2, O_2 and H_2O, which were present in the ion trap as a result of the high pressure of air in the source chamber, which led to a background pressure of $\sim 10^{-6}$ Torr in the trap. This phenomenon was useful as it reduced the intensity of background ions derived from Ar.

REFERENCES

1. Coburn, J. W.; Kay, E. *J. Appl. Phys.* 1972, *43*, 4965.
2. Kohler, K.; Coburn, J. W.; Horne, D. E.; Kay, E.; Keller, J. H. *J. Appl. Phys.* 1985, *57*, 59.
3. Coburn, J. W. *Thin Solid Films.* 1989, *171*, 65.
4. McLuckey, S. A.; Glish, G. L.; Duckworth D. C.; Marcus, R. K. *Anal. Chem.* 1992, *64*, 1606.
5. Barinaga C. J.; Koppenaal, D. W. *Proc. 41st ASMS Conf. Mass Spectrometry and Allied Topics.* San Francisco, May/June 1993, 458a.
6. *Electrical Breakdown in Gases.* J. M. Meek and J. P. Craggs, Eds.; J. Wiley & Sons. NY, 1978, pp. 215–221.
7. Sommerer, T. J.; Kushner, M. J. *J. Appl. Phys.* 1992, *71*, 1654.
8. (a) Benoit-Cattin, P.; Bernard, L. C. *J. Appl. Phys.* 1968, *39*, 5723: (b) Okamoto, Y.; Tamagawa, H. *J. Phys. Soc. Jap.* 1969, *27*, 270: (c) Okamoto, Y.; Tamagawa, H. *J. Phys. Soc. Jap.* 1970, *29*, 187.
9. May, P. W.; Field, D.; Klemperer, D. F. *J. Appl. Phys.* 1992, *71*, 3721.
10. McLuckey, S. A.; Glish, G. L. *Proc. 35th ASMS Conf. Mass Spectrometry and Allied Topics.* Denver, May 1987, 290.
11. McLuckey, S. A.; Glish, G. L.; Asano, K. G. *Anal. Chim. Acta.* 1989, *225*, 25.
12. Langford, M. L.; Todd, J. F. J. unpublished results (1991).
13. McLuckey, S. A.; Glish, G. L.; Grant, B. C. in *Proc. 3rd Symposium on Analysis and Detection of Explosives.* Mannheim, Germany, July 1989; F. Volk and H. Lehmann, Eds. Fraunhofer ICT, Pfinztal, Germany, 25.01.
14. Flurer, R. A.; Glish, G. L.; McLuckey, S. A. *J. Am. Soc. Mass Spectrom.* 1990, *1*, 217.
15. McLuckey, S. A.; Goeringer, D. E.; Glish, G. L. *J. Am. Soc. Mass Spectrom.* 1991, *2*, 11.
16. Eckenrode, B. A.; Glish, G. L.; McLuckey, S. A. *Int. J. Mass Spectrom. Ion Processes.* 1990, *99*, 151.
17. Van Berkel, G. J.; Glish, G. L.; McLuckey, S. A. *Anal. Chem.* 1990, *62*, 1284.
18. McLuckey, S. A.; Glish, G. L.; Asano, K. G. *Int. J. Mass Spectrom. Ion Processes.* 1991, *109*, 171.
19. Langford, M. L.; Todd, J. F. J. *Org. Mass Spectrom.* 1993, *28*, 773.
20. Coe, J. V.; Snodgrass, J. T.; Freidhoff, C. B.; McHugh, K. M.; Bowen, K. H. *Chem. Phys. Lett.* 1986, *124*, 274.
21. Kloto, C. E.; Compton, R. N. *J. Chem. Phys.* 1978, *69*, 1636.
22. Knapp, M.; Echt, O.; Kreisle, D.; Märk, T. D.; Recknagel, E. *Chem. Phys. Lett.* 1986, *126*, 225.
23. McLuckey, S. A., personal communication (1992).

Chapter 11

DYNAMICALLY PROGRAMMED SCANS

Anthony M. Franklin and John F. J. Todd

CONTENTS

0-8493-8251-3/95/$0.00+$.50

I. INTRODUCTION

The ion trap mass spectrometer (ITMS™) is a very versatile instrument and its operation is made relatively easy by the use of the SCAN and EXP software options. As will be apparent from the other chapters in this book, the scan function is the diagrammatic representation of the amplitude of the RF (radio frequency) drive potential applied to the trap during the acquisition of a single mass spectrum or scan. The scan function can be edited and, at the same time, a real-time display of the mass spectrum may be seen before storing these mass spectra to a data file. This data file can then be analyzed at leisure with a choice of display formats. A detailed discussion of the use of the SCAN and EXP software lies outside the scope of this review; however, a brief description later in this chapter will suffice to outline the method of dynamically programmed scanning.

II. THE SCAN AND EXP PROGRAMS

The ITMS™ software allows a number of user programs to run from within its main menu, two of which are used extensively when preparing and acquiring mass spectra, namely the SCAN and EXP programs.

The SCAN software permits a scan function to be created which can then be customized to suit the user's needs, and the editing is facilitated by the logical portrayal of the scan function being composed of a number of subdivisions linked together. These subdivisions are termed *scan tables* and each is identified by a number representing the chronological order of the tables. A very simple scan function is shown in Fig. 11.1: there are five such tables each representing a particular function and which are combined to allow ion creation, storage, and ejection together with detection. Four major variables are essential to the construction of a scan function to enable a standard ITMS™ to perform a multitude of operations: these are the amplitude of the RF drive potential, the DC (direct current) potential, the supplementary RF ("tickle") potential, and the duration of application. Within the software, these four variables are represented by seven parameters each of which, like the scan table, is represented by a number. A summary of this approach is given in the following paragraphs.

Parameter 1. The STARTING MASS represents the initial RF level for a scan table. The software calculates the RF amplitude required to move this m/z value to the point of being ejected, that is, at the co-ordinates $a_z = 0$, $q_z = 0.908$; all ions with higher m/z values are stored.

Parameter 2. The END MASS represents the final RF level for a scan table and is calculated in exactly the same way as the STARTING MASS.

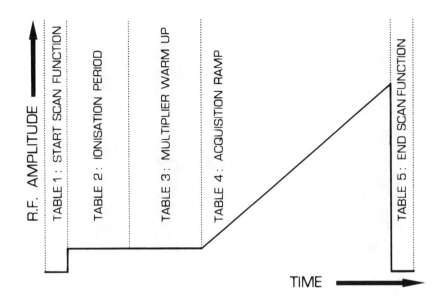

FIGURE 11.1
Diagrammatic representation of a simple scan function with reference to the scan tables.

Parameter **3**. The TIME represents the duration for which functions within a scan table operate, such as ionization time, tickle duration, and isolation period. The time is in units of microseconds and may be set up to a maximum of 100 ms.

Parameter **4**. The TICKLE MASS represents the mass/charge ratio of an ion selected for excitation. The software calculates automatically the theoretical tickle frequency from the β_z value corresponding to the a_z, q_z co-ordinates of the ion set by the STARTING MASS, and the value of the TICKLE MASS.

Parameter **5**. The DELTA value is the frequency offset from the theoretical tickle frequency determined from the TICKLE MASS. The frequency can be set within a frequency range of –30 kHz to +30 kHz and may be adjusted in increments of 1 Hz, allowing fine tuning of the resonance position when tickling an ion.

Parameter **6**. The TICKLE VOLTAGE represents the amplitude (peak–peak) of the supplementary RF field which is to excite resonantly (or tickle) ions of a particular m/z value. The TICKLE VOLTAGE has units of millivolts and can be set to a maximum value of 6000 mV.

Parameter **7**. The DC VOLTAGE allows one to set the amplitude of the direct current field, permitting the ions to be moved away from the q_z axis, that is, the $a_z = 0$ line as is required normally for ion isolation.

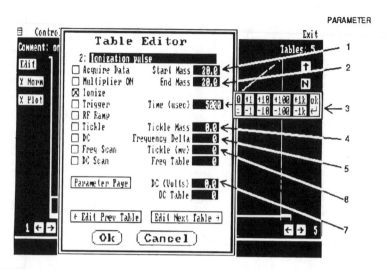

FIGURE 11.2

Example of a scan table found within the SCAN software; the parameters are labelled with numbers, as are the data values within the parameter boxes.

The DC VOLTAGE has units of volts and may be set in the range –700 to +700 V.

A typical scan table is shown in Fig. 11.2. It can be seen that there are a number of small boxes which can be employed to toggle on or off various functions such as Trigger, Ionize, RF Ramp, Tickle, and DC. Clicking the mouse onto one of the parameters on the right of the table produces a *parameter box*. The parameter box allows the parameter value to be increased or reduced by the amount shown. The increment numbers, for example 1, 10, 100, and 1000 ms, may be identified by the integers 1 through 4, whereas the decrement numbers correspond to the integers –1 through –4.

A summary of the parameters required to define the scan function shown in Fig. 11.1 is given in Table 11.1.

The edited scan function is saved as a *scan file* with the extension .SED, and this file can then be used within the EXP program in which the data can be stored to a data file as though it were a *Chromatogram*. The term chromatogram is used because the forerunner to the ITMS™, the ITD™, was used entirely for gas chromatography, the chromatographic data being recorded as a series of mass spectra acquired sequentially over a period of time. While the editing capability of the EXP program is less than that of the SCAN program, it is possible to modify the values of certain essential parameters.

The acquisition of sequential mass spectra under identical conditions is normally acceptable when one is using a chromatographic or similar

TABLE 11.1

A tabulation of the parameter values used in the scan function shown in Fig. 11.1

Scan table	Parameter 1 Start Mass/ amu	Parameter 2 End Mass/ amu	Parameter 3 time /μs	Other
1	0.0	0.0	1000	—
2	20.0	20.0	5000	Ionize on
3	20.0	20.0	5000	Multiplier on
4	20.0	650.0	See below *	Acquire on Multiplier on
5	0.0	0.0	1000	—

* The acquisition scan rate is fixed at 5550 u/s.

technique; however, when one wants to change the scan parameters while the data are being acquired during an acquisition, then modification of the scan function must be carried out manually. For example, in the investigation of ion/molecule reaction kinetics in the ion trap by systematic variation of the reaction time, either multiple acquisitions may be carried out with each data file corresponding to a different reaction time, or the reaction time may be altered manually during the acquisition. Both of these methods are time-consuming, labor intensive, and prone to human error. However, the experiment can be carried out automatically using Dynamically Programmed Scanning which is far less time consuming, simple to implement, and can eliminate human error.

III. SOFTWARE FOR DYNAMICALLY PROGRAMMED SCANS

The ITMS software, including the SCAN and EXP programs, operates in a FORTH programming environment; the software incorporates a programming option which utilizes a FORTH editor enabling one to write user programs or *key sequence* programs which can be operated within the main software. These key sequence programs can best be described as macros which mimic a series of key strokes needed for a particular operation within the ITMS software. In the original ITD™ software, a key sequence was created by pressing the relevant keys corresponding to menu-driven processes which would be stored into a file to be replayed at any time; an example might be a data processing routine for data acquired from samples injected using an auto-sampler. This early approach had many limitations. For instance, once created, a key sequence could not be edited and it was not possible to include control structures incorporating decision-making. However, with the advent of the experi-

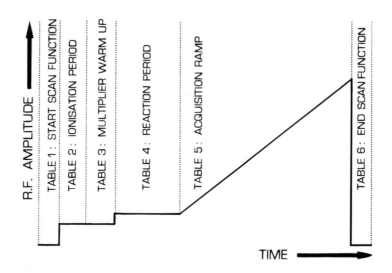

FIGURE 11.3
A simple scan function utilized in a reaction time experiment.

mental ITMS instrument, one can use the screen editor to type in high-level FORTH commands which represent key presses, and use them in the more flexible programming environment. This facility enables one to vary conditions during the operation of the EXP program and to carry out a series of experiments automatically or even semi-intelligently; we have termed this process *dynamically programmed scanning*.[1] The reactions of $CH_4^{+\cdot}$ with methane will serve as a simple illustration of an application of dynamically programmed scanning, employing the scan function shown in Fig. 11.3. In this instance, a record of the change in the intensities of products and reactants as a function of the reaction time is desired, and the following example is an extract of the key sequence program which allows the reaction time to be incremented after each acquisition of the mass spectrum.

```
⟨SUBROUTINE⟩ REACTION
                    (Text within parenthesis        )
                    (are not program syntax         )
[[ ACQUIRE-MODE-ON]] (Turns on acquire mode          )
                    (indicating all subsequent      )
                    (acquisitions are stored.       )
[[ 101]] ⟨TIMES⟩     (Begins loop and repeats 101    )
                    (times.                         )
```

```
[[ 0 CHOOSE-FILE        (Selects the datafile.        )
   1 SCAN-ACQUIRE       (Acquire one scan.           )
   10 SET-TABLE         (Select table ten.           )
   3 SET-PARAMETER      (Select parameter three, time )
SET-UP-PARAM-BOX        (Sets up parameter box.      )
4 ADD-TO-DATA-VALUE]]
                        (Add 1000 microseconds to    )
                        (time.                       )
⟨LOOP⟩                  (End loop.                   )
⟨RETURN⟩                (End of subroutine.          )
```

This program acquires a mass spectrum 101 times and, after each acquisition, the reaction time is increased by 1 ms so that the scans 1 to 101 correspond to reaction times of 0 to 100 ms. For an experiment of this kind, an essential modification would be to have an additional step before the reaction period to enable isolation of a single parent ion, ensuring a more clearly defined reaction sequence.

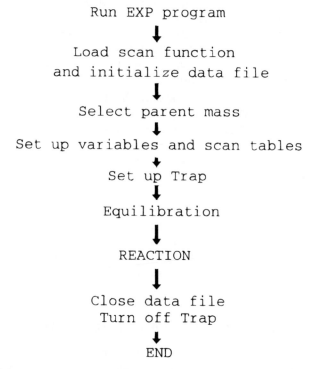

Run EXP program
↓
Load scan function
and initialize data file
↓
Select parent mass
↓
Set up variables and scan tables
↓
Set up Trap
↓
Equilibration
↓
REACTION
↓
Close data file
Turn off Trap
↓
END

FIGURE 11.4
A flow diagram showing how a key sequence program relates to a reaction time experiment.

A summarized flow diagram of such an experiment is shown in Fig. 11.4. Essentially, the program runs EXP, loads the scan function, initializes the data file, and calculates variables based upon the parent ion assigned earlier in the program. The program can then modify all the scan tables with parameters dependent on the parent mass; the trap is then set up by turning on the multiplier, RF drive potential, and filament. Scans carried out during the time allowed for the sample gas to attain equilibrium are not recorded; data are acquired subsequently during the reaction loop described in detail above.

The following example lists some of the FORTH commands used in a typical key sequence program.

```
⟨SUBROUTINE⟩ SET-UP-FILES
R EXP ⟨CR⟩              (Runs EXP represents the      )
                       (keys pressed in the manual   )
                       (operation.                   )
LOAD-SCAN-FUNCTION: SCANAME1
                       (Loads scan function          )
                       (called SCANAME1.             )
O INIT-DATAFILE: DATFILE
                       (Initializes datafile.        )
                       (called DATFILE1              )
O CHOOSE-FILE          (Selects the datafile         )
DATA-FILE-COMMENT "Methane"
                       (Inputs the file comment.     )
⟨SUBROUTINE⟩ SET-UP-TRAP
2000 SET-MULTIPLIER    (Sets the multiplier          )
                       (voltage to 2000 V            )
2 SET-FILAMENT         (Selects filament #2.         )
20 SET-MICROSCANS      (Sets number of microscans    )
                       (to 20.                       )
MULTIPLIER-ON          (Turns on multiplier.         )
FILAMENT-ON            (Turns on filament.           )
RF-ON                  (Turns on RF drive field.     )
CAL-GAS-ON             (Turns on calibration gas     )
                       (solenoid.                    )
```

IV. PROCESSING OF ITMS DATA FILES

The manner in which the ITMS data files are encoded so that the integrity of the data files is maintained can present serious problems when it is necessary to transfer data to a graph plotting package. However, there is a program called DWRITE produced by Finnigan MAT which can convert

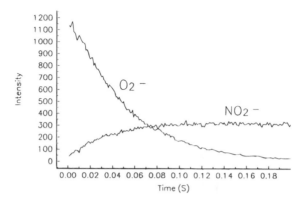

FIGURE 11.5
A plot of O_2^- intensity *versus* reaction time at $q_z = 0.3$; the plot shows a first-order exponential curve fit.

the encoded data files into a text file, although this is not in a standard ASCII format. Programs external to the ITMS software can be written to transform this text file into a format which can be read by a suitable g aph plotting package. For example, an ITMS data file containing reaction time data can be decoded by DWRITE and then converted by a simple Turbo BASIC program into an ASCII file containing a column for reaction time and columns for individual mass intensities; this data string can be read as a spreadsheet by a graph plotting package such as AXUM™ which enables manipulation of the data and presentation of the results as a graph.

V. APPLICATIONS

A. Ion Isolation and Reaction

An application of dynamically programmed scanning is the isolation and subsequent reaction of O_2^- in an atmospheric glow discharge ionization ion trap mass spectrometer.[2] From the resulting exponential ion signal intensity *versus* reaction time plot, Fig. 11.5, one can obtain the rate constant for the loss of O_2^- (which occurs through ion/molecule reactions to form NO_2^- and/or by electron detachment). A further example of the application of dynamically programmed scanning to the study of ion/molecule reactions carried out with an ITMS instrument is afforded by a gaseous mixture of silane and germane: the ions Si^+, SiH^+, SiH_2^+ and SiH_3^+ were isolated and a reaction acquisition was carried out for each ion.[3] The rate constant for the loss of parent ion and the formation of products could be calculated using a parallel reaction kinetic formula; the reaction for the Si^+ parent ion is illustrated in Fig. 11.6.

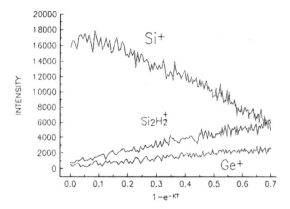

FIGURE 11.6
A kinetic plot of some of the product ions resulting from ion/molecule reactions of Si^+ with GeH_4; the rates of formation of the product ions may be calculated using a parallel reaction kinetics method.

B. Dynamically Programmed Scanning and Resonance Excitation

Resonance excitation, which is discussed in more detail in Chapter 2, allows an ion of single parent m/z value to be dissociated into smaller fragments, daughter ions, provided the frequency of the supplementary RF field applied between the end-cap electrodes is in resonance with the angular frequency of the axial secular oscillations of the parent ion. The scan function to carry out single stage resonance excitation (tickle), namely MS/MS, is shown in Fig. 11.7. The sequence of steps is ionization, ion isolation, and a tickle period, during which the tickle field is applied,

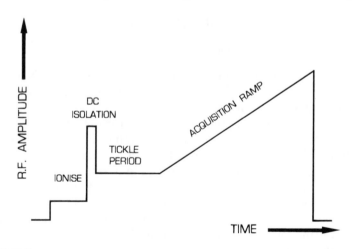

FIGURE 11.7
An MS/MS scan function representing ionization, DC isolation, and the tickle period.

after which the resulting daughter ions are scanned out of the trap and detected, and the mass spectrum is stored to a data file. The tickle frequency may be calculated theoretically (which the SCAN and EXP software does automatically once the correct *tickle mass* is set); however, it is common for the actual resonance frequency of the parent ion to be offset from the theoretical value due to variations in the helium pressure, space charge, and the inhomogeneity of the RF drive field due to the stretched geometry of the electrodes in the commercial ion trap (see Volume 1, Chapter 4).[4] Thus, under manual operation, the tickle frequency must be tuned by changing systematically the *delta* frequency offset until the parent ion becomes resonantly excited. On occasion, such tuning can be somewhat difficult to optimize successfully, but using dynamically programmed scanning, one can obtain the correct frequency quite rapidly by a slight modification of the syntax shown earlier in order to scan the tickle frequency. An example of such a key sequence program is as follows:

```
⟨SUBROUTINE⟩ DELTA
 [[ 10 SET-TABLE          (Select table ten.                      )
  5 SET-PARAMETER         (Select parameter 5, the Delta offset.  )
 SET-UP-PARAM-BOX         (Select parameter box.                  )
  -4 ADD-TO-DATA-VALUE    (Subtract 1000 μs.                      )
 ACQUIRE-MODE-ON ]]       (Turn on acquire model.                 )
 [[ 41 ]] ⟨TIMES⟩         (Loop 41 times.                         )
 [[ CHOOSE-FILE 0         (Select data file.                      )
  1 SCAN-ACQUIRE          (Acquire one scan.                      )
  10 SET-TABLE            (Select table ten and increment         )
  5 SET-PARAMETER         (the delta frequency by 100 μs.         )
 SET-UP-PARAM-BOX
  3 ADD-TO-DATA-VALUE ]]
 ⟨LOOP⟩                   (Terminate loop and end subroutine.     )
 ⟨RETURN⟩
```

Initially, the tickle frequency is decreased by 2 kHz and a mass spectrum is acquired; the process by which the tickle frequency is increased by 100 Hz and another mass spectrum is acquired, is repeated until the tickle frequency reaches a value 2 kHz higher than the original value. Thus, the total acquisition corresponds to a delta frequency scan from −2 to +2 kHz; an example of delta frequency scanning is shown in Fig. 11.8. It is seen that the loss of parent ion m/z 146 corresponds to the formation of the daughter ions m/z 111 and m/z 75.

C. Investigation of the Kinetics of Resonance Excitation

A further application of dynamically programmed scanning is the investigation of the kinetics of resonance excitation. In these experiments,[5]

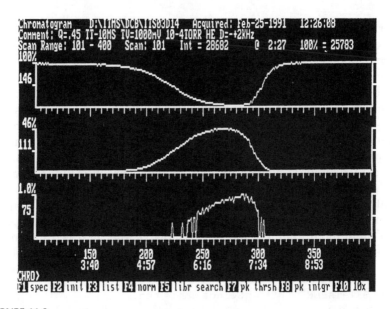

FIGURE 11.8

A mass chromatogram of 1,2 dichlorobenzene showing loss of parent ions and formation of daughter ions during the resonance period.

the aim is to follow the intensities of the parent ion, daughter ion, and total ion signals during resonant excitation as a function of the tickle period for a series of discrete tickle voltage amplitudes. This process may be automated by using the dynamically programmed scanning method summarized in Fig. 11.9. Essentially, the key sequence program is again a slight variation of the initial example and is shown below.

```
⟨SUBROUTINE⟩ TICKLE
[[ ACQUIRE-MODE-ON ]]        (Start acquiring.                          )
[[ 18 ]] ⟨TIMES⟩             (First loop 18 repetitions.               )
[[ 201 ]] ⟨TIMES⟩            (Nested loop 201 times.                   )
[[ CHOOSE-FILE 0
   1 SCAN-ACQUIRE
  10 SET-TABLE
   3 SET-PARAMETER
SET-UP-PARAM-BOX ]]
[[ 5 ]] ⟨TIMES⟩              (Second nested loop 5 times.              )
[[ 2 ADD-TO-DATA-VALUE ]]    (Increments tickle time by 100 μs.       )
⟨LOOP⟩                       (Terminate second nested loop.            )
⟨LOOP⟩                       (Terminate first nested loop.             )
[[ 10 SET-TABLE
   6 SET-PARAMETER
```

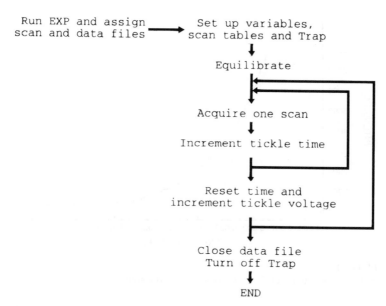

FIGURE 11.9

A flow diagram of the key sequence program corresponding to a range of tickle times being used for each tickle voltage.

```
SET-UP-PARAM-BOX
    3 ADD-TO-VALUE-DATA         (Increment tickle voltage.        )
    3 SET-PARAMETER ]]          (Re-select tickle time.           )
  [[ 10 ]] ⟨TIMES⟩
  [[ -4 ADD-TO-DATA-VALUE ]]    (Reset time to zero.              )
  ⟨LOOP⟩                        (Terminate first nested loop.     )
  ⟨LOOP⟩                        (Terminate main loop.             )
  ⟨RETURN⟩                      (terminate subroutine.            )
```

Thus, a tickle-time curve is acquired for each tickle voltage for a series of differing tickle voltages within the same data file, and once this data file has been decoded by DWRITE, a Turbo BASIC file can convert the data into separate ASCII files for each voltage or as one large ASCII file; the latter is used for plotting as a three-dimensional graph, or the former is used for individual plots for kinetic calculations on AXUM, as shown in Figs. 11.10 and 11.11, respectively.

D. Automated MS/MS/MS

Dynamically programmed scanning has the potential to carry out more sophisticated functions involving decision-making. One application

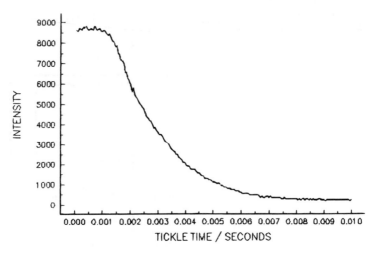

FIGURE 11.10

A plot of m/z 146 parent ion intensity *versus* tickle duration at q_z = 0.45.

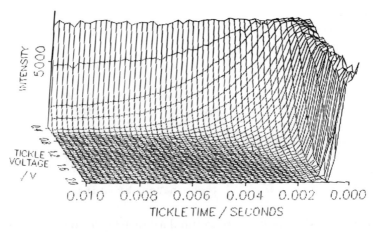

FIGURE 11.11

A three-dimensional surface plot of the signal intensity of m/z 146 from 1,2 dichlorobenzene as a function of both tickle duration and tickle voltage amplitude.

developed at the University of Kent uses semi-intelligent programs to carry out automated MS/MS/MS in an ion trap mass spectrometer. The method is based upon that employed by Cooks *et al.* on a penta-quadrupole instrument,[6] generating fixed-daughter and fixed-parent familial scans. For the fixed-parent method, a granddaughter spectrum is obtained for each daughter ion of a single parent ion; for the fixed-daughter method, a granddaughter spectrum is obtained for a single daughter ion species obtained from several parent ions. Unlike the penta-quadrupole experi-

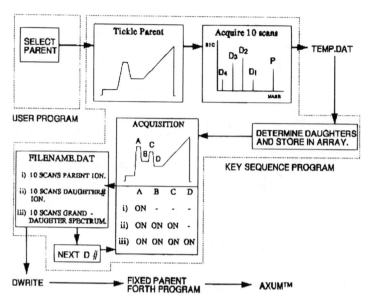

FIGURE 11.12
Flow diagram for fixed-parent ion familial scanning.

ment, in which a granddaughter spectrum is obtained for each parent or daughter m/z value whether or not an ion species exists at each m/z value, the ITMS experiment carried out with dynamically programmed scanning utilizes a semi-intelligent method to determine the possible precursor m/z values prior to analysis. The operation of a fixed-parent program is illustrated in Fig. 11.12. Initially, the parent mass is entered to the user program; the required RF and DC values are calculated for isolation and excitation of this parent ion, before loading the key sequence program. The first scan function illustrated in Fig. 11.12 is produced from these values and the parent ion is tickled for 10 scans and stored to a temporary data file. The mass spectra are averaged and all masses lower than the parent intensity and higher than a preset percentage intensity threshold are stored in an array. The program then enters a loop where (i) the DC is toggled on at point A and ten scans of the parent intensity are acquired; (ii) the tickle voltage and DC are toggled on at points B and C to tickle the isolated parent and isolate and acquire ten scans of the first daughter ion; and (iii) the tickle voltage at point D is switched on and ten scans of the granddaughter spectrum are acquired. This process is repeated for each daughter ion stored in the array, after which the data file is saved and the program ends. A similar process is used in the fixed-daughter familial scanning program shown in Fig. 11.13, but in this instance, the user program stores the fixed-daughter mass before acquiring ten scans of the primary spectrum. These mass spectra are averaged and the potential

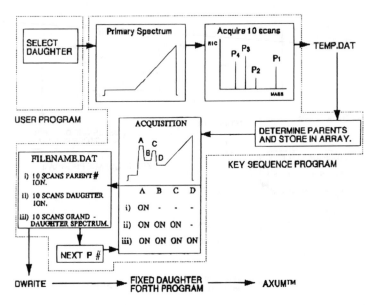

FIGURE 11.13
Flow diagram for fixed-daughter ion familial scanning.

parent masses above that of the fixed-daughter mass and above a set intensity threshold are stored in an array before acquiring MS/MS/MS spectra in the same way as in the fixed-parent method, but at this stage, only the fixed-daughter ion is isolated at step C, and the acquisition is repeated for each parent ion.

These two methods have been used successfully to obtain experimental data on various test compounds. Fig. 11.14 shows a three-dimensional mass spectral plot obtained by carrying out a fixed-parent familial scan on the parent ion m/z 314 from perfluorotributylamine. A similar plot is shown in Fig. 11.15 where a fixed-daughter familial scan was carried out on the daughter ion m/z 57 from n-decane.

E.   Mapping the Stability Diagram

A further application of the dynamically programmed scanning technique is in mapping the stability diagram and its boundary coordinates. The geometry of the commercial ion trap electrodes is "stretched" with respect to a pure quadrupolar configuration, as noted in Chapter 1. As the trapping field within the ion trap is distorted by the stretched geometry, there are changes in the boundary co-ordinates of the stability diagram; attempts have been made to use dynamically programmed scanning experiments to map these coordinates.[7] Due to the nature of the scan functions employed here, the data for the $\beta_z = 1$ boundary are rather

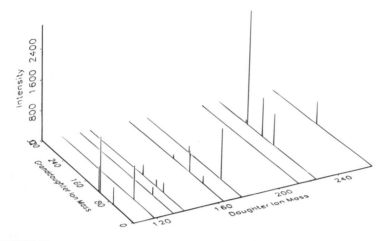

FIGURE 11.14

A three-dimensional spectral plot resulting from a fixed-parent scan of m/z 314 from perfluorotributylamine.

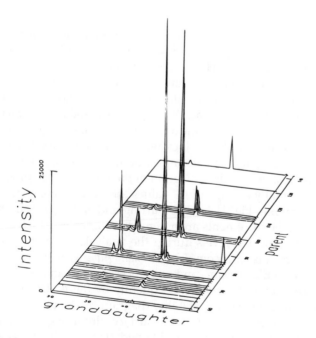

FIGURE 11.15

A three-dimensional mass spectral plot resulting from a fixed-daughter scan of m/z 57 from n-decane.

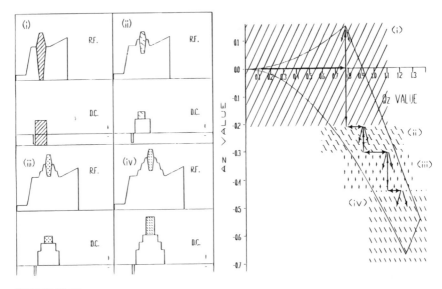

FIGURE 11.16

A diagram of the four scan functions required to map the stability diagram. The shaded regions correspond to the range of RF and DC values required to map the similarly shaded regions shown on the stability diagram.

imprecise. However, a more accurate redetermination in our laboratory[8] has utilized a more complex set of scan functions which were essentially the same as those used to characterize "black canyons" (see below).

At least four different scan functions are required to map accurately the four distinct regions of the stability diagram shown by the shaded regions in Fig. 11.16. The RF and DC amplitudes used in the scan functions are depicted on the left side of the diagram: the shaded regions correspond to the range of RF and DC values which can be mapped without ions leaving the stability region before they occupy the a_z, q_z coordinates currently under analysis. A flow diagram of the key sequence program used to scan sequentially a range of RF values after each increment of the DC voltage is shown in Fig. 11.17. The starting and ending q_z and DC values together with the increments are assigned to variables which are used to modify the RF and DC levels during the analysis stage of the program. This process results in a complex '*chromatogram*' which, after decoding with the DWRITE program, can be reprocessed into an ASCII format by a BASIC program and plotted subsequently by the AXUM program.

To summarize, the entire stability diagram can be mapped using just four scan functions as shown by the surface plot in Fig. 11.18, and from this plot, the boundary may be obtained (Fig. 11.19). Alternatively, a smaller region of the stability diagram can be mapped in greater detail to show fine structure such as the nonlinear octapolar resonance at $\beta_z =$

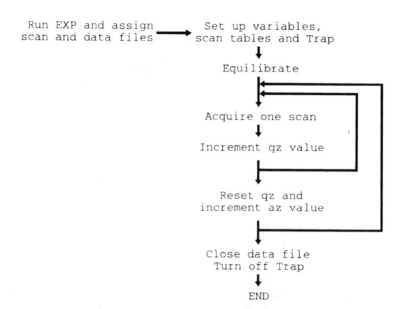

FIGURE 11.17
A flow diagram of the key sequence program which permits mapping of the stability diagram.

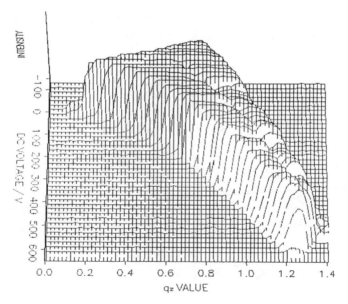

FIGURE 11.18
A surface plot of the entire stability diagram.

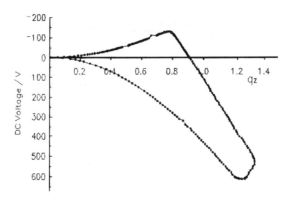

FIGURE 11.19
A plot of the boundary of the stability diagram corresponding to a signal intensity of m/z 134 from n-butylbenzene of 10% of the average level.

0.5 found at the lower apex of the stability diagram, Fig. 11.20. These experiments were carried out on m/z 134 from n-butylbenzene at a helium pressure of 10^{-4} Torr and with a residence time of 1 ms.

F.   Investigation of "Black Canyons"

Experiments carried out by Guidugli and Traldi[9] on the MS/MS of chalcone showed that upon tickling the parent ion over a series of q_z

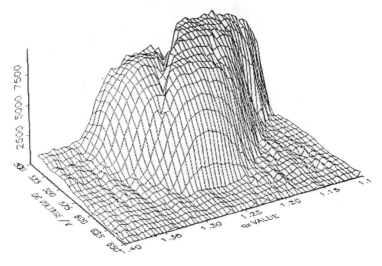

FIGURE 11.20
A three-dimensional plot of the lower apex of the stability diagram illustrating the ability of the key sequence program to map accurately a small section of the stability diagram.

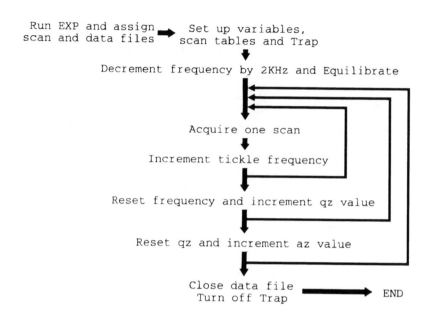

FIGURE 11.21
A flow diagram of the key sequence program which dynamically scans a small region of
the stability diagram in the search for black holes.

values, the daughter ion intensity was severely reduced when the a_z, q_z
co-ordinates of the daughter ion fell within a particular range of q_z values
around $q_z = 0.78$ ($a_z = 0$). The investigation involved spending a considerable amount of time in obtaining daughter ion intensities over a range
of a_z and q_z values and, when these intensities were plotted, the data
showed a significant drop in daughter ion intensity over a roughly circular
range of a_z, q_z values leading to the phenomenon being termed a "black
hole." Further experiments carried out in our laboratory[10] in collaboration
with the Padova group involved tickling m/z 148 from 1,2-dichloroben-
zene to produce daughter ions at m/z 111 and 113 in the intensity ratio
of 1:1. The value of q_z for the 148^+ ion during tickling was adjusted using
a dynamically programmed scan so that the intensity ratio $111^+/113^+$ could
be monitored as the a_z, q_z co-ordinates for the two daughter ions were
moved across the black hole. The results obtained validated the earlier
observations. A more complex key sequence program was then written
to map out the black hole over a wider area of the stability diagram. This
program (see Fig. 11.21 for the flow diagram) was designed to overcome
the change in the resonant frequency (which resulted from changing the
q_z value at which MS/MS occurred) by carrying out automatically, during
one acquisition, a frequency scan for both each q_z value under investigation
and for a range of q_z values for each of several a_z values. With the aid of
a Turbo BASIC program, the daughter ion intensities were obtained from

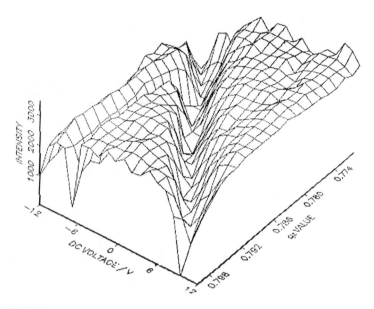

FIGURE 11.22
A three-dimensional plot of the black canyon obtained using dynamically programmed scanning.

each (delta) frequency scan at a frequency corresponding to the maximum loss of parent intensity, so that the plot shown in Fig. 11.22 was obtained. The results indicated that ion loss occurs over a narrow band of a_z,q_z co-ordinates which extended to lower regions of the stability diagram; hence, the phenomenon was renamed a "black canyon." This extension follows the line of the octapole non-linear resonance at $\beta_z = 2/3$, which is discussed in detail in Volume 1, Chapter 3.[4]

It should be noted that in the investigation of the lower regions of the stability diagram, the key sequence program was modified further on account of the considerable differences between the theoretical resonance frequencies calculated (for $a_z = 0$) by the system software and the values required at lower a_z co-ordinates. After each frequency scan, the frequency was changed by a constant amount so as to retain the parent ion axial secular frequency within the delta frequency range. A more detailed mapping of the stability diagram (sometimes using quadrupolar tickling) revealed other black canyons which can be attributed to the presence of nonlinear resonances coinciding with $\beta_z = 1/2$ and $\beta_z = 2/3$ (hexapole and octapole resonances, respectively).

VI. CONCLUSIONS

In this chapter, we have sought to show how the operation of the ion trap may be automated in order to acquire mass spectra under a range of operating conditions without repeated direct intervention by the operator. Through the normal techniques of programming in FORTH, it is possible to incorporate decision-making algorithms which confer varying degrees of machine intelligence into data gathering. Perhaps the most significant application of such power would be in the area of multiple tandem mass spectrometry (MS/MS/MS, and even higher orders), where one could envisage, for example, the combination of tailor-made scan functions for collisional dissociation with spectral library-searching for the sequencing of peptides or other biological macromolecules.

REFERENCES

1. Todd, J. F. J.; Penman, A. D.; Thorner, D. A.; Smith, R.D. *Rapid Commun. Mass Spectrom.* 1990, *4*, 108.
2. Murrell, J.; Langford, M. L.; Todd, J. F. J.; Watts, P. *Proc. 20th Meeting British Mass Spectrom. Soc.* Canterbury, U.K., September 1993, p. 271.
3. Operti, L.; Splendore, M.; Vaglio, G.A .; Franklin, A. M.; Todd, J. F. J. *Int. J. Mass Spectrom. Ion Processes,* 1994, 136, 25.
4. *Practical Aspects of Ion Trap Mass Spectrometry,* R. E. March and J. F. J. Todd, Eds., Vols. 1 and 2, "Fundamentals and Instrumentation", Modern Mass Spectrometry series, CRC Press, Boca Raton, FL, 1995.
5. Franklin, A. M.; Todd, J. F. J.; March, R. E. *Proc. 19th Meeting British Mass Spectrom. Soc.* St. Andrews, U.K., September 1992, p. 261.
6. Schwartz, J. C.; Schey, K. L.; Cooks, R. G. *Int. J. Mass Spectrom. Ion Processes.* 1990, *101,* 1.
7. Johnson, J. V.; Pedder, R. E.; Yost, R. A. *Rapid Commun. Mass Spectrom.* 1992, 6, 760.
8. Franklin, A. M.; Clarke, N. J.; Murrell, J.; Todd, J. F. J. *Proc. 20th Meeting British Mass Spectrom. Soc.* Canterbury, U.K., September 1993, p. 171.
9. Guidugli, F.; Traldi, P. *Rapid Commun. Mass Spectrom.* 1991, *5,* 343.
10. Guidugli, F.; Traldi, P.; Franklin, A. M.; Langford, M. L.; Murrell, J.; Todd, J. F. J. *Rapid Commun. Mass Spectrom.* 1992, *3,* 229.

Part 4

ENVIRONMENTAL AND BIOMEDICAL APPLICATIONS

Chapter 12

ENVIRONMENTAL ANALYSES

William L. Budde

CONTENTS

0-8493-8251-3/95/$0.00+$.50
© 1995 by CRC Press, Inc.

429

I. INTRODUCTION

As documented by Syka[1] the first commercial ion trap mass spectrometer made its debut during the mid-1980s as a gas chromatography (GC) detector, the so-called ion trap detector (ITD). Later, the more versatile and capable ion trap mass spectrometer (ITMS) was introduced. But nearly all environmental applications to date have used the GC/ITD system. Because the ITMS is of the same fundamental design as the ITD, generally no distinction will be made between these closely-related instrument systems in this chapter. When a particular environmental application depends on a feature found only on the ITMS, that fact will be noted.

The introduction of the ion trap occurred during a period of unprecedented growth in the applications of mass spectrometry for environmental analyses. Nearly all of this growth, at least in terms of the number of commercial systems sold, was in instrument systems consisting of a capillary column GC combined with a conventional single quadrupole mass spectrometer (GC/MS). Consequently, when the GC/ITD was introduced and proposed as a useful tool for environmental analyses, a competitor was already in place and its use was growing rapidly. This competitive atmosphere was not unlike the one that existed during the early 1970s when the quadrupole mass spectrometer was introduced and proposed as a substitute for the established magnetic sector and time-of-flight mass spectrometer systems.

II. MASS SPECTROMETER PERFORMANCE STANDARDS

A. Performance of the GC/MS (Quadrupole Mass Filter) System

The quadrupole mass filter GC/MS system was accepted only very slowly during the 1970s. The difficulties encountered and overcome by its proponents are analogous to the problems faced by the GC/ion trap developers during the second half of the 1980s and early 1990s. Many questions were raised about the performance of quadrupole GC/MS systems during the 1970s and many of these same questions were raised later about the ion trap systems. Several of the major performance issues of the 1970s are discussed in question-and-answer form below.

1. Stability

Were the new systems sufficiently stable over a reasonable time-frame, for example, an eight-hour work day, to permit reliable mass calibration and 100% accurate integer mass assignments? Early quadrupole systems included power supplies, amplifiers, and radio frequency (RF) modules that contained relatively unstable vacuum tube components, and these were not replaced completely in all designs until the 1980s. Also note that the effects of accumulating fractional masses (mass defects) and doubly-charged ions were not well-considered in some calibration procedures.

2. High-Mass Sensitivity

Was the transmission or detection of higher-mass ions, that is, those ions having mass/charge ratios greater than about 200 Da/charge, sufficient to give the enhanced high-mass sensitivity needed for environmental analyses? Many compounds of considerable environmental interest were known to be polychlorinated and/or polybrominated and had molecular weights in the 200 to 900 Da range. High-mass detection was relatively poor on early quadrupole systems or low-mass detection was so superior that higher-mass detection seemed relatively poor.

3. Fragmentation Patterns

Were the fragmentation patterns produced by 70 eV electron impact (EI) ionization similar to those produced by the established magnetic sector and time-of-flight mass spectrometer systems? If not, would the existing database of EI mass spectra be rendered useless? Much discussion of this issue occurred and many papers included qualifying phrases like *quadrupole spectra* and *magnetic spectra*. Many investigators were concerned with the differences in mass spectra and, later, the phrase *ion trap spectra* came into use.

4. Maintenance

What ion source and quadrupole maintenance was required to maintain good performance, and was the frequency of such maintenance reasonable and cost-effective? Quadrupole mass filters had rods that were contaminated by the discharge of filtered ions and this contamination affected performance.

B. The Role of *bis* (Perfluorophenyl) Phenylphosphine

Because of these concerns, and others, a small multi-laboratory study was organized in 1972 to compare spectra measured with quadrupole and magnetic sector GC/MS systems.[2] A test compound, *bis* (perfluorophenyl) phenylphosphine (A), also known as decafluorotriphenylphosphine (DFTPP), shown in Scheme 12.1, was selected because it had a number of favorable properties including a rich and interesting EI mass spectrum. In addition to an abundant molecular ion at 442 Da, the mass spectrum contained about 15 fragment ions fairly evenly distributed over the mass range of 50 to 442 Da.

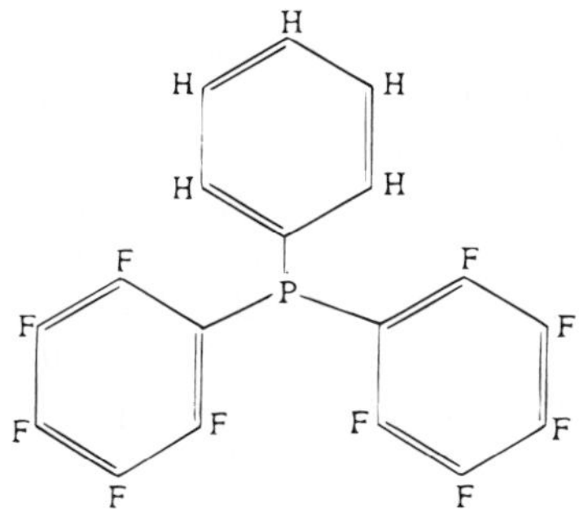

A

SCHEME 12.1

C. Multi-Laboratory Study

In the above small multi-laboratory study which was organized in 1972, the test compound was introduced into 4 magnetic sector instruments and 11 quadrupole instruments through their packed column GC inlet systems. The participating laboratories were asked to adjust or tune their instruments according to whatever procedures were used commonly in their laboratories.

The results of the study indicated clearly the need for standardization of the tuning of these GC/MS systems and for a systematic program of spectral quality control.[2] Differences in relative abundances (RAs) in spectra from all systems were significant, and even among the quadrupole systems of the same design, differences in RAs were substantial. While it was well known that users of magnetic sector instruments could affect some changes in RAs by optimizing source potentials in different ways, this study made it clear that users of early quadrupole GC/MS systems had tuning flexibility to produce just about any RA at virtually any mass. This situation was an intolerable one for instrument users who would need eventually to depend heavily on standard databases of reference spectra for compound identification. Without standardization of spectra, there was little hope that GC/MS could be employed for large-scale environmental monitoring, especially for regulatory purposes.

D. Standardization and Spectral Quality Control Proposal

In order to promote standardization and spectral quality control, compound A, (DFTPP), was proposed as a GC/MS system performance test or reference compound.[2] The proposal was that, after all calibrations and tuning adjustments were completed, the user would introduce DFTPP into the GC/MS *via* the same GC inlet that would be used for sample analysis, and verify that the instrument was performing according to specifications. The proposed specifications included a set of key DFTPP ions and corresponding RA ranges as well as checks on sensitivity, chromatographic performance, etc. The RA criteria were developed from a statistical analysis of the data obtained in the multi-laboratory study. The compound DFTPP was never intended for use as an actual calibration compound, nor has it been used as such, although it has frequently been referred to mistakenly as a calibration compound. These proposals were made with several objectives in mind:

1. The regular overall performance check would provide the needed GC/MS system quality control, a concept not popular at the time, but one that would later have its day.

2. The RA criteria would provide guidance to instrument users who could tune their equipment to give RAs in the recommended ranges. Consequently, this tuning procedure would provide some minimum degree of standardization for 70 eV EI spectra from GC/MS systems, and would permit efficient identification of compounds by comparison of the mass spectra observed with those in the historical database spectra.

3. The RA criteria would serve to guide instrument designers and manufacturers and to encourage the production of commercial GC/MS systems with the capability of producing similar EI mass spectra relatively easily.

After publication in 1975,[2] this standardization and quality control proposal lay dormant for four years except for some internal use in the U.S. Environmental Protection Agency (U.S. EPA).

E.   Clean Water Act of 1977

On December 3, 1979, the U.S. EPA proposed guidelines establishing test procedures for the analysis of pollutants.[3] These test procedures were required in order to implement provisions of the Clean Water Act of 1977. Among the proposed test procedures was U.S. EPA Method 624 for the determination of a group of 30 volatile organic compounds (VOAs), mostly organic solvents, by GC/MS. Included also in the proposal was U.S. EPA Method 625 for the determination of a group of 83 organic bases, neutral compounds, acids, and pesticides by GC/MS. These methods were pioneering in that they included, for the first time, mandatory quality control requirements together with regular verification of the performance of GC/MS systems using the performance test compound, DFTPP.

The 1979 regulatory proposals caused considerable consternation among manufacturers and GC/MS system users. After the Clean Water Act regulations became final in 1984,[4] system designers and manufacturers either had to deliver commercial products that could meet the DFTPP specifications, or risk losing a significant portion of the environmental business. Other U.S. EPA regulatory and monitoring programs followed the lead in the Clean Water Act regulations and adopted the same or very similar standardization and quality control specifications. A second test compound, 4-bromo-fluorobenzene (BFB), was introduced in 1984[4] for use in methods for determination of the VOAs. This compound was a VOA and was more convenient than DFTPP in the VOA methods, but served the same purpose.

III. THE GC/ION TRAP DETECTOR

With the delivery of the Finnigan Corporation's first commercial capillary GC/ITD systems in early 1985, several investigators began evaluations of the system with some attention to the performance issues listed above.

A. Performance of the GC/MS (Ion Trap Detector) System

The GC/ITD suffered from the usual problems of new instrumentation such as air leaks, decomposition of some analytes on hot metal surfaces, etc., but these were solved in due course. Several more fundamental performance observations are discussed, in turn, below.

1. Relative Abundances

The precision of RA measurements from repetitive injections into the capillary GC of small quantities (5 ng) of DFTPP and BFB was good, ranging from 3 to 26% relative standard deviation (RSD).[5] With the exception of a slightly excessive signal intensity of the ion of 51 Da, the GC/ITD instrument met all other BFB specifications cited in the final 1984 version of Method 624.[4,5] The RA of the ion of 51 Da was considered to be but a minor deviation from the specifications.

2. Mass Spectrum of DFTPP

The mean GC/ITD mass spectrum of DFTPP did not conform to the 1984 Method 625 specifications.[4,5] With the ion trap, the molecular ion at 442 Da/charge was often the base peak, whereas the specifications called for a base peak which corresponded to the m/z 198 fragment ion. It was not known whether this change of base peak was caused by exceptional formation and storage of the molecular ion in the ion trap or by reduced fragmentation in the ion trap, although the latter was the more likely.

3. Sensitivity

Overall system sensitivity was impressive and a recognizable complete mass spectrum of DFTPP was measured with 100 pg injected into the GC/ITD.[5]

4. Anomalies in the Mass Spectra of Polycyclic Aromatic Hydrocarbons

When polycyclic aromatic hydrocarbons (PAHs) and several other classes of compounds were introduced into the GC/IDT at levels of about 50 ng or higher, elevated abundances of $[M + 1]^+$ ions were observed in mass spectra displayed as bar graphs.[5] When the quantities injected were reduced to 10 ng or lower, near normal $[M + 1]^+$ ion abundances were observed in mass spectra displayed as bar graphs. The PAHs also gave lower than expected abundances of M^{2+} ions and unexpected significant abundances of apparently $[M^{2+} + 1]$ ions in spectra displayed as bar graphs. The unusual effects were attributed to space charging, loss of resolution, and the data acquisition algorithm used in the data system.

5. Self-Chemical Ionization

Intense $[M + 1]^+$ ions were observed also in the mass spectra of dicyclohexylamine and several methyl esters, and these ions were attributed to "self-chemical ionization" (self-CI) of the molecules by hydrogen-rich fragment ions in the ITD.[6]

B. The Necessity for Automatic Gain Control (AGC)

The above observations of self-CI impeded significantly the development of applications of the ion trap in environmental research, monitoring, and the rapidly-growing regulatory surveillance programs. Clearly, the GC/ITD had a lower-working concentration range than other types of GC/MS systems. Space charging and self-CI effects could occur with quantities of compounds well below levels that are readily separated in narrow-bore fused silica capillary GC columns. Further applications of the ion trap had to await the development of the AGC system[7] to control space charging and, in part, to reduce self-CI.

IV. ANALYTICAL APPROACHES IN ENVIRONMENTAL ANALYSES

The U.S. environmental regulatory initiatives of 1979 to 1984[3,4] also had a major impact on the analytical approaches used in environmental analyses. Two distinct approaches emerged which were characterized as the broad-spectrum and target-analyte approaches.[8]

A. Broad-Spectrum Approach

In a broad-spectrum approach, the goal is to determine a broad range of analytes that have generally similar physical or chemical properties. This traditional analysis is not guided by a predetermined list of analytes to be measured, but seeks to answer the question, "What is in the sample"? This approach is often used in environmental research and was facilitated by the development of powerful instrumental tools including the computer-controlled GC/MS systems.

B. Target-Analyte Approach

The target-analyte approach is also traditional in chemical analysis and is accepted widely in other sciences and the regulatory community. A list of the target analytes, whose concentrations are desired, is submitted to the analytical laboratory, and the methodology used by the laboratory is optimized for the target analytes. This approach was used by the U.S. EPA in its regulatory process and, after 1979, tended to dominate the environmental applications of mass spectrometry. The ultimate in target-compound analysis is the selected ion monitoring (SIM) technique widely used in mass spectrometry.

Because the target-analyte approach came to be so important in U.S. environmental regulatory programs, nearly all of the ion trap applications discussed in this chapter will emphasize that approach. But the discovery of new environmental pollutants depends on broad spectrum analyses which should be encouraged in environmental research, monitoring, exposure assessment, and other programs.

V. DRINKING WATER ANALYSES IN THE U.S.

A. Safe Drinking Water Act of 1986

The development of the AGC system solidified the performance of the GC/ITD and permitted the utilization of the ion trap in several existing analytical methods. In 1987, the U.S. EPA completed the first of several new phases of regulations concerning contaminants in the nation's drinking water supplies.[9] These regulations were required under provisions of the Safe Drinking Water Act of 1986. Although in a few earlier regulatory initiatives, analytical detection limits were used for regulatory levels, this approach was largely discredited in the 1980s. By the mid-1980s, the U.S. EPA had developed a health-risk based approach to setting maximum contaminant levels (MCLs) for regulatory purposes, and this approach was used to set the drinking water standards. Among the contaminants regulated in 1987 were 8 VOCs with MCLs set in the range of 2 to 200 μg/L. A further requirement of the 1987 regulation was regular monitoring of an additional 51 VOCs. During the last five years, drinking water MCLs were added for another 17 of the 59 VOCs. The complete current list of VOCs with monitoring requirements and MCLs is given in Table 12.1.

B. Analytical Methods for Monitoring VOCs in Drinking Water

Included with the 1987 drinking water regulation[9] was the identification of a group of analytical methods which were required for monitoring the 59 VOCs in drinking water (Table 12.1). No other method was permitted for regulatory purposes unless it could be proven statistically that the alternative method was equivalent to or better than the approved methods. Among the approved methods was U.S. EPA Method 524.2 which was basically the same as GC/MS Method 624, except that the GC requirement had been upgraded to a high-resolution capillary column, and the detection limits were lower. Because of the sensitivity required for some analytes (Table 12.1), the ion trap was a natural choice for validation with Method 524.2. The method required the GC/MS system to meet the BFB specifications, which were well within the capability of the existing GC/ITD system.[10]

Method 524.2 (and the older Method 624 as well) uses an inert gas purge of a 5- or 25-mL water sample to separate the VOCs from the aqueous phase and to entrain the volatiles in a vapor stream that is passed through a three-stage solid-phase trap. A trap consisting of equal parts of Tenax (2,6-diphenylene oxide polymer), silica gel, and coconut charcoal has proven very effective for a wide variety of compounds. The trapped VOCs are desorbed thermally during backflushing with helium, and the desorbed vapor is injected into the capillary GC/MS system for separation, identification, and quantitative analysis.[10] Instruments offered now by manufacturers have automated systems to accomplish calibration, multiple analyses, and complete reporting with minimal operator intervention.

TABLE 12.1

Volatile organic compounds regulated or monitored in drinking water supplies in the U.S.

Compound	Maximum Contaminant Level (µ/L)
Benzene	5
Bromobenzene	a
Bromochloromethane	a
Bromodichloromethane	b
Bromoform	b
Bromomethane	a
n-Butylbenzene	a
sec-Butylbenzene	a
tert-Butylbenzene	a
Carbon tetrachloride	5
Chlorobenzene	100
Chloroethane	a
Chloroform	b
Chloromethane	a
2-Chlorotoluene	a
4-Chlorotoluene	a
Dibromochloromethane	b
1,2-Dibromo-3-Chloropropane	0.2
1,2-Dibromoethane	0.05
Dibromomethane	a
1,2-Dichlorobenzene	600
1,3-Dichlorobenzene	a
1,4-Dichlorobenzene	75
Dichlorodifluoromethane	a
1,1-Dichloroethane	a
1,2-Dichloroethane	5
1,1-Dichloroethene	7
cis-1,2-Dichloroethene	70
trans-1,2-Dichloroethene	100
Dichloromethane	5
1,2-Dichloropropane	5
1,3-Dichloropropane	a
2,2-Dichloropropane	a
1,1-Dichloropropene	a
cis-1,3-Dichloropropene	c
trans-1,3-Dichloropropene	c
Ethylbenzene	700
Hexachlorobutadiene	a
Isopropylbenzene	a
4-Isopropyltoluene	a
Naphthalene	a
n-Propylbenzene	a
Styrene	100
1,1,1,2-Tetrachloroethane	a
1,1,2,2-Tetrachloroethane	a
Tetrachloroethene	5
Toluene	1000
1,2,3-Trichlorobenzene	a

TABLE 12.1

(*Continued*)

Compound	Maximum Contaminant Level (μ/L)
1,2,4-Trichlorobenzene	70
1,1,1-Trichloroethane	200
1,1,2-Trichloroethane	5
Trichloroethene	5
Trichlorofluoromethane	a
1,2,3-Trichloropropane	a
1,2,4-Trimethylbenzene	a
1,3,5-Trimethylbenzene	a
Vinyl Chloride	2
o-Xylene	d
m-Xylene	d
p-Xylene	d

[a] Monitoring requirement only.
[b] Sum of four trihalomethanes shall not exceed 100 μg/L from a 1979 regulation.
[c] Monitoring requirement only but *cis/trans* pair not distinguished in the regulation.
[d] Sum of three xylenes shall not exceed 10,000 μg/L.

C. GC/ITD Analysis of the VOCs Given in Table 12.1

The GC/ITD performed very well during repetitive measurements of the Method 524.2 target analytes at 2 and 0.2 μg/L with a 5-mL water sample.[11]

1. 2 μg/L Level

At the 2 μg/L level, the grand mean measurement accuracy for 54 compounds was 95% of the true value with a mean relative standard deviation (RSD) of 4%. Several of the VOCs were not included in the study because authentic standards were not available at the time of this validation. As m-xylene and p-xylene are not resolved chromatographically, they were reported as a single component.

2. 0.2 μg/L Level

At 0.2 μg/L with a 5-mL water sample, the grand mean measurement accuracy for 52 compounds was 95% of the true value with a mean RSD of 3%. Two compounds, chloromethane and dichlorodifluoromethane, were not detected at the lower concentration. These compounds are gases at room temperature and are difficult to detect because of their high vapor pressures and relatively poor retention on the solid-phase trap. The RSDs from repetitive determinations of dichlorodifluoromethane and another gas, bromomethane, were 25% and 27%, respectively, at 2 μg/L. This relatively high

TABLE 12.2

Volatile organic compounds added to the Method 524.2, Revision 4.0
Analyte List

Compound	Compound
Acetone	2-Hexanone
Acrylonitrile	Iodomethane
Allyl chloride	Methacrylonitrile
2-Butanone	Methyl acrylate
Carbon disulfide	Methyl methacrylate
Chloroacetonitrile	4-Methyl-2-pentanone
1-Chlorobutane	Methyl tert-butyl ether
trans-1,4-Dichlorobutene-2	Nitrobenzene
1,1-Dichloropropanone	2-Nitropropane
Diethyl ether	Pentachloroethane
Ethyl methacrylate	Propionitrile
Hexachloroethane	Tetrahydrofuran

variability is another indication of the difficulty of measuring extremely
volatile substances at very low concentrations with Method 524.2.

D.   GC/ITD Analysis of the VOCs Given in Table 12.2

In recent work, 48 additional potential analytes were evaluated for
determination by Method 524.2 using a GC/ITD.[12] By contrast with the
compounds in Table 12.1, the 48 candidates included many relatively
polar water-soluble compounds which are difficult to separate from the
water matrix. Factors considered during the evaluation included purging
efficiency, sensitivity, linearity of response, precision and accuracy of measurements, method detection limits, sample preservation and storage, and
matrix effects. This study demonstrated that 24 of the candidates could
be determined with acceptable precision, accuracy, and detection limits
of 1 μg/L or lower. The additional 24 Method 524.2 analytes are listed
in Table 12.2 and are included in Revision 4.0 of Method 524.2.[13]

E.   Summary of GC/ITD Performance with VOCs Given in Tables 12.1
and 12.2

With an analyte list of 84 target compounds, the mass spectrometric
detector is an essential part of Method 524.2. While no environmental
samples are expected to contain all of the 84 analytes, calibration mixtures
containing all analytes must be analyzed and any reasonable trade-off
between analysis time and chromatographic resolution will leave some
analytes unresolved. Under the conditions recommended for Revision
4.0, seven pairs and one group of three compounds co-eluted. However,

each member of the co-eluting pairs or group had a sufficiently different mass spectrum such that unequivocal identification of each member could be made; in addition, accurate calibration and measurement of each analyte was carried out.

F. A Multi-Laboratory Study of Method 524.2 With VOC Analytes

During 1992, the U.S. EPA and the American Society for Testing and Materials (ASTM) Committee D-19 on Water conducted a joint multi-laboratory study of Method 524.2 with 68 of the VOC analytes.[14] Over 40 volunteer laboratories participated in the study to characterize the performance of Method 524.2 in terms of accuracy, precision, and detection limits. Analyses were conducted using fortified reagent water, drinking water, groundwater, several industrial waste waters, and a simulated hazardous waste site aqueous leachate. Fortified analyte concentrations ranged from 0.2 to 80 μg/L and, generally, excellent accuracy and precision was reported. One GC/ITD system was used in the study and data from it were included in the overall statistical summary of the study results.[14]

VI. MEMBRANE MASS SPECTROMETRY

A. VOCs

Although Method 524.2 is an excellent laboratory procedure, it is not well suited to real-time or continuous monitoring of VOCs in water because a typical analysis of a so-called grab sample requires about 30 min. Although real-time monitoring is not required currently for regulatory purposes, many organizations are interested in continuous information about discharges of VOCs in waste waters, the performance of treatment processes for removal of VOCs, the monitoring of systems which may create VOCs in water, and pollution prevention through real-time control of aqueous process streams.

Continuous sampling of aqueous streams using semipermeable hydrophobic membranes has been under development for about 20 years, and mass spectrometry has been applied to continuous monitoring of the permeate from aqueous streams.[15] The VOCs are attractive analytes for this technique because many of these relatively small molecules are transported readily across commonly used cross-linked polydimethylsilicone membranes. Two membrane-mass spectrometer interfaces which have been used with the ion trap for continuous monitoring are those having the *flow-through* and the *flow-over* designs.

1. Flow-Through Design

In the flow-through design, the aqueous stream passes through the interior of a hollow fiber membrane, or over one side of a sheet membrane, and analytes permeate across the membrane and into the vacuum chamber of the mass spectrometer. High sensitivity to a variety of VOCs in the low μg/L range and rapid responses to analytes entering the system are hallmarks of the flow-through design.[16] Ion trap detection limits in the range of 1 to 100 μg/L were estimated for acetone, amyl acetate, benzene, chlorobenzene, diethyl ether, isobutyl chloride, dichloromethane, tetrahydrofuran, and toluene.

2. Flow-Over Design

In the flow-over design, the aqueous stream flows over the exterior of a hollow fiber membrane and analytes permeate into the interior of the open tube where either they diffuse to the mass spectrometer ion source, or they are entrained in a stream of carrier gas and are swept into the ion source. The flow-over design is most effective with gases and the most volatile analytes and gave excellent linear calibrations and sub-μg/L detection limits for benzene, carbon tetrachloride, 1,1-dichloroethene, and vinyl chloride.[17]

B. Other Compounds

Recent work indicates that compounds other than VOCs are amenable to membrane separations from water, and further development of these systems should allow wide application of this technology with the sensitive and inexpensive ion trap mass spectrometer.

VII. FIELD APPLICATIONS

The identification and measurement or estimation of VOCs at sampling locations, often called field measurements, are sometimes of great value. During assessment, construction, or renovation of hazardous waste sites, rapid analyses in the field may be required to provide timely information to sampling personnel, construction crews, and heavy equipment operators. The small size and relative simplicity of the ion trap has attracted several investigators to this application. Preliminary results suggest that a more rugged ion trap without a GC can be used for rapid field analyses of air, water, and soil samples.[18] The use of alternating EI and CI scans and the potential use of tandem mass spectrometry (MS/MS) should enhance the accuracy of this technique. Previously, it was demonstrated that CI/MS/MS can be applied to the determination of VOCs desorbed directly into a laboratory-based ITMS.[19] A more conventional GC/MS approach to field environmental analyses has also been reported to give good results.[20]

VIII.   VOLATILE AIR POLLUTANTS

A.   Control Standards

The Clean Air Act of 1990 requires the U.S. EPA to issue control standards to reduce emissions of 189 toxic pollutants to the ambient air from industrial and other sources. Approximately 130 of the 189 substances are VOCs and many of these 130 VOCs are listed in Tables 12.1 and 12.2. Table 12.3 contains an additional 70 potential VOC air pollutants listed in the Clean Air Act. In contrast to the relatively unreactive compounds regulated and monitored under the Safe Drinking Water Act (shown in Table 12.1), there are many other substances that are more reactive or highly reactive, such as those listed in Table 12.2 and especially Table 12.3. While some of these compounds are so reactive that, in water, they would be hydrolyzed or undergo other chemical change, they are not sufficiently reactive in air, with the result that they persist as air pollutants.

B.   Research Program for Monitoring Hazardous Air Pollutants

The Clean Air Act of 1990 also requires the U.S. EPA to conduct a research program that must include monitoring of hazardous air pollutants in a representative number of urban areas, the identification of the sources of these pollutants, and the consideration of atmospheric transformations that could increase the risks to public health. The monitoring and other requirements of the Act will require sensitive analytical methods for the sampling, separation, identification, and measurement of volatile organic compounds in air. The development of these methods has been underway for some time, and the ion trap is being applied to this analytical problem.

C.   Techniques for the Determination of VOCs in Ambient Air

Several techniques have been employed with GC/MS to determine VOCs in ambient air.[21] Method TO-1 uses an organic polymer adsorbant (Tenax) to collect VOCs prior to GC/MS; Method TO-2 uses carbon molecular sieve adsorption. Thermal desorption from these traps followed by cap-

TABLE 12.3

Selected volatile organic compounds specified in the Clean Air Act of 1990

Compound	Compound
Acetaldehyde	Ethyl carbamate
Acetonitrile	Ethylene glycol
Acetophenone	Ethylenimine
Acrolein	Ethylene oxide
Aniline	Ethylene thiourea
o-Anisidine	Formaldehyde
Benzotrichloride	Hydrazine
Benzyl chloride	Methanol
1,3-Butadiene	Methyl hydrazine
Caprolactam	Methyl isobutyl ketone
Carbonyl sulfide	Methyl isocyanate
Chloroacetic acid	4-Nitrophenol
2-Chloroacetophenone	N-Nitroso-N-methyl urea
1-Chloro-2,3-epoxypropane	N-Nitrosodimethyl amine
bis (Chloroethyl) ether	N-Nitrosomorpholine
bis (Chloromethyl) ether	Parathion
Chloromethyl methyl ether	Pentachloronitrobenzene
Chloroprene	Pentachlorophenol
o-Cresol	Phenol
m-Cresol	Phosgene
p-Cresol	1,3-Propane sulfone
Cumene	beta-Propiolactone
Diazomethane	Propionaldehyde
1,3-Dichloropropene	Propylene oxide
Diethanolamine	1,2-Propylenimine
N,N-Diethylaniline	Quinoline
Dimethylformamide	Quinone
1,1-Dimethylhydrazine	Styrene oxide
Dimethyl sulfate	2,4-Toluene diamine
4,6-Dinitro-o-cresol	2,4-Toluene diisocyanate
2,4-Dinitrophenol	o-Toluidine
2,4-Dinitrotoluene	Triethylamine
1,4-Dioxane	2,2,4-Trimethylpentane
1,2-Epoxybutane	Vinyl acetate
Ethyl acrylate	Vinyl bromide

illary column GC/MS is adequate for the determination of many nonpolar or slightly polar VOCs typical of the hydrocarbons and chlorinated hydrocarbons in Table 12.1. Method TO-14 uses a more versatile technique in which a pressurized or sub-atmospheric pressure whole air sample is collected in a stainless steel pressure vessel (canister) of about 6 L capacity. The interior surfaces of the canister are made passive by formation of a pure chrome–nickel oxide coating. The sample of ambient air is introduced into the evacuated canister through stainless steel tubing. At the laboratory, the contents of the canister are pre-concentrated in a cryogenic trap, then desorbed thermally into the capillary GC/MS for determination of the VOCs. One potential problem with cryogenic trapping of the more polar and reactive compounds listed in Tables 12.2 and 12.3 is the hydrolysis of some compounds by water collected in the canister and condensed in the trap.

These air methods are not incorporated at present into Clean Air Act regulatory requirements but they, or perhaps modified versions of them, may be included in future regulations. The mass spectral performance requirements of Method 524.2, which are based on the test compound 4-bromofluorobenzene, are included in the current version of Method TO-14. The GC/ITD is being used with these methods for ambient air monitoring and more performance data should begin to appear in the literature during the next few years.

D. VOCs in Indoor Air

One reported ion trap analysis of air samples is the detection and characterization of VOCs produced in indoor air by bacteria present on residential heating, ventilating, and air conditioning filters.[22] In this work, the compounds were swept directly from the headspace air above culture tubes into a GC/ion trap system. Several compounds including methyl thiol, methanol, ethanol, acetone, trimethylamine, dimethyldisulfide, and dimethyltrisulfide were identified. The analysis of indoor air may be a major area for growth in the application of ion trap spectrometer systems.

IX. LESS-VOLATILE ORGANIC ANALYTES

A. Sample Preparation Techniques

Only relatively few compounds of environmental interest have sufficiently high vapor pressures at ambient conditions to allow collection in a canister or inert gas purging from a water or soil sample. In order to address the wide variety of less-volatile compounds that may contaminate environmental materials, several sample collection or sample preparation techniques have been used. These techniques include: (a) traditional Soxhlet extractions or the much newer ultrasonic-assisted extractions of air filters, soils, and other solid matrices with a variety of organic solvents; (b) extractions of solid matrices with supercritical carbon dioxide or other supercritical fluids; (c) liquid–liquid extraction of water with a variety of immiscible organic solvents; and (d) liquid–solid extraction of water with solid sorbents. All of these techniques have been incorporated into a number of U.S. EPA methods which are used widely for environmental analyses.[4,10,13,21] All of these methods provide ultimately an extract which is sampled and injected into a chromatograph for separation, identification, and measurement of the components that were extracted from the environmental sample.

B. Methods for the Determination of Less-Volatile Compounds

In Table 12.4 are shown a number of less-volatile compounds that have been regulated subsequent to 1987 in drinking water in the U.S. Some of these compounds are also regulated in waste water discharges and other environmental media. As indicated previously, the drinking water maximum contaminant levels are not based on detection limits but on health risk assessments, and range over approximately seven orders of magnitude. Among the more well-known U.S. EPA GC/MS methods for determination of some of these compounds are Method 625,[4] a liquid–liquid extraction method for waste water samples, and Method 525.1,[23] a liquid–solid extraction method for drinking water samples. Both methods require GC/MS system performance to comply with DFTPP specifications.[4,23]

C. Applications of the Ion Trap

The use of an GC/ion trap with Methods 625 and 525.1 was inhibited by the DFTPP specifications because the ion trap gave a molecular ion at 442 Da that was often the base peak, whereas the specifications called for a base peak at the m/z 198 fragment ion. The DFTPP RA specifications were developed during the early 1970s[2] when perhaps only 50% of all quadrupole mass filters gave a DFTPP molecule ion RA greater than 40%. With the development of quadrupoles and the ion trap with significantly greater high-mass detection capabilities, it was incumbent on the U.S. EPA to re-examine the DFTPP specifications for RAs, and to make appropriate adjustments to take advantage of the capabilities of the newer mass spectrometers. This adjustment has been made and the ion trap can now meet the DFTPP specifications published in Method 525.1. Reports should begin to appear in the literature which describe the use of GC/ion trap systems with U.S. EPA methods for less-volatile compounds.

X. NON-VOLATILE COMPOUNDS

Fifteen of the compounds listed in Table 12.4 are not amenable to separation by GC because they do not have sufficient vapor pressures

TABLE 12.4

Other organic compounds regulated or monitored in drinking water supplies in the U.S.

Compound	Maximum Contaminant Level (µg/L)	GC/MS
Acrylamide	a	no
Alachlor	2	yes
Aldicarb	b	no
Aldicarb sulfoxide	b	no
Aldicarb sulfone	b	no
Atrazine	3	yes
Benzo[a]pyrene	0.2	yes
Carbofuran	40	no
Chlordane	2	yes
2,4-Dichlorophenoxy-acetic acid	70	no
2,2-Dichloropropionic acid	200	no
Di(2-ethylhexyl)adipate	400	yes
Di(2-ethylhexyl)phthalate	6	yes
Dinoseb	7	no
Diquat	20	no
Endothall	100	no
Endrin	2	yes
Epichlorohydrin	a	no
Glyphosate	700	no
Heptachlor	0.4	yes
Heptachlor epoxide	0.2	yes
Hexachlorobenzene	1	yes
Hexachlorocyclopentadiene	50	yes
Lindane	0.2	yes
Methoxychlor	40	yes
Oxamyl	200	no
Pentachlorophenol	1	yes (marginal)
Picloram	500	no
Polychlorobiphenyls (asdecachlorobiphenyl)	0.5	yes
Simazine	4	yes
2,3,7,8-Tetrachlorodibenzodioxin	0.00003	yes
Toxaphene	3	yes
2-(2,4,5-Trichlorophenoxy)propionic acid	50	no

a Controlled by treatment requirements.
b Future regulation expected.

at even the highest GC column oven temperatures, or they decompose at some lower temperatures. Several of these are carboxylic acids which may be converted into methyl or other esters which are sufficiently volatile for GC/MS. But ion trap-based determinations of the other non-volatiles, and many hundreds or thousands of analogous environmental contaminants, must wait for another development in ion trap technology, the liquid chromatography (LC)/ion trap interface. An LC/particle beam/mass spectrometry method (U.S. EPA Method 553) has been developed for the determination of several of these non-volatiles. Accounts of current research into the development and utilization of LC/ion trap systems are given in Chapters 5 and 6. These accounts indicate clearly a major future area of application of ion traps to environmental analyses.

REFERENCES

1. Syka, J. E. P. in *Modern Mass Spectrometry.* Vol. 1, Practical Aspects of Ion Trap Mass Spectrometry, R. E. March and J. F. J. Todd, Eds., Chap. 4, CRC press, Roca Baton, FL, 1995.
2. Eichelberger, J. W.; Harris, L. E.; Budde, W. L. *Anal. Chem.* 1975, *47*, 995–1000.
3. Federal Register, December 3, 1979.
4. Federal Register, October 26, 1984; *Code of Federal Regulations.* Title 40, Part 136.
5. Eichelberger, J. W.; Budde, W. L. *Biomed. Environ. Mass Spectrom.* 1987, *14*, 357–362.
6. Olson, E. S.; Diehl, J. W. *Anal. Chem.* 1987, *59*, 443–448.
7. Stafford, G. C.; Taylor, D. M.; Bradshaw, S. C.; Syka, J. E. P. *Proc. 35th ASMS Conf. Mass Spectrometry and Allied Topics.* Denver, May 24–29, 1987, pp. 775–776.
8. Budde, W. L.; Eichelberger, J. W. *Anal. Chem.* 1979, *51*, 567A–574A.
9. Federal Register, July 7, 1987; *Code of Federal Regulations.* Title 40, Part 141.
10. Method 524.2, Revision 3.0, in *Methods for the Determination of Organic Compounds in Drinking Water.* U.S. EPA Report EPA/600/4-88/039, December, 1988, Revised July 1991, National Technical Information Service, Springfield, VA, Order No. PB91-231480.
11. Eichelberger, J. W.; Bellar, T. A.; Donnelly, J. P.; Budde, W. L. *J. Chrom. Sci.* 1990, *28*, 460–467.
12. Munch, J. W.; Eichelberger, J. W. *J. Chrom. Sci.* 1992, *30*, 471–477.
13. Method 524.2, Revision 4.0 in *Methods for the Determination of Organic Compounds in Drinking Water.* Supplement II, U.S. EPA Report EPA/600/R-92/129, August, 1992, National Technical Information Service, Springfield, VA, Order No. PB92-207703.
14. Longbottom, J. E.; Slater, R. W.; Edgell, K. W. unpublished results.
15. Kotiaho, T.; Lauritsen, F. R.; Choudhury, T. K.; Cooks, R. G.; Tsao, G. T. *Anal. Chem.* 1991, *63*, 875A–883A.
16. Lister, A. K.; Wood, K. V.; Cooks, R. G.; Noon, K. R. *Biomed. Environ. Mass Spectrom.* 1989, *18*, 1063–1070.
17. Slivon, L. E.; Bauer, M. R.; Ho, J. S.; Budde, W. L. *Anal. Chem.* 1991, *63*, 1335–1340.
18. Wise, M. B.; Buchanan, M. V.; Thompson, C. V.; Guerin, M. R. *Proc. 40th ASMS Conf. Mass Spectrometry and Allied Topics.* Washington, D.C., May 31–June 5, 1992, pp. 501–502.
19. Wise, M. B.; Ilgner, R. H.; Buchanan, M. V. *Proc. 38th ASMS Conf. Mass Spectrometry and Allied Topics.* Tuscon, June 3–8, 1990, pp. 1481–1482.
20. Cisper, M. E.; Alarid, J. E.; Hemberger, P. H. *Proc. 40th ASMS Conf. Mass Spectrometry and Allied Topics.* Washington, D.C., May 31–June 5, 1992, pp. 503–504.
21. *Compendium of Methods for the Determination of Toxic Organic Compounds in Ambient Air.* U.S. EPA Report EPA/600/4-89/017, National Technical Information Service, Springfield, VA, Order No. PB90-116989/AS.

22. Rivers, J. C.; Pleil, J. D.; Wiener, R. W. *J. Exposure Analysis Environ. Epidemiology.* Suppl. 1, 1992, 177–188.
23. Method 525.1 in *Methods for the Determination of Organic Compounds in Drinking Water.* U.S. EPA Report EPA/600/4-88/039, December, 1988, Revised July, 1991, National Technical Information Service, Springfield, VA, Order No. PB91-231480.

Chapter 13

MULTI-RESIDUE PESTICIDE ANALYSIS

Thomas Cairns, Kin S. Chiu, David Navarro, and Emil Siegmund

CONTENTS

0-8493-8251-3/95/$0.00+$.50
© 1995 by CRC Press, Inc.

451

I. DRIVING FORCE FOR NEW APPROACH

There are a large number of multi-residue analytical procedures for the determination of pesticide residues in food. The methods in common use have adopted a simple organic partitioning procedure followed by gas chromatographic separation and final determination using a series of element sensitive detectors for P, S, N, X, etc. The establishment of such an analytical protocol usually requires a large capital investment for up to 20 gas chromatographs in addition to a significant concurrent labor overhead.

With the advent of the ion trap as a mass-selective detector operating in the full mass-scan mode, the possibility of reducing drastically the total cost of monitoring for pesticide residues deserved attention. The ability of this relatively low cost detector to obtain full mass spectral scans on picogram levels in various fruit and vegetable extracts without additional cleanup has been explored to formulate an alternative more cost-effective approach. Recent studies have optimized the gas chromatographic conditions using a 30 m DB-5 column in conjunction with a temperature-programmed injector to permit separation and detection of over 250 pesticides that have been demonstrated previously to be recovered through the Luke Procedure. Additional work performed using various stable isotopes as internal standards (various deuterated polycyclic aromatics) has provided quantification concurrent with identification and confirmation. Results on various fruit and vegetable extracts have provided strong evidence that such an approach can be a successful replacement to conventional multi-GC techniques. Work is continuing in order to evaluate this approach by rigorous comparative testing alongside the traditional element sensitive detector methods.

During the conduct of these preliminary research studies, several experimental difficulties were resolved successfully with the result that the precision and accuracy of the residue analysis results were improved. Examination of the methane chemical ionization (CI) mass spectra produced in an ion trap under automatic reaction control (ARC) conditions revealed a strong concentration dependence phenomenon even at trace (ppm) levels. The spectra produced during the elution profile of various pesticides *via* capillary gas chromatographic introduction indicated a large percentage of an electron impact (EI) spectral component superimposed on the CI spectrum. Precision and accuracy data obtained for various pesticides under such hybrid conditions (single ion *versus* total ion scans) have indicated an acceptable analytical level for trace analysis. While this duplicity of spectral character can be an important asset for confirmation of presence, it did not present a problem of obtaining acceptable precision and accuracy on quantification of pesticide residues.

A newly developed scan function and methane gas control system has permitted the production of classical CI spectra in the Finnigan Ion Trap ITS40, thus allowing direct comparison with reference spectra developed on magnetic and quadrupole instruments. This report outlines the preliminary work from this laboratory to assess the potential advantages and disadvantages of the ion trap approach to routine multi-residue pesticide analysis using a commercially available ion trap.

II. INTRODUCTION

Chemical ionization for the analysis of trace levels of pesticides by mass spectrometry/gas chromatography (GC/MS) has been well documented in the literature.[1] In particular, the selection of methane as a reagent gas has been demonstrated[2,3] to reduce potential interferences from background ions or matrix components while providing higher specificity through production of protonated molecular ions for the pesticides of interest. Therefore, it was anticipated that the same analytical approach on an ion trap would serve as a more cost-effective mechanism to regulatory pesticide analysis by providing full scan data for concurrent detection, confirmation, and quantification. The logic behind the selection of chemical ionization was a deliberate attempt to avoid saturating the ion trap under EI conditions and, hopefully, to permit the sole production of protonated molecule ions from both endogenous compounds as well as from the target pesticides.

However, during the development of such a multi-residue pesticide analytical protocol for pesticides in various fruits and vegetables,[4] it was noticed that the full scan methane CI spectra produced by the ion trap under ARC conditions were hybrid in nature. Closer examination of the spectra during typical elution profiles revealed a concentration dependence between EI and CI derived ions for the compound under investigation. This experimental observation was particularly disturbing for a number of reasons. One, the advantage of reducing ion contributions from background and matrix components through the application of CI had been effectively removed. Two, the hybrid CI spectra of chlorinated compounds could not easily be assigned a chlorine number since the molecular ion cluster was often a mixture of molecular (*via* EI) and protonated molecule ions (*via* CI). Three, the comparisons of full scan data could not be related directly to the standard pesticide library developed from previously recorded mass spectra from quadrupole and magnetic scanning instruments. While concentration dependence during GC elution has been observed previously for various drugs, the cause and effect has been attributed to thermal lability.[5]

In the case of the ion trap, however, such an explanation cannot be advanced since these pesticides have already been shown to be thermally stable under GC/MS conditions. The causes for the production of hybrid CI spectra of pesticides under ARC have been attributed to the composition of the reagent gas ionization spectrum.[6] It would seem that the ion trap can accommodate the conditions for simultaneous charge transfer (for EI ion production) and proton transfer reactions (for CI ion production).

III. EXPERIMENTAL

Pesticide standards were obtained as analytical standards from the U.S. Environmental Protection Agency repository and used without further purification. Mixtures of the various pesticides were made up at various concentration levels in methylene chloride.

Mass spectra were obtained on a Finnigan Ion Trap ITS40 equipped with a CI source, Septum Programmable Injector (SPI from Varian Associates, Inc.), Auto Sampler 200S. Standard operating conditions for gas chromatography were as follows: guard column of deactivated fused silica 2m × 0.25 mm, 30 m × 0.25 mm id, 0.25 μ DB-5 column (J & W Scientific), helium carrier gas at 28 cm³/s, operated isothermally at 60° for 1 min, then ramped at 50°/min to 180°, followed by 10°/min to 270° and finally isothermally at 270° for 13 min. Operating conditions for the programmed injector port: 40°C for 0.1 min then ramped at 175°/min to 250°C and held there isothermally for 22 min. Operating conditions for mass spectrometry: transfer line temperature of 260°C, manifold heater at 220°C, emission current 10 μA, multiplier voltage 1500 V, axial modulation amplitude 4 V$_{(0-p)}$.

IV. CHEMICAL IONIZATION OF PESTICIDES

A. Chemical Ionization in the Ion Trap

There are several major differences between the conditions under which methane CI is carried out with an ion trap and those which pertain to methane CI with a conventional quadrupole or magnetic mass spectrometer.

As with other mass spectrometers, the initial ionization step is the production of $CH_4^{+\cdot}$, CH_3^+, and $CH_2^{+\cdot}$ under EI conditions followed by ion/molecule reactions involving these primary ions with neutral molecules of methane at about 1 Torr to produce CH_5^+ (48%), $C_2H_5^+$ (41%), and $C_3H_5^+$ (6%).

In an ion trap, the accumulation of reagent ions for ionization with the analyte is controlled by the duration of the ionization time, usually in the millisecond range. With a reagent gas pressure of 10^{-5} Torr, the ion trap can produce a high concentration of reagent ions; hence, a large population of sample ions can be produced which accounts for the high sensitivity of this new technology. Optimum conditions, concurrent with the avoidance of signal saturation, are determined by a pre-scan (which is carried out with ARC) to calculate appropriate reagent gas ionization times and reagent gas reaction periods for the actual analytical scan from which observations are obtained.

In this way, both the ionization time and reaction time are altered automatically during a GC peak elution profile. For instance, longer reagent gas reaction periods may be required at lower concentrations of the analyte, for example, at the front end and tail of an elution profile. To obtain data over a capillary GC elution profile, the ionization and reaction times are selected to provide a full mass spectral scan every second, i.e., about four to five pre-scans and four to five analytical scans.

Two major types of ion/molecule reactions which take place in the trap are charge transfer and proton transfer. Additional reactions, such as condensation reactions and clustering or association reactions, are rare since the time domain of an ion trap experiment is often in the millisecond range, during which negligible collisional stabilization occurs.

1. Charge Transfer

The interaction of a positive ion such as $CH_4^{+\cdot}$ with a neutral molecule of the analyte may lead to charge transfer. The exothermicity of this charge transfer remains largely as internal energy of the radical molecule ion formed initially with subsequent fragmentation to stable carbonium ions *via* resonance stabilization.

2. Protonation

In most methane CI experiments, the ion/molecule reaction of particular interest is the proton transfer reaction from CH_5^+ to neutral molecules of the analyte. Such ion/molecule encounters are all highly efficient, as exothermic proton transfer occurs at essentially every collision. The energy transferred to such protonated molecule ions is often low and subsequent collisions with neutral methane molecule can lead to stabilization rather than fragmentation. However, the time domain within the ion trap is such that the protonated molecule ion may not experience collisions with neutral methane prior to ejection during the analytical scan.

Therefore, in an ion trap there exists the possibility of a mixture of reagent ions generated by EI as well as CI processes. For instance, the principal ions generated are the radical molecule ion, $CH_4^{+\cdot}$, and the ion/molecule reaction product ion, CH_3^+. These ions differ in their exothermicity of reaction with a given sample molecule. The EI-generated $CH_4^{+\cdot}$ is most likely to undergo charge transfer with the energy transferred *via* fragmentation to stable carbonium ions *via* resonance stabilization. On the other hand, the reaction of CH_5^+ with a molecule of the analyte takes place with a much smaller energy transfer, allowing the protonated molecule ion to be produced and stabilized (where applicable) by successive collisions with neutral methane molecules.

B. Experimentally Observed Spectral Character Under Chemical Ionization Conditions

The experimental observation of ion abundance-concentration dependence during GC elution into the ion trap for two common pesticides is illustrated in Figs 13.1 and 13.2 for chlorpyrifos (diethyl-0-3,5,6-trichloro-2-pyridyl phosphorothioate) and perthane olefin [1-chloro-2,2-bis(p-ethyl-phenyl) ethylene], respectively. In Fig. 13.1, the four mass spectral scans represent the sequential data collected over the elution of a capillary GC peak of chlorpyrifos. The first clear observation is that the spectral character changes dramatically from low concentration to high concentration of the compound in the ion trap. The protonated molecule ion for chlorpyrifos is identified as the cluster at m/z 350, 352, and 354, representing a triple chlorine atom-containing structure. However, this cluster is the base peak only at the apex of the elution profile. At low concentrations, the base peak is m/z 153 with strong fragment ions at m/z 322/324 and m/z 198/200.

An interpretation of this dependence could be based on the balance between the two major reactions occurring in the trap, charge transfer and protonation. At higher concentrations of the analyte, the ARC will impose shorter ionization and reaction times automatically to maintain a reasonable number of ions in the trap and avoid saturation; at lower concentrations of the analyte, the reverse is true. Effectively, automatic reaction control has introduced two concurrent reaction mechanisms to the experiment. One, at the highest analyte concentrations, the source pressure will be such that CI reactions should predominate *via* proton transfer. However, shorter reagent ionization times should favor charge exchange at high sample concentrations. To decipher which reaction dominates at high concentration of the sample, the mass spectral data were examined in detail in an attempt to provide guidelines for identifying those processes which were operative.

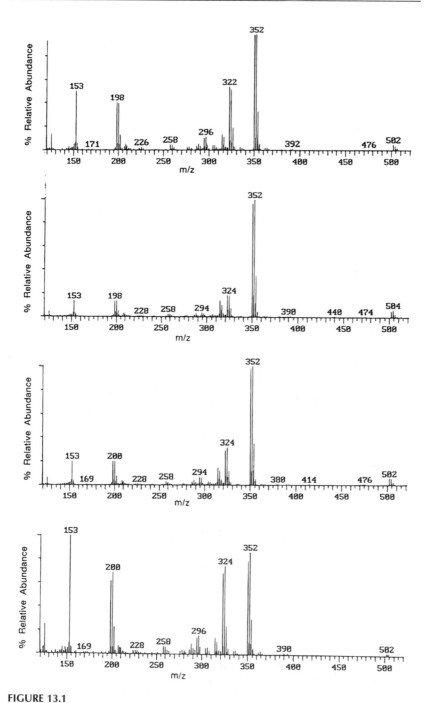

FIGURE 13.1

Spectral data collected for chlorpyrifos under ARC-CI conditions; the top spectrum represents start of elution profile and the bottom spectrum the end of the elution profile.

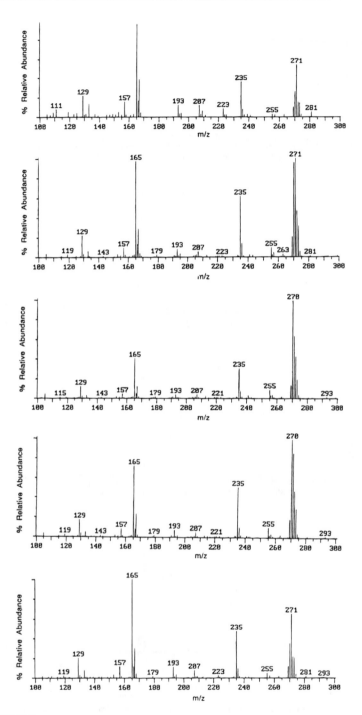

FIGURE 13.2

Spectral data collected for perthane olefin under ARC-CI conditions; the top spectrum represents start of elution profile and the bottom spectrum the end of the elution profile.

1. Chlorpyrifos

Clearly, the cluster of ions at m/z 350 is produced *via* proton transfer (molecular weight for chlorpyrifos is 349) and illustrates little if any concentration dependence over the elution profile. Furthermore, the ion abundance ratios for a triple chlorine atom-cluster are within the theoretical ratios. All other major ions (m/z 153, 198, 320) illustrate a strong concentration dependence, almost inversely proportional to the sample concentration. Therefore, it would appear that ions which were found to be concentration dependent over the elution profile could be labeled as generated *via* charge transfer with $CH_4^{+\cdot}$ followed by subsequent fragmentation. Examination of the CI spectrum for chlorpyrifos under quadrupole conditions revealed that the base peak was the protonated molecule cluster at m/z 350 with the major fragment ion at m/z 153 (41% relative abundance). The spectrum showed no ion cluster at m/z 322. It would appear, therefore, that all four spectra taken over the concentration elution profile should be classified as hybrid CI since they always contain elements of CI and EI character.

Since the ARC programming reduces the ionization and reaction times at the highest concentration level, it must be assumed that the resultant spectra reflect the composition of the reagent gas spectrum. The fragmentation observed from CI-generated ions such as the protonated molecule ion could be caused by energetic reactions (charge transfer or EI-like), resulting in fragment ions of the same masses expected from purely CI mechanisms. This mechanism would explain the experimental observation that the relative intensities of many CI fragment ions remain concentration dependent since they are produced in EI-like reactions.

2. Perthane Olefin

In the case of perthane olefin (Fig. 13.2), the five spectra recorded over the GC capillary elution profile illustrate a more serious situation where a relatively intense EI-like contribution is superimposed upon the CI spectrum. Examination of the molecular ion region indicates that there is strong competition between the CI and EI components of the spectra. The protonated molecule ion appears at m/z 271 while the EI-like molecular radical cation produced *via* charge transfer appears at m/z 270. Due to the severity of the overlap of those ionization processes, the isotopic contribution due to one Cl^{37} in the CI mass spectrum, at m/z 273, is not discernible. Indeed, the two overlapping clusters at m/z 270 (EI-like) and 271 (CI) have obscured completely the identification of a one-chlorine containing ion. Such competing ionization processes routinely make the appropriate assignment of a chlorine number to structure almost impossible. From an overall examination of these spectra for perthane olefin, the radical molecule ion at m/z 270 is the most abundant at high analyte concentration, that is, at the apex of the elution profile. This experimental observation would seem to indicate that the shorter ionization and reaction times favor EI-like reactions.

In summary, four main features have been proved to occur simultaneously during CI conditions in the ion trap. One, charge transfer reactions can occur *via* the $CH_4^{+\bullet}$ to produce EI-like ions. Two, protonation can occur *via* ion/molecule reactions with CH_5^+ and $C_2H_5^+$. Three, fragmentation of ions produced *via* protonation can take place because of lack of stability due to insufficient collisions with neutral methane before ejection or *via* charge transfer in collision with $CH_4^{+\bullet}$. Four, the resultant spectra are, usually, predominantly CI in nature with an EI-like component superimposed. Those ions which are concentration dependent can be attributed to an EI mechanism favored at higher analyte concentrations.

C. Proposed Cause for Hybrid CI Spectra

Given that the experimental evidence supports two main reaction mechanisms occurring under methane CI in the ion trap controlled by ARC, that is, charge transfer and protonation, the main cause must reside in the reagent gas spectrum generated just before the reaction period and the duration of the reaction period. While the reagent gas spectrum is optimized to favor m/z 17 and m/z 29 (CH_5^+ and $C_2H_5^+$) by allowing adjustment of gas pressure to minimize the primary EI ions at m/z 15 and m/z 16 by collision with neutral methane, there is a delay of several milliseconds before the acquisition of the actual CI spectrum. This delay must allow re-introduction of the primary EI ions responsible for charge transfer reactions. While an increase in the CI gas pressure can reduce these competing ions from appearing in the reagent gas spectrum, the additional time required to allow the reaction to enhance the occurrences of m/z 17, 29, and 41 is not practical under capillary GC conditions where four to five full scans are required per peak.

D. Mathematical Model for Observed Ion Ratios (EI/CI)

A rapid recognition of the production mechanism of such hybrid ions has proved to be a helpful diagnostic tool for their ultimate prevention. The intensity of an EI-like ion in the spectrum of a compound, I_{EI}, is proportional to ion production rate, α_1, sample concentration, C, and the ionization time t_{ion}:

$$I_{EI} = C\alpha_1 t_{ion} \tag{13.1}$$

The intensity of the ion peak is the area as detected by the electron multiplier and as viewed in the profile mode. However, during AGC data acquisition, this intensity is normalized or scaled as follows to give the abundance (that is, digital counts) recorded in the spectrum as A_{EI}:

$$A_{EI} = I_{EI}/t_{ion} = \alpha_1 C \qquad (13.2)$$

The quantity, A_{EI}, is proportional, therefore, to the sample concentration. The factor α_1 increases with emission current, multiplier voltage, and RF storage level. The extent of ionization depends also on the efficiency of injecting electrons into the trap, that is, the filament alignment can play an important role. In this respect, α_1 may change dramatically when switching to a different filament or replacing the filament assembly.

In the CI mode, the scaling operation is more complicated because the intensity of a CI reagent ion is proportional to both the initial reagent ion intensity and the reaction time. A first-order approximation can be assumed provided that not all reagent ions are consumed. In order for this condition to hold true, it is desirable to optimize the reagent ion storage level so as to store the largest number of reagent ions possible. However, a higher storage level in standard ARC-CI will lead to more efficient storage of analyte EI ions. As far as the maximum reagent ionization time is concerned, the proportional relationship with reagent ion intensities is true only for ionization times up to about 1 to 1.5 ms; once the trap is full, no more reagent ions can be accommodated.

The intensity of a CI ion, I_{CI}, is given by

$$I_{CI} = \alpha_2 C t_{ion} t_{rea} \qquad (13.3)$$

where α_2 is the CI ion production rate, t_{ion} is the ionization time, and t_{rea} is the reaction time. In the Finnigan ITS40 CI software, the reaction time is always a constant multiple of the ionization time, so that:

$$t_{rea} = \alpha_R t_{ion} \qquad (13.4)$$

where α_R is the constant multiple. The reagent ion intensity is proportional, therefore, to the square of the ionization time.

$$I_{CI} = \alpha_2 \alpha_R C t_{ion}^2 \qquad (13.5)$$

α_2, the CI ion production rate, increases with emission current, multiplier voltage, and RF storage level, and also with reagent gas pressure. From Eq. (13.5), the intensities reported in the spectrum are proportional to analyte concentration. In the spectrum recorded by the ARC-C! acquisition, the normalized intensity is given by

$$A_{CI} = I_{CI}/t_{ion}^2 = \alpha_2 \alpha_R C \tag{13.6}$$

Consider the case of analyzing the same compound at different times t(1) and t(2) at which the analyte is present at two different concentrations C(1) and C(2), respectively, that is, at different times during the elution profile of the analyte. Under ARC control, there will be different ionization times, t_{ion} (1) and t_{ion} (2). When t(1) corresponds to a time shortly after the leading edge of the peak, and t(2) to the peak maximum, then C(1) < C(2) and t_{ion} (1) > t_{ion} (2). Therefore, the reported intensities for two different ions in the spectrum of the analyte, one EI-like and the other a CI ion will be, for the CI ion:

$$A_{CI}(1) = I_{CI}(1)/t_{ion}^2(1) = \alpha_2 \alpha_R C(1) \tag{13.7}$$

and

$$A_{CI}(2) = \alpha_2 \alpha_R C(2) \tag{13.8}$$

For the EI-like ion, however, the normalized intensity A_{EI} is not proportional to the concentration only because the data acquisition uses the CI scaling process:

$$A_{EI}(1) = I_{EI}(1)/\alpha_R t_{ion}^2(1) = \alpha_1 C(1)/\alpha_R t_{ion}(1) \tag{13.9}$$

and

$$A_{EI}(2) = \alpha_1 C(2)/\alpha_R t_{ion}(2) \tag{13.10}$$

The equations derived above can now be used to calculate the ratio, R, of the EI-like ion to the CI ion at two different concentrations.

$$R(1) = \frac{A_{EI}(1)}{A_{CI}(1)} = \frac{\alpha_1 C(1)}{\alpha_R t_{ion}(1)\alpha_2 \alpha_R C(1)} = \frac{\alpha_1}{\alpha_2 \alpha_R^2 t_{ion}(1)} \tag{13.11}$$

and

$$R(2) = \frac{\alpha_1}{\alpha_2 \alpha_R^2 t_{ion}(2)} \tag{13.12}$$

Therefore the ratio of the observed ion abundances, r, is predicted mathematically to be:

$$r = \frac{R(2)}{R(1)} = \frac{t_{ion}(1)}{t_{ion}(2)} \qquad (13.13)$$

For example, if during the elution of an analyte the ionization time decreases from t_{ion} (1) = 1 ms on the side of the elution profile to t_{ion} (2) = 300 μs at the apex of the elution, the ratio of the suspected EI-like ion to the known CI ion increases by a factor of 3.33. This effect will not be strongly visible if the EI-like ion is initially of low relative abundance, for example, 3% of the CI ion. However, if the EI-like ion is about 30% of the CI ion (the base peak), the EI-like ion might well become the base peak at higher concentrations. This model, therefore, provides for easy recognition of the presence of EI-like ions in the CI spectrum, that is, inversely proportional to the ionization time.

E. Modified ARC-CI Scan Function

With the realization that the various experimental approaches outlined above would not improve the CI spectral character, attention was focused on the main cause for this phenomenon, that is, the presence of undesired ions formed during the reagent gas ionization period, m/z 15 for example. Based on the mathematical mode described above, a modification to the scan function, made at the source code level, allows for the rejection of the EI-like ions. This new ARC-CI function is a protected Finnigan proprietary procedure, and was developed to resolve the observed duplicity of spectral character.

This modified scan function has offered two main advantages leading to pure CI spectra in an ion trap. One, the time allocated for the RF ramp from ionization to reaction level provides extra time for the collision reactions forming the principal CI reagent ions. Two, the reagent ion spectrum is close to that expected classically when the reaction period starts, and any subsequent contributions from unexpected reactions (for example, charge transfer) are reduced or eliminated.

V. ANALYTICAL APPROACH TO PESTICIDE ANALYSIS

At the present time, there are approximately 250 pesticides that have been extracted and recovered experimentally, *via* the Luke Method, from a wide variety of fruits and vegetables.[7,8] The utility of the ion trap would be as a replacement detection system capable of automatic confirmation and quantification in one analytical step.

A. Gas Chromatographic Conditions

Since the analytical protocol involved injecting 1 μL of a raw, uncleaned, extract representing 3.6 μg of product, the optimal experimental parameters for the gas chromatographic separation were determined to involve a temperature program of about 25 min run on a DB-5 capillary column (Fig. 13.3). Additionally, the use of a Septum Programmable Injector (SPI) equipped with an insert and guard column was employed to observe early eluting compounds as well as to protect the column from damage from extracts.

B. Precision and Accuracy of the Ion Trap Technology

The first question to be addressed was the experimental ability of the commercial ion trap to detect and quantify all 250 target compounds within acceptable precision and accuracy at the residue levels required. Rather than prepare a complex standard reference mixture of over 250 pesticides, the analytical approach adopted by this laboratory was to divide the target compounds into 11 different mixtures using a temperature-programmed GC run to permit ample separation of compounds to calibrate the ion trap for an eventual automatic identification approach. These precision data on the standard reference mixtures have been published in detail elsewhere.[9]

For illustration of the construction of this target compound database on the ion trap, the reference mixture 1 containing 31 pesticides commonly encountered in residue work and 6 deuterated internal standards was examined under this analytical protocol (Fig. 13.4). The data from five consecutive injections are illustrated in Table 13.1. It was demonstrated that the ion trap had the capability to detect the target pesticides in this mixture at residue levels with a RSD of less than 10% with a correlation coefficient of 0.995 or greater. While these data represented a standard reference mixture injected at 1 ng per compound, the digital counts representing each compound were of sufficient size to extrapolate to lower detection levels in the range of 10 to 40 pg injected on column. This experiment established clearly the ability of the ion trap to detect and quantify the target compounds under investigation.

Having established the potential ability of the ion trap to analyze for the target compounds, a detailed analytical protocol was established to address detection, quantification, and confirmation in a pre-programmed automatic fashion. Linearity of response for each target compound was established using one or more ions at three different concentration levels; the slope of the response was a direct measure of the sensitivity of detection to the elected internal standard (usually a fused ring deuterated aromatic compound such as acenaphthene-d_{10}) selected for the compound.

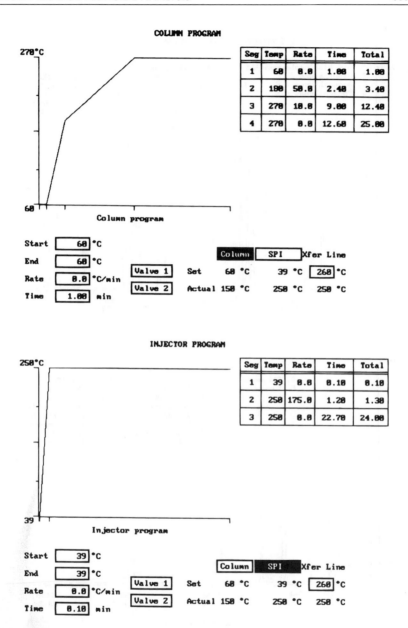

COLUMN PROGRAM

Seg	Temp	Rate	Time	Total
1	60	0.0	1.00	1.00
2	100	50.0	2.40	3.40
3	270	10.0	9.00	12.40
4	270	0.0	12.60	25.00

Column program

Start	60	°C
End	60	°C
Rate	0.0	°C/min
Time	1.00	min

Column	SPI	Xfer Line
Valve 1	Set	60 °C	39 °C	260 °C
Valve 2	Actual 150 °C	250 °C	250 °C

INJECTOR PROGRAM

Seg	Temp	Rate	Time	Total
1	39	0.0	0.10	0.10
2	250	175.0	1.20	1.30
3	250	0.0	22.70	24.00

Injector program

Start	39	°C
End	39	°C
Rate	0.0	°C/min
Time	0.10	min

Column	SPI	Xfer Line
Valve 1	Set	60 °C	39 °C	260 °C
Valve 2	Actual 150 °C	250 °C	250 °C

FIGURE 13.3

Experimental details of temperature programming for both the SPI and the GC column adopted for the multi-residue pesticide analysis.

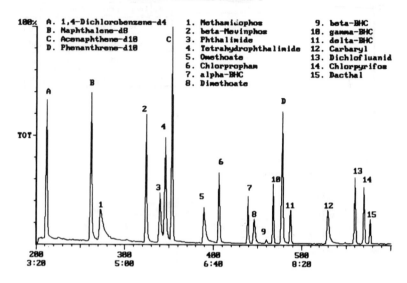

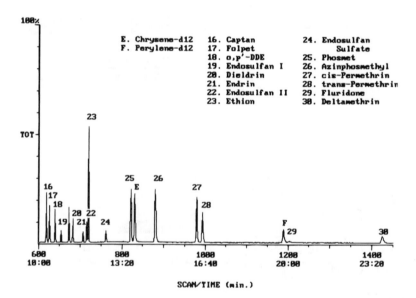

FIGURE 13.4

Total ion chromatogram for standard reference mixture 1 containing 31 pesticides and 6 deuterated internal standards (A–F).

TABLE 13.1

Analytical data for standard reference mixture 1 containing 30 pesticides and six deuterated internal standards

Pesticide	GC			Detection/Quantification					MS Confirmation				
	Scan No.	Rt Min.	Quan. Ions	Int. Std.	Resp. Fact.	M	SD	%RSD	Five Strongest Ions % Relative Abundance[a]				
1,4-Dichlorobenzene-d$_4$	212	3:32	151	A	1.00	133	14	10.5	151	153 (65)	150 (45)	152 (36)	155 (10)
Naphthalene-d$_8$	262	4:22	137	B	1.00	250	11	4.4	137	136 (58)	138 (10)	135 (4)	134 (7)
Methamidophos	274	4:34	142, 94, 125	C	0.88	753	41	5.4	142	94 (49)	125 (15)	141 (8)	95 (5)
beta-Mevinphos	324	5:24	193	C	1.08	994	56	5.6	193	127 (24)	192 (14)	194 (7)	225 (4)
Phthalimide	341	5:41	148, 130, 147	C	0.70	834	33	3.9	148	130 (17)	149 (9)	147 (8)	146 (1)
Tetrahydrophthalimide	347	5:47	152, 81	C	0.99	627	20	3.2	152	81 (90)	151 (17)	79 (10)	153 (8)
Acenaphthene-d$_{10}$	353	5:53	165	C	1.00	233	13	5.6	165	164 (63)	166 (13)	162 (12)	163 (2)
Omethoate	390	6:30	183, 196, 214	C	0.68	513	47	9.2	183	196 (36)	214 (32)	155 (16)	156 (15)
Chlorpropham	407	6:47	172	C	0.58	377	13	3.4	172	174 (33)	213 (13)	154 (13)	173 (11)
alpha-BHC	439	7:19	183, 181	C	0.26	151	7	4.6	183	181 (99)	219 (88)	217 (67)	221 (41)

[a]The abundance is relative to that of the ion species listed first.

TABLE 13.1
(Continued)

Pesticide	GC Scan No.	RtMin.	Quan. Ions	Int. Std.	Resp. Fact.	M	SD	%RSD	MS Confirmation — Five Strongest Ions % Relative Abundance[a]				
Dimethoate	447	7:27	199	C	0.39	302	9	3.0	199	88	87	171	198
			88							50	20	16	15
beta-BHC	461	7:41	183	C	0.06	28	0.7	2.5	183	181	219	217	221
			219							91	81	63	39
gamma-BHC	468	7:48	183	C	0.4	173	3	1.7	183	181	219	217	221
			181							97	88	66	37
			219										
Phenanthrene-d$_{10}$	478	7:58	189	D	1.00	194	6	3.1	189	188	190	187	186
										69	15	9	1
delta-BHC	487	8:07	181	D	0.37	160	7	4.4	181	183	219	217	221
			183							96	88	68	39
			219										
Carbaryl	530	8:50	145	D	0.79	571	27	4.7	145	144	173	146	202
										23	13	10	4
Dichlofluanid	560	9:20	224	D	0.36	204	9	4.4	224	110	226	126	167
			110							67	64	59	44
			126										
Chlorpyrifos	571	9:31	153	D	0.28	269	7	2.6	153	350	352	198	200
										74	69	60	59
Dacthal	577	9:37	333	D	0.33	351	15	4.3	333	331	335	301	299
			301							76	47	28	23

[a] The abundance is relative to that of the ion species listed first.

TABLE 13.1
(Continued)

Pesticide	GC Scan No.	GC RtMin.	Quan. Ions	Int. Std.	Resp. Fact.	M	SD	%RSD	MS Confirmation — Five Strongest Ions % Relative Abundance[a]				
Captan	619	10:19	264	D	0.22	180	16	8.9	264	236/69	238/63	134/44	79/33
Folpet	626	10:26	260, 261	D	0.50	436	20	4.6	260	262/66	264/12	105/11	261/11
o,p' DDE	640	10:40	281, 283	D	0.24	197	10	5.1	281	283/95	285/27	318/15	282/15
Endosulfan I	654	10:54	71, 243, 277	D	0.03	26	1	3.8	71	69/80	243/79	277/67	241/67
Dieldrin	682	11:22	79, 243	D	0.09	41	2	4.9	79	243/46	245/44	279/36	263/35
Endrin	707	11:47	108, 245, 113	D	0.03	24	0.9	3.8	108	79/56	243/42	245/41	113/38
Endosulfan II	716	11:56	71, 277, 243	D	0.07	52	2	3.8	71	277/75	279/53	241/47	243/44
Ethion	720	12:00	199, 231, 171	D	1.10	644	21	3.3	199	231/36	171/22	153/16	201/8
Endosulfan sulfate	762	12:42	325, 289	D	0.10	90	4	4.4	325	327/80	289/75	291/60	323/51

TABLE 13.1
(*Continued*)

Pesticide	GC Scan No.	GC RtMin.	Quan. Ions	Int. Std.	Resp. Fact.	M	SD	%RSD	MS Confirmation — Five Strongest Ions / % Relative Abundance
Phosmet	822	13:42	160	D	1.69	715	31	4.3	160; 161 (10); 477 (5); 286 (4); 318 (3)
Chrysene-d_{12}	831	13:51	241	E	1.00	155	5	3.2	241; 240 (60); 242 (18); 239 (8); 269 (6)
Azinphosmethyl	881	14:41	160, 132	E	1.72	532	22	4.1	160; 132 (92); 159 (11); 133 (8); 104 (7)
cis-Permethrin	979	16:19	183, 211	D	0.84	579	18	3.1	183; 211 (16); 184 (14); 365 (7); 163 (5)
trans-Permethrin	994	16:34	183, 211	D	0.61	474	20	4.2	183; 211 (15); 184 (14); 163 (5); 365 (5)
Perylene-d_{12}	1189	19:49	265	F	1.00	88	2	2.3	265; 264 (73); 266 (20); 263 (7); 267 (1)
Fluridone	1204	23:45	330, 331	E	0.16	191	18	9.4	330; 331 (26); 310 (23); 358 (17); 329 (16)
Deltamethrin	1425	23:45	208, 281	E	0.41	168	13	7.7	208; 281 (45); 181 (24); 279 (19); 283 (18)

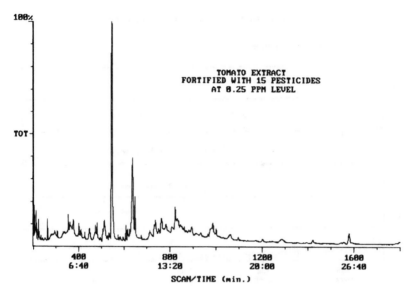

FIGURE 13.5
Total ion chromatogram for a tomato extract which has been spiked with 15 representative pesticides at 0.25 ppm.

Response factors for standard reference mixture 1 were calculated for each compound as illustrated in Table 13.1; these response factors were a measure of the sensitivity of detection of each compound relative to the chosen internal standard for quantification. The programmed system has been set up to search automatically for each target compound based on the ion selected, followed by confirmation of presence based on a fit using a minimum of four to five ions and, finally, quantification using the internal deuterated standard.

C.   Residue Analysis

Fig. 13.5 illustrates a typical profile for a tomato extract fortified with 15 pesticides at 0.25 ppm. All the major peaks represent endogenous compounds from the tomato, either intact, thermally-stable molecules or compounds resulting from pyrolysis in the injection port of thermally-unstable molecules. Pesticides at the ppm level would not be visible on the total ion chromatogram. However, when the data system re-examines the total ion chromatogram using the preselected ions for pesticide detection within the correct retention window, the pesticides are clearly identified (Fig. 13.6). For instance, in the case of *trans*-permethrin (Fig. 13.7), the ions selected for identification were m/z 183 and m/z 211. Once these ions were detected, the area of the combined ion profile was calculated,

TOMATO EXTRACT RECOVERY STUDY AT 0.25 PPM

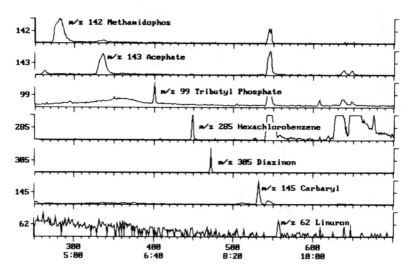

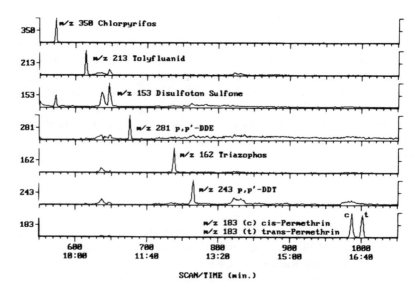

SCAN/TIME (min.)

FIGURE 13.6

Series of single ion chromatograms of the spiked tomato extract to illustrate the detection
of the 15 pesticides.

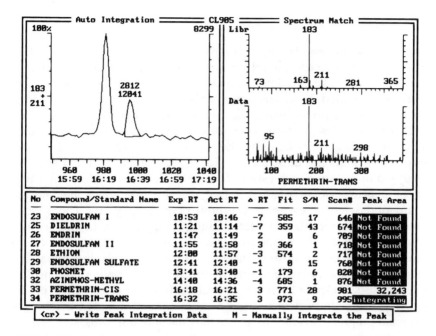

FIGURE 13.7
Typical screen output from data system to illustrate the detection of *trans*-permethrin: auto integration depicts the ion chromatogram (m/z 183 and m/z 211) used for detection; spectrum matching of extract spectrum with reference spectrum for *trans*-permethrin; and results section showing the finding of *cis*-permethrin and *trans*-permethrin together with deviation from expected retention time, spectral fit, and signal/noise ratio.

provided the library spectrum matched the data or experimental spectrum within the minimum fit.

Rejection of detection of a particular compound is dependent upon three main criteria. One, the retention time must be within 10 to 20 seconds of the retention of the standard reference material. However, the retention time of certain compounds such as methamidophos, acephate, and captan may shift with the sample matrix. Therefore, the search window for these compounds must be expanded more than the allowable 20 seconds. Two, the signal/noise ratio must be greater than 5:1. Three, the spectral match or fit must conform with a preselected operator value, usually greater than 800. In this manner, the operator or laboratory manager has previously imbedded into the data system the necessary quality assurance data to ensure confident results.

The process of searching for 250 target compounds using the 11 reference mixtures can be performed concurrently while the next analysis is being performed. The time taken for this sweep of 11 reference files normally takes 8 to 10 min while the analysis time is 25 min.

D. Comparison With Conventional GC

Since the process of determination of recoveries at 2 or 3 concentration levels for 250 target compounds in over 3000 different fruit and vegetable matrices would take a lifetime of experimental work, an alternative mechanism was adopted. In this new mechanism, laboratory pesticide residue analyses are performed using the conventional multi-GC approach[9] on over 5000 samples per year. Therefore, it was decided to run a parallel study on a daily basis between the conventional approach and the ion trap. The results are illustrated in Table 13.2 for the first 100 samples analyzed in the comparison study. Clearly, the ion trap method was capable of detecting violative samples, that is, residues in which the detected compounds were in excess of the limit permitted by law. There were, however, a few exceptions in determining compounds with no tolerance level permitted at all. The ion trap failed in some of those cases since the limit of detection was insufficient to see ultratrace levels. In general, the quantification by the ion trap was within 20% of the values obtained by the various GC element sensitive detectors. In a few cases, however, the ion trap was able to detect target compounds not detected by the gas chromatographic systems. In particular, the ion trap was sensitive at detecting nitrogen-containing compounds, a known weakness in the GC approach. However, the results from this small number of samples have pointed out a weakness in the ion trap method. There now exists the need to develop a quick sample cleanup procedure to allow the detection of ultra-trace levels for those compounds where no tolerance exists.

VI. CONCLUSIONS

From this preliminary study of the potential use of an ion trap for pesticide residue analysis, several important facts have emerged. One, under automatic reaction control for methane CI, the resultant spectra can exhibit duplicity of character in that an EI component is often superimposed on the CI spectrum. Such a condition might introduce a condition whereby identification using well-established spectral databases may require scrutiny before acceptance. Two, the precision and accuracy of the ion trap for trace levels of pesticides is well within acceptance ranges, that is, RSD of less than 10% with a correlation coefficient of 0.995 or greater. Furthermore, the use of a single ion area measurement (provided the ion selected is a true CI-generated ion) is acceptable for trace level quantification.

The sensitivity data generated for the selected 250 target compounds have revealed that full mass scan data can be collected for the majority of compounds at the 0.1 ppm level. There were about six compounds, mostly

TABLE 13.2

Comparison data: conventional multi gas chromatographic detection *versus* ion trap

Commodity	Pesticides Detected	GC Element Sensitive	Ion Trap MS	Violative Sample
		Levels Found (ppm)		
Anaheim peppers	Dacthal	.08		
	Chlorpyrifos	.40	.33	Yes
	Endosulfan I	.04		
	Endosulfan II	.05		
	Endosulfan Sulfate	.01		
Baby yellow squash	Dimethoate	.16	.40	Yes
Baby yellow squash	ND	ND		
Butternut squash	Chlorpyrifos	.05	.06	Yes
	Endosulfan I	.01		
	Endosulfan II	.01		
	Endosulfan Sulfate	.04		
Butternut squash	Thiabendazole	.05		Yes
Butternut squash	Thiabendazole	.20	.25	Yes
Butternut squash	Endosulfan Sulfate	.08		
Butternut squash	Thiabendazole	.35	.56	Yes
Cabbage	Permethrin	.13	.09 (trans) .07 (cis)	
Caribe peppers	Dacthal	.07		
Caribe peppers	Dacthal	.18		
	Endosulfan I	.07		
	Endosulfan II	.09		
	Endosulfan Sulfate	.01		
Carrots	ND	ND		
Carrots	Botran	.08		
	Chlorthalonil	.03		
Cauliflower	ND	ND		
Cauliflower	ND	ND		
Cauliflower	Endosulfan I	.01		
Cauliflower	ND	ND		
Cauliflower	ND	ND		
Celery	Acephate	.31	.14	
	Chlorthalonil		.14	
	Methamidosphos		.03	
	Endosulfan I	.01		
	Endosulfan II	.01		
	Endosulfan Sulfate	.01		
Celery	Chlorthanonil	1.0	.99	
	Permethrin		.03 (cis) .01 (trans)	
Celery	Dyrene	.50	.53	
Celery	Permethrin	.08		
	Chlorthanonil	.19	.86	
	Dyrene	.22		

TABLE 13.2

(*Continued*)

Commodity	Pesticides Detected	Levels Found (ppm) GC Element Sensitive	Ion Trap MS	Violative Sample
Cherry tomato	ND	ND		
Cucumber	Endosulfan I	.05		
	Endosulfan II	.04		
	Endosulfan Sulfate	.08		
Grapefruit	Imazalil	1.9	1.4	
Grapes	ND	ND		
Grapes	Iprodione	.30	.77	
Red grapes	Iprodione	.40		
Green beans	Endosulfan I	.05		
	Endosulfan II	.05		
	Endosulfan Sulfate	.06		
	Captan	5.3		
Green beans	Dacthal	.16		
	Captan		.62	
	Tetrahydropthalim		.56	
Lemons	Imazalil	2.4	1.4	
Nappa	ND	ND		
Nappa	ND	ND		
Raspberry	Captan	.25	.50	
Red leaf lettuce	ND	ND		
Tomato	ND	ND		
Watermelons	Acephate	.28	.32	Yes
Watermelons	Chlorthalonil	.07	.05	
Yellow summer squash	ND	ND		
Peaches	Iprodione	4.1	2.4	
Peaches	Iprodione	5.1	3.2	
Celery	Chlorthalonil	.4	.13	
	Acephate	.14		
Red grapes	Iprodione	.60	.54	
	Captan	.29	.40	
	Tetrahydropth.		.08	
Pears	Beta-phosdrin	.03		
Pears	Iprodione	.36	.08	Yes
Plums	ND	ND		
Yellow squash	Endosulfan I	.1		
	Endosulfan II	.03		
	Endosulfan Sulfate	.08		
Sweet apple	Captan	.8		
Green bean	Methamidophos	.04	.08	Yes
	Endosulfan I	.15		
	Endosulfan II	.12		

TABLE 13.2

(Continued)

Commodity	Pesticides Detected	Levels Found (ppm) GC Element Sensitive	Ion Trap MS	Violative Sample
Opa squash	Methamidophos	**	.04	Yes
Japanese cabbage	Methamidophos	.04		
Carrots	Dacthal	.03		
Kale	Permethrin	.15	.16 (cis) .08 (trans)	Yes
Bittermelon		ND	ND	
Peaches	Iprodione	2.0	3.0	
Red grapes	Captan	.5	.92	
Red grapes	Captan		.31	
Plums		ND	ND	
Plums	Iprodione	.01		
Mokwa squash	Methamidophos	.13	.16	Yes
Celery	Chlorthalonil	.10	.13	
	Diazinon	.10	.13	
Cabbage	ND	ND		
Celery	Chlorthanonil	.15	.12	
Passion fruit		ND	ND	

sulfur-containing pesticides, for which the ion trapping technique showed insufficient sensitivity.

The residue data gathered for comparison of the conventional multi-GC approach *versus* the ion trap, would indicate that the analytical approach is sound, but sample cleanup would improve both the detection limits and quantification data. Work is proceeding to investigate a simple, sample cleanup that would provide greater sensitivity for all target compounds.

REFERENCES

1. Cairns, T.; Siegmund, E. G.; Stamp, J. J. *Mass Spectrom. Rev.* 1989, *8*, 93.
2. Cairns, T.; Siegmund, E. G.; Krick, F. *J. Agric. Food Chem.* 1987, *35*, 433.
3. Mattern, G. C.; Louis, J. B.; Rosen, J. D. *J. Assoc. Off. Anal. Chem.* 1991, *74*, 982.
4. Cairns, T.; Chiu, K. S.; Siegmund, E. G. *Rapid Commun. Mass Spectrom.* 1992, *5*, 331.
5. Cairns, T.; Siegmund, E. G.; Stamp, J. J.; Skelly, J. *Biomed. Mass Spectrom.* 1983, *10*, 203.
6. Cairns, T.; Chiu, K. S.; Siegmund, E. G. *Rapid Commun. Mass Spectrom.* 1992, *6*, 449.
7. Luke, M. A.; Froberg, J. E.; Masumoto, H. T. *J. Assoc. Off. Anal. Chem.* 1975, *58*, 1020.
8. Luke, M. A.; Masumoto, H. T.; Cairns, T.; Hundley, H. K. *J. Assoc. Off. Anal. Chem.* 1988, *71*, 415.
9. Cairns, T.; Chiu, K. S.; Navarro, D.; Siegmund, E. G. *Rapid Commun. Mass Spectrom.* 1993, *7*, 1976.

Chapter 14

ELECTROSPRAY/ION TRAP MASS SPECTROMETRY: APPLICATIONS TO TRACE ANALYSIS

Hung-Yu Lin and Robert D. Voyksner

CONTENTS

0-8493-8251-3/95/$0.00+$.50

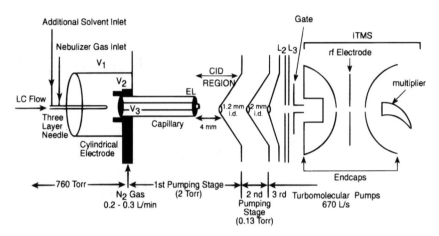

FIGURE 14.1
Schematic of Analytica electrospray-ITMS systems described in the text.

I. INTRODUCTION

Electrospray is an ionization method which can efficiently desorb ions formed in solution into the gas phase for mass analysis.[1-4] Ion currents generated by electrospray are typically stronger than currents generated by other desorption ionization techniques.[5] The fundamental aspect of electrospray is the formation of ions by ion evaporation ionization.[4,6,7] This ionization process is initiated by applying a high electric field to the sample spraying needle to produce micron-sized charged droplets. As the solvent is vaporized continuously by a heated nitrogen bath gas, the radius of the charged droplets decreases and hence the field density increases. Droplets with excess charge undergo further fission, since coulombic repulsion exceeds the forces of cohesion (Rayleigh stability). However, at sufficiently small droplet diameter, a field strength in excess of 10^8 V/cm^2 can be achieved, resulting in direct ion evaporation of analyte from the solution bringing it into the gas phase.

The ability to produce singly- or multiply-charged ions directly from analytes in solution and bring them into the gas phase with high efficiency without heat was a vital breakthrough in the on-line LC/MS analysis of ppb levels of non-volatile and thermally labile compounds. Furthermore, the ability to form multiply-charged ions in electrospray nearly eliminates the need for mass analyzers to extend beyond 4000 Da to analyze biopolymers of over 100,000 Da.[8] It is possible, therefore, to determine the molecular weight of small proteins accurately (<0.02% mass error) with a low resolution mass spectrometer.[8-11]

Ions formed by electrospray or other atmospheric pressure ionization (API) techniques need to be transported from atmospheric pressure to the vacuum of the MS. One transport design (Fig. 14.1) uses a capillary and a series of skimmers to reduce pressure sequentially and to move ions into the mass spectrometer. An important aspect of this and most API interfaces is the ability to generate collision-induced decomposition (CID) spectra to obtain structural information.[12–15] This is of vital importance with a desorption ionization technique such as electrospray, since the ion evaporation ionization process is very soft, generating only molecular adduct ions. Through control of the capillary exit potential, the internal energy of the ions can be controlled. At increased capillary voltages, ions are accelerated for multiple collisions with the drying gas (air or nitrogen) between the capillary and skimmer (~1 Torr pressure), resulting in increased internal energy. If the capillary voltage is of sufficient magnitude, enough internal energy can be imparted into the ion to break the covalent bonds between atoms to generate product ions (fragments) which can provide structural information about the compound. This CID process is very efficient due to minimal loss in ion current from scattering and due to the excellent conversion efficiency of molecular ion species into product ions. Electrospray transport CID can input over 16 eV of internal energy into a molecule,[12] which is sufficient to generate product ions from the most stable organic molecules.

Tandem MS (MS/MS)[16–18] can be used to provide structural information from ions generated by soft ionization processes but does not compete with CID in the transport region in terms of cost, complexity, and sensitivity. However, tandem MS provides a CID spectrum specific for the selected ions, an important advantage when dealing with mixtures and co-eluting components in LC/MS. This mass-selection is not possible in the electrospray transport region; therefore, the CID spectrum obtained is a composite for all components of the mixture. A tandem MS approach that rivals electrospray transport CID for efficiency is available with ion trap technology.

Ion trap mass spectrometry (ITMS)[19–22] is well suited for coupling with electrospray for achieving the low limits of detection and maximizing specificity through MS^n capabilities.[23–25] The quadrupole ion trap contains two end-cap electrodes and a ring electrode. The end-cap electrodes are normally at ground potential while an RF voltage is applied to the ring electrode. This generates a three-dimensional quadrupole field to trap ions. A high background pressure (about 1 to 2 mTorr of helium) is necessary to dampen ions collisionally injected into the trap to obtain the maximum trapping efficiency.[26] In this way, the ion kinetic energy distribution is reduced and the mass resolution is enhanced. Ions are trapped or isolated by the application of an RF or RF/DC field as described by the Mathieu stability diagram.[20] The mass-dependent trapping parameter q_z of the ion increases with RF voltage, and the ion will be ejected when the q_z reaches 0.908.

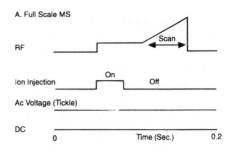

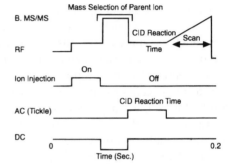

FIGURE 14.2
The ITMS scan functions for electrospray: (A) full scan mode of operation, and (B) electrospray MS/MS mode of operation.

II. OPTIMIZATION OF COUPLING OF ELECTROSPRAY TO THE ITMS

The utility of electrospray combined with ITMS has been explored by several investigators to achieve additional sensitivity and/or specificity in a potentially less costly package.[27-29] The approach pursued in our laboratory employs the use of a commercially available electrospray interface that was coupled to an ITMS using minimal hardware modifications (Fig. 14.1).[30] Briefly, the electrospray interface was mounted on the 10-in. flange opposite the RF electronics. The trap was rotated to 180° and moved to the opposite end of the vacuum chamber. This required the retapping of the RF coil to rebalance the RF power. The electrospray interface was axially coupled to the existing entrance end-cap electrode of the ITMS (filament assembly was removed) by a tubular lens which also served as the gate (Fig. 14.1). Also, the vacuum pumping speed on the ITMS was increased from 170 to 690 L/s to handle gas flow thorough the capillary and skimmers.

The ITMS scan function (Fig. 14.2) controlled the various voltages and frequencies in the trap and interface to acquire a mass spectrum (or MS/MS spectrum) in an analogous manner, as in EI operation. Ions were introduced into the ion trap by a short pulse of voltage applied to the gate electrode. The gating voltage was provided by a condition circuit which controllably allowed up to ± 100 V to the tubular gate lens. This circuit was driven by the EI gate pulse. The circuit inverted the polarity of the EI gate pulse, allowing for a maximum injection of ions into the trap at −40 V, while ions were not injected into the trap at 20 V. By ramping the RF amplitude, ions of consecutively increasing mass-to-charge ratio can be ejected and detected (mass-selective instability scan).[31] Since there is a high sampling duty cycle (compared to a quadrupole instrument) and minimal losses of transmission from ion optics, ion traps can achieve superior sensitivity over quadrupole and sector instruments.

The ion trap is currently the least expensive mass spectrometer on the market which can be operated in the MS^n mode to generate additional structural information from desorption ionization techniques. Mass isolation is accomplished by a combination of RF and DC voltages. A suitable AC tickle voltage (V_{p-p}) is selected to excite resonantly the ion of interest for CID with the helium bath gas (Fig. 14.2(B)). Since the CID process occurs inside the RF trapping field at a high pressure (~1 mTorr), there are minimal losses and the resulting product ion currents are significantly higher than those obtained by triple quadrupole and sector mass spectrometers.[32–34]

The limited mass range of the commercial ITMS (10 to 650 Da) can be easily extended for electrospray usage through resonance ejection. By scanning the RF amplitude, ions can be brought into resonance with a supplementary RF signal applied to the end-cap electrodes. As the ions come into resonance with the supplementary signal, they may absorb sufficient energy to be ejected from the ion trap at a lower value of q_z than would ordinarily be required. This method was used to extend the detectable mass range of the ITMS to more than 50,000 Da.[35–37]

Due to the multiple charging phenomenon observed with electrospray, only 4× mass extension (up to 2,600 Da) was necessary to detect ions generated from peptide samples analyzed by the electrospray-ITMS system. This small increase in mass range also minimizes the disadvantages of mass range extension on the current commercial instrument, resulting in under sampling of the signal, leading to poor mass resolution and signal intensity. A 4× increase in mass range correlates to 1/4× the number of data points per Da that define a peak.

While these hardware modifications and scan functions to control the trap were relatively straightforward, operating conditions needed to be optimized to achieve the best sensitivity. Most of this optimization involved the ion trap operation since ion transmission for the electrospray interface was virtually independent of the type of mass analyzer (trap or quadrupole). The parameters that were evaluated included end-cap aperture diameter, ion injection time (gate), ion energy, bath gas pressure, q_z value during trapping, and de-solvation time. The diameter of the ion entrance aperture in the end-cap was varied from 1 to 6 mm (Fig. 14.3(A)). It was determined that a 3 to 4 mm end-cap opening was optimal for ion

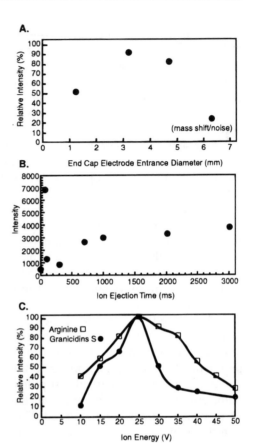

FIGURE 14.3

Optimization of electrospray-ITMS parameters to achieve maximum sensitivity for the ions of arginine (MW 174) and gramicidin S (MW 1141). (A) Total ion current level for arginine and gramicidin S with various end-cap aperture diameters. (B) Total ion current level for arginine and gramicidin S at different ion injection times. (C) Ion current for arginine and gramicidin S for various ion energies (controlled by skimmer 1 in the electrospray source). (D) Ion current for arginine and gramicidin S for various helium pressures in the ion trap (uncorrected ion gauge readings). (E) Ion current for arginine and gramicidin S for different q_z trapping values during ion injection. (F) Total ion current for arginine and gramicidin S using different reaction times (de-solvation times) after ion injection into the ITMS.

injection. Larger end-cap electrode openings resulted in a decrease in signal levels, possibly due to inhomogeneities in the trapping field.

Ion injection time into the ion trap indicated 50 to 70 ms gating was optimal for loading the trap with ions (Fig. 14.3(B)). Longer ion injection times appeared to decrease signal levels, as well as decreasing mass resolution, due to space charging. It was found that −40 V on the tubular lens coupled to the entrance end-cap was optimal for transmitting ions into the trap while +20 V effectively prevented ions from entering.

The ion energy (set by skimmer 1 on the electrospray interface) for injecting ions into the trap was mass dependent (Fig. 14.3(C)). Arginine showed a broad optimum while gramicidin S exhibited a narrow optimum between 20 to 30 V. These patterns were most likely due to ion transmission and collisional dampening efficiencies of ions with different masses.

Helium pressure played a major role in optimizing signal level and the qualitative appearance of the spectra (Fig. 14.3(D)). In general, higher

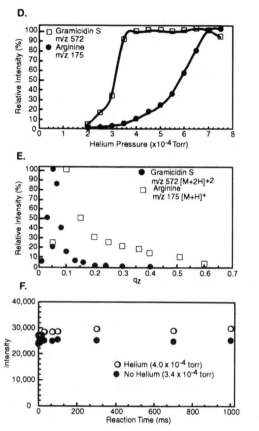

D.

E.

F.

FIGURE 14.3
Continued

helium pressures were optimal for trapping ions. However, there is a mass dependence in which lower molecular weight ions are trapped at lower pressures, while higher molecular weight compounds (gramicidin S) are more efficiently trapped at higher operating pressures. This trend can be explained in terms of the inability of helium to dampen collisionally high-mass ions relative to low-mass ions. Therefore, the qualitative appearance of the mass spectra for arginine and gramicidin S could be drastically different, depending on the helium trapping pressure. Higher pressure showed a higher relative intensity for gramicidin S $[M + 2H]^{+2}$ ion compared to the intensity of arginine. Low pressure showed primarily $[M + H]^{+}$ for arginine and a weak signal for the $[M + 2H]^{+2}$ ion of gramicidin S. To enhance trapping of high-mass ions, pulsing helium into the trap during ion injection was investigated. In this way, high helium pressure could dampen the ions, then a one second pump-out time reduced the pressure before mass analysis was initiated. The pump-out time reduced the pressure by a factor of three to four, minimizing multiplier noise and mass spectral peak broadening.[38]

The q_z value for ion injection was mass dependent with arginine optimizing near a q_z value of 0.15 and gramicidin S near a q_z value of 0.05 (Fig. 14.3(E)).

The period of time after ionization that the ions were stored in the trap (de-solvation time) was not a critical parameter. This stems from the fact that the ions are de-solvated and de-clustered in the electrospray transport region on the described system *versus* within the trap on using collisional warming for other electrospray-ITMS configurations.[27–29] De-solvation reaction times from 0 to 1000 ms, with and without helium bath gas, resulted in less than 10% signal intensity variance (Fig. 14.3(F)).

III. APPLICATION OF ELECTROSPRAY-ITMS TO ENVIRONMENTAL ANALYSIS

There are growing concerns among the people of the world about environmental quality. Air, food, and water are rightly scrutinized for contaminants that can produce health risks. Most methods resulting from legislation to clean our environment have utilized gas chromatography/mass spectrometry (GC/MS) and LC. While GC/MS techniques have been the "staple" for measuring volatiles and semivolatiles, there are major shortcomings when monitoring non-volatiles that comprise water, soil, and sludge pollutants. Often, only a fraction of these pollutants can be detected by GC/MS. The unaccounted mass is summarized as non-volatile, thermally unstable, polar or high molecular weight material that requires an alternative detection strategy, namely LC/MS.

Pesticides and azo dyes are two classes of compounds of great environmental interest that require the capabilities of LC/MS for specific and sensitive detection. Therefore, the electrospray ITMS system was evaluated for the detection of these two classes of compounds. Of particular importance was the question of when CID in the electrospray transport or CID within the ITMS should be employed. The complexity in setting the ion resonance frequency, tickle voltage, and activation time for CID in the ITMS made the latter method much more time-consuming and difficult compared to CID in the electrospray transport region. However, while CID in the electrospray transport only required changing one voltage, the technique did not resolve co-eluting peaks or components of different m/z values in a mixture, resulting in a CID spectrum of all components in the electrospray interface at that particular time. For example, the electrospray mass spectrum of aldicarb sulfone (MW 222) only generates an $[M + H]^+$ ion with no collisional activation in the electrospray

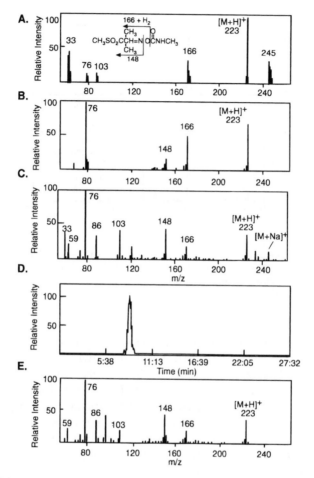

FIGURE 14.4
Electrospray-ITMS spectra for aldicarb sulfone (MW 222). (A) Flow injection with a low capillary voltage (60 V) to minimize CID. (C) Flow injection with a high capillary voltage (140 V) to generate CID product ions. (B) Flow injection with a low capillary voltage (60 V) and selecting m/z 223 to generate the MS/MS spectrum (q_z 0.3, tickle voltage 1.5 V_{p-p}, tickle time 80 ms). (D) Capillary column chromatography of aldicarb sulfone C_{18} 0.32 × 150 mm column, 5 μm particles, gradient of 10 to 70% acetonitrile (1% acetic acid) in 15 min at a flow rate of 6 μL/min. (E) The mass spectrum of the compound eluted at 8 min was acquired with a capillary voltage of 140 V.

or ITMS (Fig. 14.4(A)). Increasing the voltage in the electrospray transport region (capillary voltage on the Analytica source or repeller voltage on a Vestec source) increased the collisional energy, resulting in CID of the [M + H]⁺ ions as well as product ions (Fig. 14.4(C)). Further increasing these voltages can result in over 16 eV of internal energy imported into the ion, producing numerous bond cleavages (0.5 to 1.5 eV are needed to break

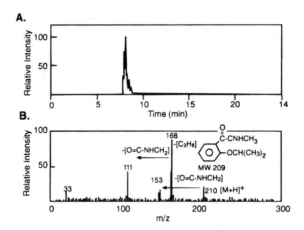

FIGURE 14.5
Capillary LC/electrospray-ITMS analysis of 60 pg of propoxur. (A) Total ion chromatogram using separation conditions described in Figure 14.4D. (B) Electrospray-ITMS mass spectrum using CID in the electrospray transport region (30 V using a repeller on Vestec source) to generate product ions for confirmation.

a covalent bond).[12] Likewise, the ITMS can mass-select the [M + H]+ ion at m/z 223 for collisional activation inside the trap to generate the CID spectrum shown in Fig. 14.4(B). The spectrum shows the same structurally relevant product ions, but has less background than the spectrum shown in Fig. 14.4(C). Since these samples were flow-injected into the system, the background could be reduced through on-line LC/electrospray-ITMS analysis of aldicarb sulfone employing electrospray transport CID (Fig. 14.4(D)). The added specificity offered by chromatography resulted in a clean electrospray transport CID spectrum, comparable to the MS/MS spectrum obtained by the ITMS.

The sensitivity of the electrospray-ITMS system was sufficient to detect 60 pg of propoxur using CID in the electrospray transport region under LC/MS conditions (Fig. 14.5). The signal/noise obtained was about two to three times superior to a quadrupole under the same full scan conditions. Furthermore, the relatively clean aqueous samples did not result with any interferences in the confirmation ions for propoxur at m/z 210, 168, and 111 (Fig. 14.5).

A number of azo and anthraquinone dyes were analyzed to compare the CID spectra generated in the electrospray transport region and the MS/MS spectra obtained in the ITMS.[30] Both CID techniques showed structurally significant product ions (Figs. 14.6 and 14.7). Often the ion trap provided less product ions than CID in the electrospray transport due to the wider range of internal energy that can be imparted into the ions by CID in the electrospray transport. Product ions were primarily observed from the cleavage of the azo bond (N=N) and the N-C bonds,

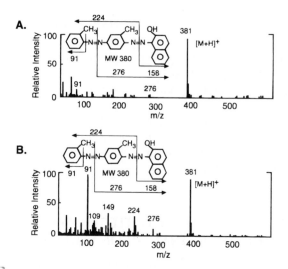

FIGURE 14.6
Electrospray-ITMS determination of 300 pg of Solvent Red 24. (A) CID spectrum in electrospray transport region (30 V repeller). (B) MS/MS spectrum of the [M + H]⁺ ion at m/z 381 in the ITMS (q_z 0.2, tickle voltage 1.0 V, tickle time 50 ms).

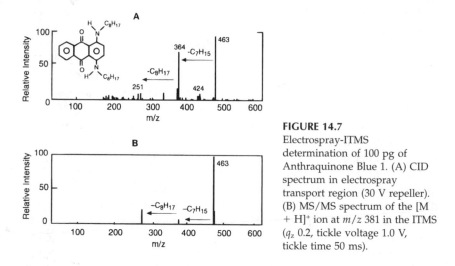

FIGURE 14.7
Electrospray-ITMS determination of 100 pg of Anthraquinone Blue 1. (A) CID spectrum in electrospray transport region (30 V repeller). (B) MS/MS spectrum of the [M + H]⁺ ion at m/z 381 in the ITMS (q_z 0.2, tickle voltage 1.0 V, tickle time 50 ms).

as well as from the loss of OH and alkyl substituent groups. Clearly the MS/MS spectra in Fig. 14.6(B) and Fig. 14.7(B) exhibited ions only due to the compound of interest, while the CID in the electrospray transport showed other low intensity ions due to impurities in the sample and solvents.

These examples show that the ability to isolate selectively the ions of interest for MS/MS in the ion trap is often preferred, especially during flow-injection analysis. However, the MS/MS capability of the ion trap depends on many variables including tickle voltage (V_{p-p}), tickle frequency (which can vary with space charging), collision time, and the q_z value used for trapping. All these parameters must be optimized to acquire MS/MS spectra with high signal/noise ratio. On the other hand, CID spectra generated within the electrospray transport region are easily obtained. Voltage on the capillary represents the only parameter which needs to be set. However, this method is useful only on relatively pure samples or samples showing minimal background, placing more of an emphasis on LC separation.

IV. APPLICATION OF LC/ELECTROSPRAY-ITMS FOR THE DETERMINATION OF NEUROPEPTIDES

Neuropeptides are involved in multiple forms of intercellular communication in the control of a variety of biologically important functions such as reproduction, metabolism, and other activities. It is important, therefore, to analyze these compounds and their metabolites both qualitatively and quantitatively. However, the trace level of neuropeptides in biological fluids and the complexity of the matrices make it very time-consuming to purify these compounds for routine analysis.

The mass spectra of the neuropeptides obtained from our laboratory were acquired using resonance ejection to extend the m/z range to 2600 Da. Flow-injection or infusion analysis of α-melanocyte stimulating hormone (α-MSH), α-endorphin, and β-endorphin resulted in mass spectra that showed several multiply-charged ions of the compound (Fig. 14.8(A)–(C)) that could be used to calculate the molecular weight of each peptide. Since electrospray is a soft ionization technique, no significant fragmentation was expected. It was noticed that the multiple charge envelope was at higher m/z values using electrospray on the ion trap rather than a quadrupole mass analyzer. It is postulated that the RF trapping field and the high pressure dampening gas used to assist in the trapping of ions can result in charge stripping. The success of electrospray to ionize peptides led to the evaluation of on-line LC/electrospray-ITMS for their detection in biological fluids.

In order to increase analysis speed it was decided to evaluate the performance of perfusion capillary LC.[39,40] Perfusion capillary columns are packed with porous particles to maximize mass transfer while minimizing flow resistance. Therefore, minimal height equivalent per theoretical plate (HETP) values are obtained at very high solvent linear velocities. Separations of neuropeptides that took 30 to 45 min on a C_{18} capillary column (150 × 0.32 mm) could now be accomplished in 4 to 5 minutes on a 150 × 0.32 mm perfusion column. Analysis speed and detection sensitivity

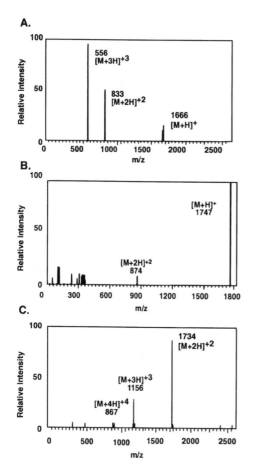

FIGURE 14.8
Flow injection electrospray-ITMS
mass spectra of 10 pmol of the
following neuropeptides (130 V on
capillary). (A) α-MSH (MW 1665).
(B) α-endorphin (MW 1746). (C) β-
endorphin (MW 3465).

were demonstrated in the LC/electrospray-ITMS analysis of neuropeptides spiked in serum. The total ion current (TIC) chromatogram (Fig.
14.9) clearly distinguishes the three neuropeptides spiked into serum at
subpicomole level. The ITMS mass spectrum could confirm their identity
based on the multiply-charged ions detected.

Perfusion LC/electrospray-MS was also used to determine hydrolysis
degradation products of dynorphin A 1-13. The TIC chromatogram indicated the presence of one degradation product (Fig. 14.10). The electrospray mass spectrum indicated correctly that the molecular weight of
dynorphin A 1-13 was 1602 and the degradation product was 1474. Based
on this information and the sequence for dynorphin A 1-13 it was postulated, and later confirmed with a standard, that this degradation product
was dynorphin A 1-12 (loss of lysine from dynorphin 1-13). Furthermore,
the signal for dynorphin 1-12 is estimated to represent 80 to 100 fmol

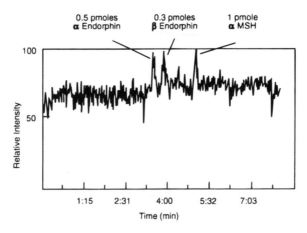

FIGURE 14.9
Perfusion LC/electrospray-ITMS total ion current chromatogram for the determination of α-MSH, α-endorphin, and β-endorphin spiked in serum at the 0.3 to 1.0 pmol level. Separation conditions: Poros II R/H 0.32 × 150 mm perfusion column, gradient of 0 to 60% acetonitrile in 5 min (0.1% trifluoroacetic acid) at a flow rate of 35 μL/min.

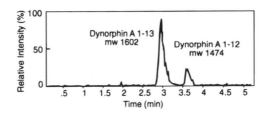

FIGURE 14.10
Perfusion LC/electrospray-ITMS total ion chromatogram for the determination of the hydrolysis product formed from dynorphin A 1–13 (1 pmol injected). Separation conditions are described in Fig. 14.9.

injected into the perfusion column, indicating the sensitivity that can be achieved with the perfusion LC/electrospray-ITMS combination.

V. CONCLUSIONS AND OUTLOOK

Electrospray occupies an important niche in mass spectrometry, particularly for the biochemists and protein chemists. Electrospray is also continually undergoing development to become more versatile with chromatography. Techniques including ion spray,[41] ultrasonic nebulization,[42] and atmospheric pressure chemical ionization (APCI)[43] increase the range of solvents, buffers, and flow rates acceptable to electrospray MS, making it a preferred LC/MS technique for the foreseeable future.

Likewise, the ITMS has significant advantages stemming from (1) a high ionization duty cycle resulting in excellent sensitivity, (2) the capability of MS^n to generate structural information, (3) the ability to explore ion/molecule reactions, (4) the capability to conduct high-mass analysis and high resolution mass analysis.[19-21] However, current instrumentation lags in its capabilities and versatility to achieve all these goals easily. Researchers must rely on their expertise to modify the hardware and software either to simplify the operation or to achieve a certain mode of operation to perform the desired analysis such as LC/MS. Future versions of ion traps should address these concerns, changing the trap from a novelty to a tool that can reliably accomplish a job.

ACKNOWLEDGMENTS

The authors are grateful for support from the National Institute on Drug Abuse, Grant No. 5-R01 DA06315 and the U.S. Environmental Protection Agency, Co-operative Agreement No. CR-819555.

REFERENCES

1. Whitehouse, C. M.; Dreyes, R. N.; Yamashita, M.; Fenn, J. B. *Anal. Chem.* 1985, *57*, 675.
2. Fenn, J. B.; Mann, M.; Mong, C. K.; Wong, S. F.; Whitehouse, C. M. *Mass Spectrom. Rev.* 1990, *9*, 37.
3. Matthias, M. *Org. Mass Spectrom.* 1990, *25*, 575.
4. Fenn, J. B. *J. Am. Soc. Mass Spectrom.* 1993, *4*, 524.
5. Voyksner, R. D. *Pesticide Chemistry—Advances in International Research, Development and Legislation.* H. Frehse, Ed. Vol. XIV, VCH, New York 1991, 383.
6. Iribane, J.; Thompson, B. *J. Chem. Phys.* 1976, *64*, 2287.
7. Thompson, B.; Iribane, J. *J. Chem. Phys.* 1979, *71*, 4451.
8. Loo, J. A.; Edmonds, C. G.; Barinaga, C. J.; Udseth, H. R. *Anal. Chem.* 1990, *62*, 882.
9. Chowdhury, S. K.; Katta, V.; Chait, B. T. *Rapid Commun. Mass Spectrom.* 1990, *4*, 81.
10. Smith, R. D.; Light-Wahl, K. J.; Winger, B. E.; Loo, J. A. *Org. Mass Spectrom.* 1992, *27*, 811.
11. Smith, R. D.; Loo, J. A.; Ogorzalek, R. R.; Busman, M.; Udseth, H. R. *Mass Spectrom. Rev.* 1991, *10*, 359.
12. Voyksner, R. D.; Pack, T. *Rapid Commun. Mass Spectrom.* 1991, *5*, 263.
13. Duffin, K. L.; Wach, T.; Henion, J. D. *Anal. Chem.* 1992, *64*, 61.
14. Katta, V.; Chowdhury, S. K.; Chait, B. T. *Anal. Chem.* 1991, *63*, 174.
15. Smith, R. D.; Loo, J. A.; Barinaga, C. J.; Edmonds, C. G.; Udseth, H. R. *J. Am. Soc. Mass Spectrom.* 1990, *1*, 53.
16. Cooks, R. G. *Anal. Chem.* 1985, *57*, 823A.
17. Kondrat, R.; Cooks, R. G. *Anal. Chem.* 1978, *50*, 81A.
18. Yost, R. A.; Enke, C. G. *Anal. Chem.* 1979, *51*, 1251A.
19. Cox, K. A.; Williams, J. D.; Cooks, R. G.; Kaiser, R. E. *Biol. Mass Spectrom.* 1992, *21*, 226.
20. Nourse, B. D.; Cooks, R. G. *Anal. Chem.* 1990, *228*, 1.

21. Todd, J. F. J.; Penman, A. D. *Int. J. Mass Spectrom. Ion Processes.* 1991, *105* 1.
22. March, R. E.; Hughes, R. J. *Quadrupole Storage Mass Spectrometry.* Chemical Analysis Series, Vol. 102, John Wiley & Sons, New York, 1989.
23. Kaiser, R. E., Jr.; Cooks, R. G.; Syka, J. E. P.; Stafford, G. C., Jr. *Rapid Commun. Mass Spectrom.* 1990, *4,* 30.
24. Stafford, G. C., Jr.; Kelley, P. E.; Syka, J. E. P.; Reynolds, W. E.; Todd, J. F. J. *Int. J. Mass Spectrom. Ion Processes.* 1984, *60,* 85.
25. Johnson, J. V.; Yost, R. A.; Kelley, P. E.; Bradford, D. C. *Anal. Chem.* 1990, *62,* 2162.
26. Louris, J. N.; Amy, J. W.; Ridley, T. Y.; Cooks, R. G. *Int. J. Mass Spectrom. Ion Processes.* 1987, *88,* 97.
27. McLuckey, S. A.; Van Berkel, G. J.; Glish, G. L.; Huang, E. C.; Henion, J. D. *Anal. Chem.* 1991, *63,* 375.
28. Van Berkel, G. J.; Glish, G. L.; McLuckey, S. A. *Anal. Chem.* 1990, *62,* 1284.
29. Van Berkel, G. J.; McLuckey, S. A.; Glish, G. L. *Anal. Chem.* 1991, *63,* 1098.
30. Lin, H. Y.; Voyksner, R. D. *Anal. Chem.* 1993, *65,* 451.
31. Syka, J. E. P.; Louris, J. N.; Kelley, P. E.; Stafford, G. C.; Reynolds, W. E. U.S. Patent 4,736,101, 1988.
32. Kaiser, Jr., R. E., Cooks, R. G., Syka, J. E. P.; Stafford, Jr., G. C. *Rapid Commun. Mass Spectrom.* 1990, *4,* 30.
33. Stafford, Jr., G. C.; Kelley, P. E.; Syka, J. E. P.; Reynolds, W. E.; Todd, J. F. J. *Int. J. Mass Spectrom. Ion Processes.* 1984, *60,* 85.
34. Johnson, J. V.; Yost, R. A.; Kelley, P. E.; Bradford, D. C. *Anal. Chem.* 1990, *62,* 2162.
35. Hemberger, P. H.; Moss, J. D.; Kaiser, R. E.; Louris, J. N.; Amy, J. W.; Cooks, R. G.; Syka, J. E. P.; Stafford, G. C., Jr. *Proc. 37th ASMS Conf. Mass Spectrometry and Allied Topics.* Miami Beach, May 21–26, 1989, p. 60.
36. Kaiser, R. E., Jr.; Cooks, R. G.; Moss, J.; Hemberger, P. H. *Rapid Commun. Mass Spectrom.* 1989, *3,* 50.
37. Kaiser, R. E.; Louris, J. N.; Amy, J. W.; Cooks, R. G. *Rapid Commun. Mass Spectrom.* 1989, *3,* 225.
38. Lin, H. Y.; Voyksner, R. D. *Proc. 41st ASMS Conf. Mass Spectrometry and Applied Topics.* San Francisco, May 31–June 4, 1993, p. 461a.
39. Afeyan, N. B.; Fulton, S. P.; Regnier, F. E. *J. Chromat.* 1991, *544,* 267.
40. Liapis, A. I.; McCoy, M. A. *J. Chromat.* 1992, *599,* 87.
41. Bruins, A. P.; Covey, T. R.; Henion, J. D. *Anal. Chem.* 1987, *59,* 2642.
42. Whitehouse, C. M.; Shen, S.; Fenn, J. B. *Abstracts 40th ASMS Conf. Mass Spectrometry and Applied Topics.* Washington, DC, May 31–June 5, 1992, p. 69 (no extended abstract).
43. Covey, T. R.; Lee, E. D.; Bruins, A. P.; Henion, J. D. *Anal. Chem.* 1986, *58,* 1451A.

Author Index

CHEMICAL INDEX

SUBJECT INDEX

A

A, coefficient, 260
a_r, definition, 37, 259
a_u, definition, 11, 259
a_x, definition, 11
a_y, definition, 11
a_z, definition, 11, 37, 259
a_z, q_z, coordinate, 20, 419
a_z, q_z space, 11, 36
Accuracy, 464
Accurate mass assignment, 122
ACE, *see* Alternating chemical
    ionization/electron
    ionization
Adducts, 22
Adduct formation, 80, 195, 244, 248
AGC, *see* Automatic gain control
Alternating chemical ionization/
    electron ionization, 24, 136,
    206, 242
Analytical scan, 19, 59, 107, 124
Analyzer, 4
Anglescan, 105
Angular momentum, 10
Anions, *see* Negative ions
Anode, 384
Anode fall region, 383, 384
Apex isolation, 65
APCI, *see* Atmospheric pressure
    chemical ionization
API, *see* Atmospheric pressure
    ionization
Appearance energy, 53
Appearance potential, 284
ARB, *see* Arbitrary waveform
    generator
Arbitrary waveform generator, 153
ARC, *see* Automatic reaction control

ASGDI/ion trap system, 97, 389, 391
ASGDI source, *see* Atmospheric
    sampling glow discharge
    ionization
Atmosphere-borne materials, 25
Atmospheric pressure chemical
    ionization, 216, 492
Atmospheric pressure ionization, 216,
    220, 225, 480
Atmospheric sampling glow
    discharge ionization, 97,
    157, 395
Automated MS/MS, 77
Automated MS/MS/MS, 416
Automatic gain control:
    16, 17, 127, 142, 150, 172, 173, 191,
    436
    EI mass spectrum, 92
    ion trap response, 92
    mode of operation, 17, 48, 66
Automatic ionization control, *see*
    Automatic gain control and
    Automatic reaction control
Automatic reaction control, 22, 127,
    244, 452, 456
Auxiliary RF potential, 38, 261
Axial:
    component of motion, 11
    kinetic energy, 261
    modulation, 19, 24, 38, 47, 103, 132,
    298, 300, 392
    potential well depth, 267
    secular frequency, 37, 39
    stability, 37
"Avalanche" effect, 381
AXUM™ software, 411
Azo dye analysis, 486

507